'Chee takes us on an encounter-as-detour, tracking spatial stories imbued with affect and haunted by precarious power relations. Undoing the discipline's instrumental ambitions and revealing the fallacy of its purported autonomy, she asks: If we become better attuned to the micro-politics of radically open-ended affective encounters, can our knowledge practices in architecture be transformed?'

——— Hélène Frichot, Professor of Architecture and Philosophy,
Melbourne School of Design, University of Melbourne

'Lilian Chee's *Architecture and Affect* is an exceptionally important contribution to the architectural field. It asks a simple but most profound question: how can we account for affective responses to architecture, so often dismissed as incidental yet so vital to lived experience? Chee draws on a series of encounters with spaces in Singapore in order to build a frame through which to incorporate affect into architectural histories, theories, and practices. These encounters immerse readers in residual spaces where scholars still-too-rarely go, from mass housing to cemeteries, and dwell on everyday practices of inhabitation and care, from feeding stray cats to exhuming tombs. A noted filmmaker, architectural designer and feminist theorist, Chee's original and rooted portrait of architectural affect reflects years of intimate engagement with her sites across many registers. The result is itself a beautiful "monument" – to squinting from blind spots, being captivated by subjects, and never ignoring the tiger under the billiard table.'

——— Barbara Penner, Professor in Architectural Humanities,
Bartlett School of Architecture, UCL

Architecture and Affect is motivated by two questions: Why does dismissed affective evidence trouble us? What would it mean for architecture to assemble such discrepant evidence into its discourse?

Arguing that the persistent refrains of lived affect dwell in architecture, this book traces such refrains to a concept of architecture wedged in the middle ground—jammed amidst life, things and events. Rather than being aloof from its surrounds, architecture-in-the-midst challenges an autonomous epistemology. Beyond accounting for the vivid but excluded, this book develops a frame and a disposition for thinking critically about, speculatively through, and being grounded by, encounter. Examining affect through a constellation of spaces in contemporary Singapore, it details architecture's uneasy but inextricable relationship with key subjects relegated to the incommensurate, the peripheral, the scenic and the decorative. The outcome is a politicized architectural discourse simultaneously grounded and speculative; bridging depth and intuition, thinking and feeling.

Lilian Chee is Associate Professor of Architectural Theory and Design at the National University of Singapore, where she co-leads the Research by Design Cluster. Her research revolves around architectural representation, affect theory, feminist politics, and creative practice methods. Her works include the award-winning essay film *03-FLATS* (2014), the documentary *Objects for Thriving* (2022), and a co-edited book *Remote Practices* (2022). She leads a Social Sciences Research Council funded project about home-based labour. She writes on affect, architectural representation and domesticity.

ROUTLEDGE RESEARCH IN ARCHITECTURE

The *Routledge Research in Architecture* series provides the reader with the latest scholarship in the field of architecture. The series publishes research from across the globe and covers areas as diverse as architectural history and theory, technology, digital architecture, structures, materials, details, design, monographs of architects, interior design and much more. By making these studies available to the worldwide academic community, the series aims to promote quality architectural research.

Exteriorless Architecture: Form, Space and Urbanities of Neoliberalism
Stefano Corbo

Transgressive Design Strategies for Utopian Cities: Theories, Methodologies and Cases in Architecture and Urbanism
Bertug Ozarisoy, Hasim Altan

Architecture and Affect: Precarious Spaces
Lilian Chee

Modernism in Late-Mao China: Architecture for Foreign Affairs in Beijing, Guangzhou and Overseas, 1969-1976
Ke Song

The Spatialities of Radio Astronomy
Guy Trangoš

The Ambiguous Legacy of Socialist Modernist Architecture in Central and Eastern Europe
Mariusz E. Sokołowicz, Aleksandra Nowakowska, Błażej Ciarkowski

Architecture, Ritual and Cosmology in China
The Buildings of the Order of the Dong
Xuemei Li

For more information about this series, please visit: https://www.routledge.com/Routledge-Research-in-Architecture/book-series/RRARCH

Architecture and Affect:
Precarious Spaces

Lilian Chee

Routledge
Taylor & Francis Group
LONDON AND NEW YORK

First published 2023
by Routledge
4 Park Square, Milton Park, Abingdon, Oxon OX14 4RN

and by Routledge
605 Third Avenue, New York, NY 10158

Routledge is an imprint of the Taylor & Francis Group, an informa business

British Library Cataloguing-in-Publication Data
A catalogue record for this book is available from the British Library

ISBN: 978-1-4724-5463-8 (hbk)
ISBN: 978-1-032-40754-8 (pbk)
ISBN: 978-1-315-60456-5 (ebk)

DOI: 10.4324/9781315604565

Cover image by Simryn Gill

Typeset in Gill Sans Nova & Palatino Linotype
by Lin Derong

For Product Safety Concerns and Information please contact our EU representative: GPSR@taylorandfrancis.com
Taylor & Francis Verlag GmbH, Kaufingerstraße 24, 80331 München, Germany.

for Joan, for George

Table of Contents

IMAGE CREDITS

COVER

Photo: Simryn Gill, 2006.

PREFACE

0	Model: Lilian Chee, 2006.

CHAPTER 1

Opening	Drawing: Lin Derong, 2019.
1.1	Image: Raffles Hotel, *Straits Times* 1902.
1.2	Photo: National Museum of Singapore and National Heritage Board, 1890s. Accession No. 1994-04802.
1.3	Photo: Jeremy Seah, 2019.
1.4	Image: Raffles Hotel Collection, c. 1902.
1.5	Image: *The Singapore and Straits Directory*, 1909.
1.6	Image: Raffles Hotel Singapore, c. 1905.
1.7	Drawing: Lin Derong, 2019.
1.8	Drawing: Lin Derong, 2021.
1.9	Image: Heinrich Leutemann, collection of National Museum of Singapore and National Heritage Board, c. 1865. Image Accession No: 2000-06610.
1.10	Image: *Guide to the Zoological Collections of the Raffles Museum*, Singapore, 1908.
1.11	Image: Kelly Chopard and Patrick Yee, Landmark Books, 1996.
1.12	Image: Kathy Creamer, Little Pink Dog Books, 2017.
Closing	Drawing: Lin Derong, 2019.

CHAPTER 2

Drawings: Lilian Chee and Toby Fong, 2020.

CHAPTER 3

Opening	Drawing: Lin Derong, 2020.
3.1	Photo: Lin Derong, 2016.
3.2a	Drawing: Lin Derong, 2019.
3.2b	Drawing: Lin Derong, 2019.
3.3	Photo: Lin Derong, 2016.
3.4	Photo: Finbarr Fallon, 2019.
3.5	Photo: Liew Yuqi, 2015.
3.6	Photo: Simryn Gill's *Guide to the Murals*, 2006, 15 x 21 cm, Lin Derong, 2016.
3.7	Photo: Lilian Chee, 2015.
3.8	Photo: Simryn Gill's *Guide to the Murals*, 2006, 15 x 21 cm, Lin Derong, 2016.
3.9	Photo: Simryn Gill's *Guide to the Murals*, 2006, 15 x 21 cm, Lin Derong, 2016.

3.10	Photo: Simryn Gill's *Guide to the Murals*, 2006, 15 x 21 cm, Lin Derong, 2016.
3.11	Photos: Simryn Gill, 2006.
3.12	Photos: Simryn Gill, 2006.
3.13	Photos: Simryn Gill, 2006.
Closing a Closing b	Drawing: Lin Derong, 2019.

CHAPTER 4

Opening	Drawing: Lin Derong, 2019.
4.1	Drawing: Lin Derong, 2019.
4.2	Photo: Lilian Chee, 2017.
4.3a	Drawing: Lin Derong, 2019.
4.3b	Drawing: Lin Derong, 2019.
4.4	Photo: Lin Derong, 2017.
4.5	Drawing: Lin Derong, 2019.
4.6	Photo: Lin Derong, 2017.
4.7	Photo: Lin Derong, 2017.
4.8	Photo: Lin Derong, 2020.
4.9	Photo: Jeremy Seah, 2019.
4.10	Photo: Jeremy Seah, 2019.
Closing	Drawing: Lin Derong, 2019.

CHAPTER 5

Drawings: Lilian Chee and Lin Derong, 2020.

CHAPTER 6

Opening	Drawing: Lin Derong, 2020.
6.1	Film still: *03-FLATS*, Lei Yuan Bin (dir.), Lilian Chee (PI, concept), 2014.
6.2	Film still: *03-FLATS*, Lei Yuan Bin (dir.), Lilian Chee (PI, concept), 2014.
6.3	Film still: *03-FLATS*, Lei Yuan Bin (dir.), Lilian Chee (PI, concept), 2014.
6.4	Film still: *03-FLATS*, Lei Yuan Bin (dir.), Lilian Chee (PI, concept), 2014.
6.5	Film still: *03-FLATS*, Lei Yuan Bin (dir.), Lilian Chee (PI, concept), 2014.
6.6	Film still: *03-FLATS*, Lei Yuan Bin (dir.), Lilian Chee (PI, concept), 2014.
6.7	Film still: *03-FLATS*, Lei Yuan Bin (dir.), Lilian Chee (PI, concept), 2014.
6.8	Drawing: Lilian Chee, 2014.
6.9	Drawing: Lilian Chee, 2014.
6.10	Drawing: Lilian Chee, 2014.
Closing	Drawing: Lin Derong, 2020.

CHAPTER 7

Opening	Drawing: Lin Derong, 2019.
7.1	Photo: Finbarr Fallon, 2018.
7.2	Drawing: Lin Derong, 2019.
7.3	Drawing: Lin Derong, 2019.
7.4	Photo: Lin Derong, 2018.
7.5	Photos: Katie Chavez, 2015; Laura Freeman (www.expatadventureinsingapore.com), 2011; Hou Chia Yi, 2014; Claire Leow, 2013; Lin Derong, 2016, 2018, 2019; Peter Pak (Rojak Librarian), 2013, 2018; RememberSingapore.org, 2011; Rosalind Tan, 2012; Calvin Teo, 2011, 2018. Collage: Lin Derong, 2019.
7.6	Photo: Lin Derong, 2018.
7.7	Photo: Lin Derong, 2018.
7.8	Photo: Lin Derong, 2018.
7.9	Photo: Wong Zi Hao, 2012.
7.10	Photo: Lin Derong, 2018.
7.11	Photo: Lin Derong, 2020.
7.12	Photo: Lin Derong, 2019.
Closing	Drawing: Lin Derong, 2020.

CHAPTER 8

Models: Lilian Chee and Wong Zi Hao, 2020.

CHAPTER 9

Opening	Drawing: Lin Derong, 2021.
9.1	Photo: Lin Derong, 2022.
9.2	Drawing: Lin Derong, 2019.
9.3	Photo: Lilian Chee, 2021.
9.4	Photo: Chi Yin Sim/Magnum Photos, 2017.
9.5	Photo: Chi Yin Sim/Magnum Photos, 2017.
9.6	Video Still: Charles Lim Yi Yong, 2015.
9.7	Photo: Charles Lim Yi Yong, 2015.
9.8	Photo: Luis D'Orey/Reuters, 2003.
9.9	Video Still: Charles Lim Yi Yong, 2015.
9.10	Photo: Charles Lim Yi Yong, 2008.
9.11	Video Still: Charles Lim Yi Yong, 2008.
9.12	Video Still: Charles Lim Yi Yong, 2008.
9.13	Drawing: Wong Zi Hao, 2020.
Closing	Drawing: Lin Derong, 2022.

LIST OF ABBREVIATIONS

API	Asia Paranormal Investigators
FMS	Federal Malayan States
FMSR	Federated Malay States Railway
HDB	Housing and Development Board
KTM	Keretapi Tanah Melayu
LEEDR	Low Effort Energy Demand Reduction
PAP	People's Action Party
POA	Points of Agreement (On Malayan Railway Land in Singapore between the Government of Malaysia and the Government of Singapore)
SCDF	Singapore Civil Defence Force
SIAJ	Singapore Institute of Architects Journal
SIT	Singapore Improvement Trust
UMNO	United Malays National Organisation
URA	(Singapore's) Urban Redevelopment Authority

PREFACE

Squinting from a Blind Spot

0 *Stripes*, wooden table, 100% Cotton print from *Liberty's* 'Indian Stripes' collection, glue.

IT BEGAN WITH AN ANECDOTE OF A TIGER IN A BILLIARD ROOM (Fig. 0).[1] This was the first of my many affect troubles, occurring decades ago during my doctoral studies. Back then, though I did not use the term, it engrossed and frustrated me in equal measure. Every piece of writing was a painful process of trying to find the right words, the right arguments, the right theories to talk about this—affect—a still unnamed term that had not yet landed properly in architectural discourse though it was always central to the discipline's creative practice. My writing was persistently sabotaged by detours and slowness; a constant looping back to retrace missed nuances, or prematurely dismissed material, and the frustration of crucial evidence not fitting an overarching theory. In retrospect, these roadblocks were hallmarks of working through affect.

Getting to this book involved constant figuring out of why affect sticks, what it means, how it is produced, how it circulates, how it works and why it is so difficult to pin down. For a long time, hunting down affect's unresolved and circuitous definitions became my obsession. This quest for clarity was incapacitated by affect's complex range. The thematic diversity, relevance and application of affect spans a multitude of fields ranging from the sciences to the humanities. Concepts associated with affect read like a litany of mixed metaphors: 'assemblage, flow, turbulence, emergence, becoming, compossibility, relationality, the machinic, the inventive, the event, the virtual, temporality, autopoiesis, heterogeneity and the informational'.[2] I realised that any attempt at a general definition of affect theory was futile. Instead, affect comes into its own when it is played out as 'a dimension of an event'[3]—demonstrating in that arrangement changefulness and a sense of vitality, of being connected to and being in the world, especially when one is 'drawn in by a situation, captured by its eventfulness, rather than you capturing it'.[4] Sticking to Stripes, the tiger, I was not just shown another route into the architectural history and theory of the billiard room, I was also shown another way of seeing and thinking.

'Affect' derives from the Latin noun *affectus* (something having been affected or influenced) and its verb *afficere* (to affect or to be affected by). Affect is a potential carried by an event or an encounter. Affect, in this book, refers to the visceral and experiential exchanges that happen between subjects in encounters with other subjects or objects in a spatial setting. Architecture, pertains here to a written discourse about the built environment and its altered landscapes. While quite a few studies in architecture speak about how the architecture itself, by way of design, singularly produce affects (which loosely map to 'affect as atmosphere'), affect, as it is tracked in this book, is produced and transmitted relationally,[5] in the joint configuration that architecture finds itself—populated by things, humans and non-humans.

To place architecture next to affect is to ask whether there is a new way to think the political[6] in the former, specifically here how architectural histories and theories might be wrested from their biases to engage subjects in the social imaginary excluded from architecture's archival and published records. Reciprocally, to think with architecture is to accord affect theory what it misses—specificity of physical concreteness and materiality. This pairing produces a hinge: An infrastructure of thinking and disposition for writing about architecture as a subject mediated by, and emerging through, affective encounters, *and reciprocally*, an infrastructure of thinking and disposition for writing about affect as a subject mediated by, and emerging through, architectural encounters.[7]

There are many obstacles in this pairing. My approach to affect in architecture is unified by attunement to the body and the embodied. Yet, this familiar analytic category also started to break down since 'body' consequently included animals, non-humans, 'species bodies, psychic bodies, machinic bodies and other-worldly bodies'.[8] Similarly, to write architecture following the trail of affective evidence means the research is immersive and 'deterritorialising', that is to say, one is dislodged from certainties and limits—physical, intellectual, imaginative—and compelled to venture beyond a 'cognitive communication of meaning'.[9] While commonplace to creative work in the design studio, the power of deterritorialisation greatly diminishes with the relentless justification demanded in academic writing. To attempt a reading of architecture through diverse corporealities requires what media and cultural theorist Anna Gibbs describes as a psycho-somatic shift. Gibbs argues that the affective genre of fictocriticism challenges bodily limits and assumptions, such that to be able to partake in that kind of writing one must be willing 'to lend one's body to the words of another, to be—albeit temporarily—possessed by alien affects and to put ourselves at risk of being transformed by them'.[10] This is the risk of being implicated. Architecture through affect is implicated in the world—a position at odds with its assumed autonomy, of architecture being made for the sake of itself.

Similarly, affect is not without its own blindspots. During the late twentieth-century 'turn to affect', regulated conceptions of 'form' and 'representation' were superseded by more fluid notions of sensation, materiality and embodiment. Film theorist Eugenie Brinkema argues that affect is repeatedly (ill) defined as 'what undoes, what unsettles, that thing [one] cannot name, what remains resistant, far away (haunting, and so ever beautiful); indefinable, it is said to be what cannot be written', with affect's fleetingness traded down for its generic workings through 'pressures, forces, intensities'.[11] Yet, in order to access the locations where affect precipitates particularly for aesthetic constructs such as literature, art, film and architecture, specific forms and genres of representation require negotiation. Arguing for close readings of 'the formal dimension of affect' to allow for the density of details and specificity as well as to recover 'the minor, inconsequential, secret, atomic',[12] Brinkema asks that we 'let affect press back on theory'. Thus, affect might be treated 'as a problematic of structure, form, and aesthetics'[13] with affective encounters subject to theoretical scrutiny, and vice versa.

Affect moves us; affect is concerned with troublesome things that stick to us. My reading of affect is influenced by feminist philosophy, where 'affect' operates as an aestheticised and politicised analytic frame accentuating a populated, embodied and relational construction of architectural space. 'The point of thinking with affect is to think through our implication in relational fields, and the potential we might find there'.[14] Philosopher Brian Massumi argues that in this way, in the self-consciousness of being implicated, affect makes demands on ethics—there is a consequence to how we choose to think and act in response to each situation, 'Ethics is about how we inhabit uncertainty, together'.[15] To attend to the affective encounter, as this book demonstrates, is to also attend to the ethical.

Architecture and Affect is motivated by two questions: Why do dismissed affective leftovers persistently haunt, or transmit, as troubling experience or imaginaries? And, what would it mean for architecture to assemble such discrepant evidence into its discourse? Beyond accounting for the vivid but excluded, I develop a frame and a disposition for thinking critically about, speculatively through, and being grounded by, encounter. I seek to return architecture to a location where the relations of production—historical, political, biological, social, ecological, financial and cultural—are re-enacted, in other words, to return architectural discourse to where worlding happens. Lastly, I recognise architecture as a site where 'structures of feeling'[16] reside; that is, architecture's capacity to confront the symptoms of a neoliberal, neocapitalist, posthuman and neoimperialist world rife with anxiety, helplessness, resignation, agitation, shame and anger. A record of architecture, after affect, could offer insights into how people belong, how they experience attachment, manage anxiety, hold up hope and achieve security. These trajectories saturate the relationships, histories and theories in *Architecture and Affect*.•

ACKNOWLEDGEMENTS

This book was seeded long ago when I was still a student in London. Compelled to focus on Singapore-based fieldwork then, I experienced firsthand academic resistance to affective encounter. This formidable occasion spurred on *Architecture and Affect*. Over one-and-a-half decades, I started on and finished other projects, putting this one on the backburner. My determination to pull the unwieldy affective threads together into a book was really only sustained (and many times revived) by conversations, emails, Whatsapp messages, year-end greetings, and recently, zoom calls, with wonderful colleagues from all over the world.

I am deeply grateful to Barbara Penner and Jane Rendell who saw the beginnings of this project—the intransigent tiger—and did not shut it down. Jane's ethical position continues to guide mine, while Barbara's intellectual generosity and friendship fuel my commitment to an academic's ofttimes treacherous course. I am indebted to Jane M. Jacobs, a trusted colleague, invaluable mentor, and honest critic. Anoma Pieris remains a benchmark of what a researcher and academic should be. These four persons will be aware of the turning points when their counsel held me steady. They are model teachers and academics, whose dispositions I strive to mirror and carry into my own professional life.

Various others have contributed by their friendship and collegiality: the *Insurgent Domesticies* crew under Anooradha Iyer Siddiqi; conversations at the Asia Research Institute with Chua Beng-Huat, Mike Douglass, and Tim Bunnell; *The Good Life* Public Lecture Series, Berlage Institute with Salomon Frausto; numerous presentations at the Future Cities Lab-ETH Singapore Centre and ETH Zurich hosted by Stephen Cairns, Phillip Ursprung, Marc Angélil, Charlotte Malterre-Barthes and Ben Leclair; Murray Fraser at the *World of Architectural History* conference; Lloyd Pratt, Nancy Armstrong, Ankhee Mukerji and Helen Small at the *NOVEL* Colloquium in St. John's College, Oxford; Hélène Frichot and Helen Runting at the University of Melbourne for the High-Rise Colloquium; Ahmad Mashadi and the curators at NUS Museum; Ute Meta Bauer at NTU CCA, Gulsum Baydar, Iain Borden, Lori Brown, Karen Burns, Hugh Campbell, Emma Cheatle, Finbarr Fallon, Jennifer Ferng, Colin Fournier, Penelope Haralambidou, Hilde Heynen, Jonathan Hill, Ho Puay Peng, Erik G. L'Heureux, Thomas Kong, Constance Lau, Andrew Leach, Lee Kah-Wee, CJ Lim, the late William S.W. Lim, Lesley Lokko, Matthew Mindrup, Ong Ker-Shing, Natalie Pang, Julieanna Preston, Peg Rawes, Charles Rice, Eunice Seng, Shabbir Hussain Mustafa, Chi Yin Sim, Naomi Stead, Igea Troiani, Katie Lloyd Thomas, and Audrey Yue, and my colleagues at Architecture, College of Design and Engineering (CDE) at the National University of Singapore (NUS). Lei Yuan Bin's partnership in the making of *03-FLATS* sharpened my conviction about architecture's interdisciplinary potential. Simryn Gill and Charles Lim's works and friendship push me to think, write, and make things shimmer through affect.

A disposition towards affect influenced my intellectual perspectives and shaped my teaching and research agendas at the Architecture, NUS. NUS provided me with two MOE Academic Research Tier 1 grants which allowed me to work on the essay film *03-FLATS* and develop research for this book. Another Humanities and Social Sciences Seed grant enabled *Situating Domesticities*, realising my longstanding interest in domesticity and establishing an international collective of researchers working collaboratively in the field. The inter-institutional University of Sydney-NUS Partnership Collaboration Awards brought international voices to discuss architecture as a remote practice. I was given a half-year sabbatical for research, spent at FCL-ETH Singapore and The Bartlett School of Architecture, UCL, for which I remain grateful. Undergraduate, masters and postgraduate students from Architecture, CDE NUS have listened to and worked with many of the theoretical themes and subjects that make up this book in lectures and design studios. The students shared incredible energy, enthusiasm, and creativity. I learnt much from teaching and working with them.

I am grateful to the patience and trust of publishers and editors who commissioned and worked on this book. Val Rose, who first saw the faint outline and took a gamble on it. From Routledge, Grace Harrison, Fran Ford and Caroline Church whose generosity enabled me to finish what I started, and the assistance of Aoife McGrath, Trudy Varcianna and Varun Gopal on the editorial work. To Alanna Donaldson for turning the manuscript into print with speed and precision, thank you. I am indebted to Barbara Penner, a generous intelocutor who was especially alive to the ambit of this project.

The research, the drawings, the thinking, the editing—these suffer without support. Lin Derong's drawings and photographs make the affective encounter come alive in these pages; he understood what I was sensing and thinking even before these became words on a page. His creative labour animates *Architecture and Affect*. Tan Yi-Ern Samuel's intelligent feedback, robust counter arguments and finely tuned edits sharpened the chapters considerably. Nicholas Lua, Serene Ng, Jeremy Seah, Rachel Sim, Sherilyn Yeo and Anthea Phua all worked tirelessly through different stages of the project. The making of *The Third Archives* relied on the generosity, collaboration and talent of three accomplished designers—Lin Derong, Toby Fong and Wong Zi Hao.

My family were not exempt from dinner table confessions of affect anxiety. Peter Sim is still my first and last reader, and always, my best friend. Ailian Chee is my kick-ass reality checker, and source of sustenance. Without them, this book would not be. I am grateful to my mother Ching Choo and my aunt Kim in Penang for their fervent prayers and unreserved belief. My helper Cecil Asong, my righthand at home for almost a decade now, made it possible for me to write, on and off, in relative solitude. Ginger, Mojo, Felix, Horatio, Luna, and Monty's exuberant influences kept me going; their optimism for life bounce off these pages.

Amidst all these, Joan Sim and George Sim came into our lives. Being with them daily teaches me why affect matters. I dedicate this book to them.

•••

A brief section of the Introduction was presented at the *Remote Practices* symposium, NUS-University of Sydney, October 2020, and previously published as 'Amidst…: Afterimage, Affect, Architecture', in *Remote Practices: Architecture at a Distance*, edited by Matthew Mindrup and Lilian Chee, 137–143. London: Lund Humphries, 2022.

A version of Chapter 1 was presented as 'Domesticating the Tropics: Ruler, ruled and the unruly' at the *Architectures of the Novel*, St John's College, University of Oxford, UK, June 2019. An earlier version of a section of Chapter 1 was previously published as 'Under the billiard table: Animality, anecdote and the tiger's subversive significance at the Raffles Hotel', in *Singapore Journal of Tropical Geography* 32, no. 3 (2011): 350–364. © 2011 Lilian Chee. *Singapore Journal of Tropical Geography* © 2011 Department of Geography, National University of Singapore and Blackwell Publishing Asia Pty Ltd.

An earlier version of Chapter 3 was previously published as 'After the Last Train: Narrating the Tentative Monument through Simryn Gill's Tanjong Pagar Railway Station', in *Proceedings of the Society of Architectural Historians, Australia and New Zealand: 33, Gold*, edited by AnnMarie Brennan and Philip Goad, 122–132. Melbourne: SAHANZ, 2016. An earlier version of the chapter was presented at the Centre for Contemporary Art (CCA-NTU) Singapore, Public Programmes *Detour* Series in conjunction with *Simryn Gill: Hugging the Shore* exhibition, April 2015.

Parts of Chapter 4 were presented at the *High-Rise Colloquium*, Melbourne Design Week, March 2022. An earlier version of this chapter was previously published as 'Keeping cats, hoarding things: domestic situations in the public spaces of the Singaporean housing block', in *The Journal of Architecture* 22, no. 6 (2017): 1041–1065. © The Journal of Architecture, reprinted by permission of Taylor & Francis Ltd, http://www.tandfonline.com on behalf of The Journal of Architecture.

Earlier versions and parts of Chapter 6 were presented at: *Behind Closed Doors: The secret life of home in Singapore*, ARI Asia Trends, The National Library Singapore, April 2018; *Spaces of Transition: Globalisation, transnationalism and urban change in the Asia-Pacific*, University of Melbourne, July 2016; Yale-NUS Urban Speaker Series, March 2016; Asia Research Institute-FASS Cities Cluster of the Faculty of Arts and Social Sciences, NUS, May 2015; Bartlett School of

Architecture, UCL, London, February 2015; *The Good Life* Public Lecture Series, Berlage Institute, Delft University, Delft, February 2015. College of Design and Engineering (CDE)

Earlier versions of Chapter 7 were presented at: *A World of Architectural History* international conference, Bartlett, UCL, London, November 2018; *Urban Nature* Seminar Programme, FCL-ETH, Singapore, June 2016.

Earlier versions and parts of Chapter 9 were presented at: Society of Architectural and Urban Historians of Asia *Interdisciplinarity, critique and risk: New directions in urban and architectural histories in and of Asia,* August 2020; SAHANZ *Distance Looks Back,* University of Sydney, July 2019; *Production Sites: Resituating the Culture of Architectural Knowledge,* University of Sydney, Australia, October 2015.

INTRODUCTION

Knowing Otherwise

Architecture after affect

I. Groping for meaning: *Interpreting encounters*

> Affect's potential is in pointing out the non-questions and non-answers at play in any theory. Treating affect as a form is another way of demanding that we read for and speculate on these non-answers.[1]

SOMEWHERE IN ANTHROPOLOGIST MICHAEL TAUSSIG'S FIELDWORK NOTES, there is a drawing of a woman sewing herself and a man, into a white nylon potato bag.[2] They are seen at the entrance to a highway tunnel in Medellin, Colombia as Taussig's taxi sped past the scene. It turns out that the figures creating a makeshift home were displaced peasants involved in the region's unravelling territorial conflict with the paramilitary.

Titled *I Swear I Saw This*, the drawing is a vivid afterimage. It behaves like a talisman, a piece of magical evidence resulting from Taussig's vital witnessing. The drawing conveys a range of possibilities from the concrete to the speculative. It provokes questions. It discloses contrarian routes into unpicking familiar historical knowledge. Yet this compelling 'evidence' is problematic. Placed within a prevailing research paradigm that demands distance between purposeful knowledge and mere coincidence, the inchoate questions prompted by two figures disappearing into a nylon bag amidst a traffic slipstream are curtailed by the encounter's happenstance and transience. 'How can that encountered connectability be conveyed so that it does not allow itself to be weighed down by the history of a place, or contained by its identity, or tethered to the official belongings attached to these?'[3]

Taussig's after-event drawing details an in-process situation, capturing its urgency as well as the anthropologist's uncertainty and disbelief. The drawing pauses at the point where something is about to, or has just, happened. It holds potential—one can ask of it questions of what, why and how. Yet the dilemma here is that the academic's curiosity may not, as Taussig tells us, be settled with his transcription of the poignant event, or translation of this drawing, into words. Why not? Taussig retorts that academic discourse or 'agribusiness writing'—'a mode of production ... that conceals the means of production'[4]—is beholden to questions which arrive at the self-same conclusions because academic discourse 'cannot estrange the known, that with which it works, its itselfness'.[5]

The point he makes is that something 'other' happens if we recount occurrences without embellishment or prior assumptions, trying to make sense of what we encounter but cannot fathom, and so sticking to 'only what we *know*, without adding anything, and the satisfaction we are trying to get from the explanation comes of itself'.[6] The discourse deriving from the Medellin encounter, in other words, produces a different kind of knowing if the evidence of the couple in that potato bag is closely explicated. What is known here leaves more unknown, and likely even more, unknowable. The drawing seizes 'a seeing that doubts itself', capturing a tentative but tantalizing moment of knowing.[7] Taussig differentiates 'knowing' as 'alive', tenuous, it is what compels him to 'grope for meaning when you least expect it in a taxi rushing into a tunnel'.[8] He holds this kind of knowing apart from knowledge, which would likely be an 'inert record', conferring the knower with seeming certainty.[9] Taussig values the drawing's spontaneity—its ability to preserve the immediacy of the encounter, to point to where things are still unfolding. This potentiality of the incomplete moment of encounter, if formative to the academic process of 'writing-up', could influence the trajectory of the academic argument, more in alignment with the tenuous scaffolds of such contingent knowledge: 'This is not a plea for exact reproduction of the fieldwork notebook', Taussig adds, 'but rather a plea for following its furtive forms and mix of private and public in what can only be called, as in cinema, a "dissolve" or "fade out" that captures ephemeral realities, the check and bluff of life'.[10]

Furtive and implicated, the afterimage presents an insistent 'third meaning',[11] a concept borrowed from literary theorist Roland Barthes' reading of the film image. Barthes argues that apart from what is given as information and what is symbolic, there is, in arresting visual moments, a 'third meaning' that is 'evident, erratic and obstinate'.[12] Comparing the third meaning's 'obtuse' character to the first and second's 'obvious' meanings—symbolic and cultural—Barthes claims that the obtuse meaning extends beyond the analytical, finding itself 'on the side of the carnival'.[13] The third meaning is transgressive and unexpected: it penetrates into the spectator; it presses against the other two meanings to make them quiver, hesitate. Yet this obtuse third does not readily offer a meaning of its own. It compels a search, a chase, or as Taussig says, it imposes a tendency to grope around. Requiring intimate investment in the encounter, the risk of such investment is a reconfigured territory of discourse as one moves from the distant yet familiar perspective towards a near yet unknown one:

> The obtuse meaning is a signifier without a signified, hence the difficulty in naming it. My reading remains suspended between the image and its description, between definition and approximation. If the obtuse meaning cannot be described, that is because, in contrast to the obvious meaning, it does not copy anything—how do you describe something that does not represent anything? ... [I]n short, what the obtuse meaning disturbs, sterilises, is ... criticism.[14]

The third meaning, Taussig insists, is not a meaning at all. It is really a gap created in the struggle of interpretation.[15] The third meaning occurs where the seams of discursive interpretation begin to come apart. In straining to know, one must negotiate the divided realms of knowledge production—the collective and the individual; the abstract and the particular; the analytical and the intuited—because the third meaning envelopes, exerts, infuses, '... you are, as it were, inside it'.[16] The researcher is emplaced in relation to her subject, and this relationship becomes epistemological. So, although it hit him 'first and foremost', Barthes cannot make up for it in explanation, description, or identification: 'I cannot name that which pricks me'.[17] Locating this obtuse meaning in the film still, on which Barthes also assigns a 'filmic' quality, the third meaning draws us in because it triggers a sense of something astray, it points to a 'permutational framework of unfolding'—allowing the reader to connect the still beyond its located frame and sequence. The still, Barthes emphasises, is not a subproduct of the film. It is 'the inside of the fragment', the fragment here is akin to a 'quotation'—the still provides a 'second text' that is related to, but not the same as the film.[18] The filmic quality of the still encapsulates in one instant everything that is important to say. And unlike the film which follows sequential time, the still can be perceived in one's own time, it can be read in context with the film or in context with one's life. It can be read in passing or in slow burn. It can be scrutinised attentively or seen in distraction.

Lingering on as afterimage and talisman, the drawing and the still hold residues of affect. The affective encounter punctures the hierarchical order of a 'proper and automatic relationship between thinking subject, concept, and world'.[19] It compels before one might emote why the encounter cannot be ignored, of 'this-worldly things with the unimperative allure of happenstance that softly, almost surreptitiously, draw(s) one in'.[20] The scene pulls one into a field of connections and relationships. These must be examined in reverse and criss-cross order, from aspects closest to the encounter leading up to its furthest points. The encounter inscribes its own geography; it indicates a constellation of possibly unremarked relations, their possible histories *and* projective prospects. Taussig's drawing beckons with its obtuse third meaning. Its allure lies in the connection made between the strangeness of the couple sewn into bag-as-home being a residue of the forest, which when replaced by the freeway and the paramilitaries, was obliterated from historical consciousness. Provoked by gestures and residues deposited in the sites, figures, and images of encounter, thought movements track the opaque relations of power, their invisible effects, blind spots and repressions.

Engaging the encounter in discourse without losing its immediacy is germane. The assumption here is that architecture is encountered circuitously, through people and things that are caught in its frame. Such knowing is mediated by peripheral subjects/objects, modes of occupancy and experience which sit outside architecture's disciplinary frame or its orbits of dissemination. Such knowing takes seriously the pull of serendipitous encounters. The accommodation of happenstance as a trajectory of inquiry requires what philosopher Brian Massumi calls a sense of an 'in-each-otherness'.[21] Massumi explains that we can only be affected by 'difference' if our concept of 'distinction'—how we see things as being the same or different from what we already know—is transformed from 'extensive distinction', the tendency to diminish differences between an 'inside' and an 'outside', towards a more nuanced and fine-grain 'modal distinction' which registers process, emergence and becoming.[22] The former privileges stability and identity while the latter is invested in action, in the happening of an event, thus being open to the logic of 'interfusion, intrication, envelopment'.[23] This is not to say that boundaries collapse and merge, rather the opposite happens: differences are intensified, they become more pervasive. As the 'boundary condition for the event' approaches the remarkable, the boundary will capture tensions or 'intensities'; it will hold tendencies and potentials.[24] The boundary condition for the event—the periphery as it were—constitute actions, events and processes which do not belong within, but intersect with, the subjects involved in that event.

Thus, the sense of 'in-each-otherness' focuses on relations between. The theory and practice of in-each-otherness is immanent: theory and practice are implicated in and thus, made through, the unfolding relations between subjects. Massumi calls this phenomenon an 'occurrent emergence'.[25] He emphasises that the anticipation of, and attention to, the contours of events are critical. An ethics of encounter—how the researcher is affected by what she discovers and how she affects that which she discovers through writing, making, thinking—compels her to care for those she encounters. Framing the practice of architecture as a 'creative ecology' connected to 'environment-worlds' binding anthropocenic relations, material politics, and radical action/thought, architectural theorist and philosopher Hélène Frichot likewise emphasises 'creativity' as constituting open-mindedness and generosity.[26] Acknowledging the concept-tools exchanged and learned from the extra-disciplinary practices, Frichot advocates a 'pedagogy of concept' which turns the discipline's peripheral practices into creative resources for architecture's praxis, to repair a depleted, posthuman landscape.

Attuning to emergent relations is not without risk. Anthropologist Kathleen Stewart describes the researcher's own analytic practice as an act of creative and mindful shadowing, where she must be willing to follow instinctually, sometimes even flying blind, to catch what is about to come:

> ...a kind of haptic description in which the analyst discovers her object of analysis by writing out its inhabited elements in a space and time. ... [t]o outline what [Nigel] Thrift... calls a geography of what happens—a speculative topography of the everyday sensibilities now consequential to living through things.[27]

This book is likewise concerned with a disposition that encounters architecture—its object of analysis—through fine-grained, tactile, passionate and embodied discovery. It pays close attention to architecture's surroundings—its speculative topography or boundary conditions—comprising compelling 'oblique events and background noises',[28] those aspects which draw someone towards, and into it, as the source of evidence and theory. It argues for a radical scholarship which questions the value of information from the institutionally designated, placing the former in dialogue with the socially encountered, and by conferring the latter the status of architectural evidence. It acknowledges the contradictorily imaginative and situated character of encountered knowledge as critical to embodying architectural history within a lived contemporary milieu. Yet, it should be said that the politics of encounter are not uncomplicated. The encounter throws together many threads, which at first glance seem disparate, not having a point and often, not making an argument especially in the distant and objective manner demanded of academic discourse. There is also the problem of the researcher's proximity, her closeness to the subject, of being implicated in the knowledge she produces. These issues are broached in the forthcoming chapters.

What is the epistemological outcome of allowing a two-way relation between the lived encounter with other histories/stories-in-the-making, and the building's sanctioned histories? Locating 'architecture' in the middle-ground, such that it comes into focus only through a situation, the book is constructed around this sense of 'in-each-otherness'. At every turn, it negotiates the encounter and writes its way towards an immanent discourse. It seeks to mend the spilt faced by someone who is constructing architectural histories and theories, influenced by the creative aspects of design, and invested in the lived historical present. It makes intelligible and critical to discourse, the knowing of architecture from intimate and intuitive encounters with subjects and objects drawn to architecture from elsewhere, and equally, to admit the manifold forms of such encounters as architecture's evidence. *Architecture and Affect* recuperates the immediacy of coming to architecture. It takes the affective encounter as key to architectural epistemology—thus reclaiming the distracted, instinctive and impulsive ways of encountering architecture as instrumental and relevant to architecture's discourse, its histories and its theories.

II. Detecting architectural histories: *Experience, intuition, evidence*

> ... is it not the ultimate betrayal to render stories as 'information' and not as stories? ... The next step in this betrayal is the instant translation of the story into a fact, or what is called 'data', and along with that the storyteller is translated as an 'informant; once these steps have been achieved... the philosophical character of the knowing is changed. The reach and imagination of the story is lost.[29]

> ... place is something that throws itself together in moments, things, in aesthetic sensations and affective changes.[30]

Can architectural discourse attentively follow untrained, spontaneous, ordinary modes of knowing? What becomes of architectural knowledge when a building, a space, or a landscape is approached, as it often happens in lived encounters, through a compelling detour—an event, or another subject located outside the discipline? How can architectural scholarship critically account for this detour, giving this mode of knowing a discursive space in-between, as opposed to being divorced from, architecture's traditional objects and their supporting archives? Attuning to such affective, lived encounters, how does knowing and knowledge of architecture change?

These questions—how we know what we know, why we know what we know and thus, what knowledge is produced or suppressed—preoccupy *Architecture and Affect*. They are couched within the subdisciplinary space of architectural history where a particular empiricist epistemology of designated building-related archives (not coincidentally serving as evidence) that scaffold traditional subjects of architecture (such as design lineages, architects' histories and aspirations, built environment histories) hold court. The object of study in this strand of architectural history is self-evident. Its empirical tradition is bound to what the discipline designates as legible and legitimate. Its subjects, argument, narratives and evidence retrace the architect's path and the built environment's development. Architectural history of this more conventional kind views scholarship as doing the work of expanding knowledge and/or reclaiming an archive. It approaches the object of architecture as distant, specialist and uncontaminated by instinct or affinity.

In contrast, architectural historians Barbara Penner and Charles Rice ask a fundamental question which matters to this book: What exactly is 'architectural' in the subdiscipline of architectural history? The question reveals an inherent assumption that 'architectural' presupposes particular subjects of study, while also not fully querying whether 'architectural' extends to specificity of methodologies and their accompanying forms of evidence. Penner and Rice opine that if 'the architectural object is no longer simply "there", waiting for the architectural historian to uncover it', then there must be an examination into other means and modes of evidence from which architecture can be 'detected'. They further entertain the possibility of whether it would be desirable to have an architectural history without architecture as its focus, detecting architecture instead through the evidence at hand:

> Does focussing on evidence itself as an object of inquiry open historical narrative to less expert, more exploratory accounts and perceptions? ... [T]hinking about detection in this way maintains the elusive positioning of architecture, its 'background effect', and makes this elusiveness central to the inquiry. This, we believe, is the real challenge of the shift of the last 20 years. It does not mean architectural historians should be interested in things other than architecture, but that their interest in architecture would be shifted by the alternate possibilities for what might constitute evidence, and the modes of detecting that they would imply.[31]

Architecture is visceral, visual, tangible and in many cases, physically accessible. The capacity of architecture to accommodate simultaneous temporal arcs of past, present and future in one space, offer possibilities of discovering, engaging and producing counter- or partial archives and narratives, thus changing the implication of the archive and its narrative(s) as perpetually becoming. Rather than annotate architecture's passing, Penner and Rice's prescient call for architectural history to embrace an active mode of 'detecting' transforms its potentials. The switch emphasises the role of architectural history as a methodology to do the work of thinking, making, viewing, understanding and intervening. This engagement goes beyond the architectural object. Architectural history in an active mode may be relevant to other disciplines: film, literature, visual art, cultural studies, geography, anthropology and ethnography, amongst others. Furthermore, the emphasis on the 'architectural' aspect of architectural history resonates with the visceral, visual, sensorial and psychological training in architectural design. The latter compels attunement to what is 'in the air': disciplinary skill to read patterns in-formation, and vicarious experience through another. These learned disciplinary attunements similarly challenge the disinterested distance touted by the conventional threads of architectural history.

Nevertheless, the admission of the visceral and the experiential within the realms of knowledge production need to be problematised. Can an embodied experience be both singular and shared? In *The Evidence of Experience*, feminist historian Joan W. Scott recounts an autobiographical narrative of one man's life-changing experience in a bathhouse in the 1960s.[32] The author Samuel Delany, a gay black man, speaks of his first public encounter with an 'undulating mass of naked male bodies' dimly lit in haunting blue light and 'spread wall to wall'.[33] Scott emphasises that one possible reading of this scene is authorised of Delany's 'vision' and his first-hand 'experience':

> Knowledge is gained through vision; vision is the direct apprehension of a world of *transparent* objects. In this conceptualization, the visible is privileged, writing is put at its service. Seeing is the origin of knowing. Writing is reproduction, transmission – the communication of knowledge gained through (visual, visceral) experience.[34]

Scott argues that the evidence of experience is problematic in orthodox historical research. Experience here is 'uncontestable' and 'originary' while remaining safely within the limits of a method which privileges the historian's 'vision' as a reliable source of evidence.[35] The concern lies in the authoritative and foundationalist discourse that experience might encourage, where the latter ignores the historian's 'position or situatedness' and its effects on the types of knowledge they produce.[36]

Yet Scott recognises that experience is too much a part of everyday life to be dismissed. As such, the construction of experience is 'that which we seek to explain'.[37] Through it, we understand how unevenness in class, race, gender alter experience, and thus knowledge. Scott advocates a reflexive attitude towards experience where its construction, coupled with recognition of how vision and language are shaped by privilege or paucity, should be accounted.[38] Echoing Gayatri Spivak, she challenges historians to problematise 'concepts and identities as historical events in need of explanation'.[39] Rather than producing an authoritative knower, experience may be taken as energetic occurrences or event-based encounters which set up relationships and opportunities for relating to others.[40] This experience of encounter is indeterminate since relationships are being enacted through detours and chance. They allow us to pick up situations-in-the-making. Cultural theorist Lauren Berlant points out that a situation is an event in-the-making that enters our consciousness:

> A situation is a state of things in which something that will perhaps matter is unfolding amid the usual activity of life. It is

> a state of animated and animating suspension which forces itself on consciousness, that produces a sense of the emergence of something in the present that may become an event.[41]

When we gravitate towards a situation, we intuit that something is unfolding in our midst. Or if the subject that galvanises our detour is historical, we intuit that whatever is past still persists, demanding our present attention and action. For Berlant and Massumi, the unfolding of a situation-soon-to-be-event approximates Scott's unforeclosed experience—these experiences require explanation; they are discursive rather than conclusive. The situation suspends 'contemporary historicity as one lives it'.[42]

Untrained modes of knowing have a lineage in affect theory. Berlant advances the role of intuition in 'sensual data-gathering', arguing that intuition, rather than being autonomic, is embodied and learnt through memory and life experiences—where 'affect meets history'.[43] Comparing the role of affect theory to ideology theory, she argues that a focus on affect gives insights to historical situations beyond an understanding of the disciplinary controls enacted by orthodox institutions and practices. Affective evidence squares up what is sensed against what is known or recognisable, it traces how a lived situation is negotiated between norms and individual desire:

> Laws, norms, and events shape imaginaries, but in the middle of the reproduction of life people make up modes of being and responding to the world that altogether constitute what gets called 'visceral response' and intuitive intelligence.[44]

Berlant asks what a historicism which takes affect seriously might have to consider outside, and in addition, to 'history's proper evidence' since this history is a history of the present, which is being lived through as much as it is being observed and written. In this uncertain condition where things at the cusp of becoming-event, and becoming-historical, she argues that intuitive intelligence and being attuned is a critical contact zone 'between affects and their historical contexts of activity'.[45] This commitment to knowing through strategic inexactness recalls Taussig's description of a groping around to make sense of what we encounter. Differentiated from 'technical empiricism', affective epistemologies grow from 'concrete', 'vernacular' and 'sensual' knowledge.[46]

The work that Berlant describes is not so much a proverbial return to context, but an attentiveness to the exact shape of the eventful encounter, to come close to its saturated details, and to situate such encounters as part of a historical continuum. Responding to affect's emphasis on immediacy and its relationship to historical framing, Massumi and philosopher-artist Erin Manning argue that affect prioritises the event 'as the primary unit of the real'.[47] Following philosopher Alfred Whitehead's idea of 're-enaction' or what Massumi calls 'reactivation', Massumi and Manning argue that we learn to sense from the residue of past events; picking up historical and contextual connections intuitively when encountering a present event. This process articulates the micropolitical act of thinking-feeling which operates in a tentative space of connections, transitions and co-presence.[48] Massumi describes micropolitics as abductive and enactive, in that it derives its momentum and reasoning from being in the in-between:

> The feeling of the transitional encounter is not 'raw' feeling. It is imbued with an immediate understanding of what is under way, what might be coming—and what we are becoming. This is enactive understanding: it is one with action. It is what I call a thinking-feeling. ... [I]t is clear that the affective thinking-feeling is not the thinking or feeling or a particular object—or a particular subject. It pertains more directly to the event, what passes in-between objects and subjects... [49]

Through the thinking-feeling of an affective event, the present relation to the past is 'direct, unmediated' and singular.[50] Massumi calls this present-past relation 'immediation': 'Immediation is the past bumping against the future in the present'.[51] An attention to affect thus allows the slowing down and precise tracking of a 'situation', a 'happening', 'getting the drift of things' which may eventually be registered as an event.[52] Affect allows us to trace the happening's resonance to the historical. The attunement to the affective situation creates the possibility of inhabiting the past in an embodied historical present. Historicisation attentive to affect is dialogical.

The identification of evidence in *Architecture and Affect* is key. Typically inadmissible to architectural discourse, evidence such as the anecdote and the situation, must be detected. Affective evidence produces micro-tears through the known and prompts a rethinking of the materials around which discourse is typically constructed. The affective encounter re-conceptualises the commonly held definition of 'primary source' material as one previously locked away in archives since 'primary source' here is *primary to*, such encounters, and thus accessible. It critiques the nature of architectural evidence couched in influential interests and identities. In place of such convenience, the work here engages the messiness of affective encounters where agency, experience and an embodied historical present coalesce.

The notion of affect being autonomous describes affect's resistance to being captured fully or expended completely in language, thought or representation.[53] Affect thus, leaves a surplus of unrealised potentials in the remaindered. Category leftovers and residues direct architectural discourse in the forthcoming chapters. Discrepant evidence surrounding a tiger, a ruin, hoarded objects, street cats, single women, ghosts and failing infrastructure reside outside the norms of architectural history. Discrepancies are tactical; they produce counter-narratives and train activism on 'the sly'.[54] Discrepant evidence question not just the object of architectural history but also its *modus operandi*—asking what is privileged as evidence, why, and where such alternate forms of evidence might lead us.

If affect, following the Spinozist definition, is the capacity to affect and to be affected, this book echoes Massumi's contention that power structures and ideologies are secondary and tertiary effects of affective encounters.[55] Where there is power to affect, power accumulates, and where power accumulates, ideology is expressed. In admitting the primacy of these affective encounters and their concomitant evidence into the rational framework of architectural discourse, this book shifts architectural history's disposition from one of habitual distance mediated by disciplinary norms, to one of critical proximity unsettled by the affective encounter.

III. Theorizing architectural discourse: *Advenience, structures of feeling, the broken middle, immanent critique*

> If the founding violence of the archive is obliteration, the founding truth of speculative and close narrative forms is that there is more, we might call it life, interiority, vision, imagination, desire that exceeds archival documentation and that this more is a legitimate subject of history and scholarly writing.[56]

Explaining the attractive force of the image, Roland Barthes introduces the term 'advenience' to describe the pull exerted on him by the image, '... it seemed that the best word to designate (temporarily) the attraction of certain photographs extend upon me was *advenience* or even *adventure*. This picture *advenes*, that one doesn't'.[57] 'Advenience', later superseded by *punctum*, derives from the Latin word *adventus* (Christian 'advent'), outlining a nascent aesthetics and politics. 'Advenience' means to come to, to reach or to arrive at, to be called by, a poignant something. Of interest here is Barthes' description of the draw or force of the image, or in the case of this book, the lure of the image as evidence.

In his reading of Barthes, political and cultural theorist Davide Panagia locates advenience within the realm of sensorial aesthetics, where the image or visual object wields a disruptive force that disorganises the mind and body: 'By "advenience" I refer to the capacity of things to stand forth and affront the spectator and, through this bodying forth, to strike at one's perceptual milieu'.[58]

Yet advenience 'emerges in the interstice between sensation and reference',[59] forcing us to regard its appearance (or its arrival) but also leaving us unable to explain why it haunts. Panagia describes this arresting experience as necessarily elusive to our cognitive grasp, thus fuelling a curiosity or desire to scratch at the impression which sticks. Advenience holds us to that event. We recognise it as unsettling, agitating, distracting; we allow that which advenes to intrude upon our space and our bodies, and in beholding it, the encounter points us towards knowing otherwise:

> ... the advenience of an appearance is a political event not because it is meaningful but because it acts on our perceptual competencies and invites a turning of our attentions and a reconfiguration of those correspondences that mediate our worldly interactions.[60]

Barthes emphasises the photograph's retention of an 'irreducible' affect such that on seeing it, the viewer is assailed by emotion: 'desire, repulsion, nostalgia, euphoria'.[61] 'I see, I feel, hence I notice, I observe, and I think'.[62] Thinking in feeling. The focus of the image is given by the 'studium' which captures sociocultural and/or political themes.[63] The studium's meaning is unequivocal and economical—these exist without duality, indirection or disturbance.[64] In contrast, the 'punctum' which gives the photograph its advenience, is an unintentional detail, operating in excess of the 'studium'. The puntum is disruptive and subversive—it makes the viewer doubt, it beckons thought.[65] The 'punctum' functions as 'a supplement ... at once inevitable and delightful', escaping the control of the photographer who cannot '*not* photograph the partial object at the same time as the total object'.[66]

The punctum in the photograph, or image, exerts an embodied relation to its reader; it wounds, pricks, or stings them. They may discover additional meanings behind that image.[67] It is the reader who adds to the 'punctum' by making sense of it, but it is not something she construes since the 'punctum' is co-existent with the 'studium'. 'It is what I add to', Barthes says, '... and *what is nonetheless already there*'.[68] For Jacques Derrida, the punctum has a contrapuntal relation to the studium—it introduces a 'polyphonic rhythm' of other possible meanings through an inclusion of 'the voice of the other', releasing what is obscured or repressed.[69] The punctum's discursive formation constitutes a 'blind field',[70] unlike the all-seeing, authoritative historical premise set up by the studium. Hence, to engage with its latent meanings, the punctum shifts the reader's perceptions by referencing a 'secondary, untimely object', which 'manages to half conceal, delay or distract'[71] our habitual modes of knowing.

The punctum is a desirous surplus element which fractures status quo meaning. It detours by pulling in stories, subjects, contradictions and discrepancies otherwise dismissed from normative perspectives. The punctum provokes the third or obtuse meaning. But Barthes is adamant that the obtuse meanings of the punctum, just like its signifier, is contingent upon encounter, 'My reading remains suspended between the image and its description, between definition and approximation'.[72] For the structuralist-turned-poststructuralist Barthes, the punctum and its obtuse meanings 'outplay meaning', seeding instead a provocation to produce a circumstance-dependent and uncertain 'counter-narrative' that contrasts with the immovable studium.[73] In his opening section to the methods later set out in *Camera Lucida*, Barthes declares his intentionality to theorise photography but at the same time preserve affect, keep alive the 'pathos' of *images* and most importantly, explore photography not as a question or a theme 'but as a wound'.[74] The outcome of this 'wound' for which Barthes lays out an approximate anatomy, but also calls a 'compromise' to the 'logic' of method, is affect.

Affect in *Architecture and Affect*, is a worldly, embodied, relational intensity—a pre-conscious sensing of potential and non-cognitive sense of 'something's happening' borne by encounter. Affect is 'what sticks, or what sustains or preserves the connection between ideas, values and objects'.[75] Affect is not autonomous to architecture, it is not contained in architecture alone because it needs to transmit between things, people, spaces. Sensed when architecture is eventful and framed in-the-midst-of the event, affect is carried by architecture in the latter's relation to something else. 'To experience an object [or architecture] as being affective or sensational is to be directed not only toward an object', argues feminist theorist Sara Ahmed, 'but to "whatever" is around that object which includes what is behind the object, the conditions for its arrival'.[76] This relational configuration of in-between-ness and beside-ness emphasises the impinging experience, or citing Barthes, the advenience of something, someone, somewhere commanding attention, and magnetizing us into its ambit. The affective encounter transpires in how the detail in a photograph sets Barthes off on an adventure, and how the couple in the potato bag sends Taussig scrambling for meaning. In both interpretations, affect relies on the scaffolds of a relational capture-surrender, for its structure and its form. This is also how I approach affect in this book.

Affect is not inert. Leaving its 'refrains' on human and non-human bodies which in turn intersect with landscapes and institutions, affect accrues in atmospheres which are co-compositional and co-constitutive of competing subjects whose presences overlap across competing temporalities.[77] Anthropologists Shanti Sumartojo and Sarah Pink argue that rather than coherent or staged, atmosphere emerges through the ways we live, act, and react to being in the world.[78] It follows that atmospheres are political—they trace where regions of power are located, and atmospheres are often operationalised as powerful instruments of public opinion, whose powers are especially potent when factual evidence is lacking.[79] Atmospheres of fear, solidarity, security, and exclusion at national borders, for example, incite acts of self-surveillance, censure and judgement on what is right or wrong. Yet, I find the vagueness of atmosphere compounding further the vagueness of affect. This inadequacy led me towards other affective theorisations of a shared milieu.

'Structures of feeling' was first introduced within literary and film criticism in Raymond Williams' *A Preface to Film* (1954). Williams argued that the intricate formations of taste and feeling were important cultural indicators of a historical milieu. Structures of feeling is dialectical—it bridges the personal and the historical. The concept articulates Williams' ongoing focus on how to capture the tension between the fixed and formal monoliths encompassed by terms like 'society' and 'culture', contrasting these with the receding forms of lived experiences in an everyday present context.[80] It captures the dualities of thinking felt, or feelings thought, derived from a specific milieu—of a place, time, people—where and when a particular culture unfolds. Detecting the lived residues of an age in art and in the novel, 'structures of feeling' acknowledges the significance of 'pattern(s) of impulses, restraints, tones'[81] within the hegemonic, and the need to separate details of the textured particular from the broad erasing strokes of a general milieu.

Williams outlines the importance of maintaining 'an unease, a stress, a displacement, a latency' between one's present and the milieu studied.[82] Significantly, the most effective site to test the concept, Williams points, is in discourse, since ideology is put under pressure when confronted with the 'discordant elements in exemplary personal experience'.[83] The strategic vagueness of Williams' theory has the effect of delaying conclusive analysis, insisting instead on thick descriptions of unfolding lived and felt relations. Committed to 'radical empiricism', cultural theorist Ben Highmore argues that Williams' corporeally inflected concept is grounded in our spontaneous acts of world-making.[84] 'Structures of feeling' loosens structuralist logic with the interruption of experience such that the often separate categories of experience and meaning converge in analysis. 'The suggestive, provisional, even vague quality of the formulation is in fact its virtue. It enables Williams to access an area of uncertainty, interest and inarticulacy...'[85] Resonating with Scott's call to historicise experience, 'structures of feeling' implies an emerging cultural armature onto which fleeting experiences and feelings might be attached, bringing ephemeral but recurring moments into public and discursive consciousness. Together with Barthes' figuration of the wound, William's schema folds the question of relationality: Where are we located in relation to our structures of knowledge?

Psychoanalytic readings see affect as irreducible to language, and thus glimpsed only in the spaces of non-correspondence or what literary theorist Isobel Armstrong calls 'the broken middle', a space held between representation/ language (writing, drawing, film, sculpture) and affect.[86] Although the understanding of affect referenced in this study is not undergirded by psychoanalytic theory, I want to hold onto Armstrong's concept of the broken middle because the middle also sets out the epistemological bearings for this project. The broken middle—where interpretation fails—is where Armstrong locates 'cognitive affect' or 'thinking affect', a disposition that goes back-and-forth between analytical and creative realms. 'Affect precedes representation in creation; it follows representation in reception'.[87] If we think of architectural history and theory as the practice of creation while our encounters with architecture are points of reception, affect becomes the mediating in-between (dis)connecting architecture with its discourse. The significance of the broken middle is because it holds together 'elements of detour and conflict', these being tied to how things are perceived and meanings are made.[88] In other words, the broken middle gives us insights into how we perceive:

> Affect itself is an ambiguous, alternating force. It moves between the destruction of representation, opening up an abyss in consciousness by violently breaking the barrier of repression, and appropriating, thieving, representation. It belongs to a chain of discourse and breaks it: it alternates between being bound and unbound, attached to signification and rupturing it. … The concealing and revealing, exposing and masking process which belongs to affect is structurally tied to the possibility of meaning.[89]

Armstrong's project seeks an intellectual model for figuring out the epistemic possibility of a thinking affect. She wants to know what affect can do for the critique of aesthetic work (in her case, for literature, art and film), whilst recognising the limits of discursive language and its ideologies. Beyond reading the work with a precision towards feeling-thinking, she is asking where this close reading might carry us. Her ambition is that cognitive affect moves us beyond the frame of the aesthetic work, such that what we experience but still have no words to describe—'the prosody of the gap, the blank space, articulation through the pause, the moment of void'—might be, though uncertain and discomforting, a space where discourse can tarry a little longer. To write in this gap is to create 'the conditions of "epistemophilia" ([Melanie] Klein's term), a longing to know and discover, a primal curiosity which is both a drive and a mental experience, where learning is indivisible from the passions'.[90] Recall Taussig's couple in the potato bag, and Barthes' urge towards adventure because an image advenes. Passion—the curiosity and hunger to know-with, to know-through, to be led by rather than to be in control of knowledge. In other words, to start in the middle of—for Taussig the potato bag at the mouth of the tunnel, for Barthes the wound rendered by the image.

To think and write amidst things, people, the world around us; to move back and forth between architecture and something else, and something other without the assurance of control—this is a paradigmatic shift for disciplinary knowledge which compels the forthright staking of disciplinary boundaries. For Stewart, to stay 'in the middle of things' requires one to be both 'speculative and concrete', looking inwards but also out and forwards, to intertwine thought and self, subject and world.[91] Being in the middle of, capitulates to relationality. To be distinguished from another location—the margins populated by polemicists and dissenters—the middle signals company, connection and affiliation, as opposed to animosity and separation associated with the margins. To be in the middle and close to, rather than to be distanced and overlooking. To think, theorise, write amidst others, means to work the ruins of critique where subjects may be unrecognisable or unremarkable to the discipline. Often this middle requires intervention, creativity and invention to bring subjects into architecture's orbit. And because the middle is where things are unravelling, it is also a space of the more-than, a position that tends 'to what surpasses the frame, *the something else* and *the what-might-be*'.[92]

To do research 'amidst' means anticipating outcomes that are more 'fractious, multiplicitous and unpredictable'[93] because the affective world is not already 'finished'; it is not the academic version of a readymade space ripe for conclusive analysis. Composed of multiple residual situations which may coalesce to compose interconnected events—big or small, momentous or banal, in 'the world which affect proposed',[94] analytic critique needs to be slowed down, routed to swerve away from the script which seeks a big underlying structure, avoiding trite conclusions and 'bottom-line arguments' simply because these underlying explanations could demolish the very thing which piques our curiosity, which could also be the thing that lures us out of our disciplinary blind spot.[95] Stewart uses the term 'attuned' frequently. To be attuned is to be attentive, to be willing to adapt, accommodate, adjust, acclimatise, rather than foreclose, or master. Contrary to academic ambition, the outcome of affective thinking-feeling is not authoritative knowledge. Rather, its purpose is 'to fashion some form of address … adequate to [the] form [of the encounter]'.[96] The method is situationally attuned, thus again, really an anti-method, where one is obliged to create a clearing for addressing 'the complex and uncertain objects that fascinate, [that] … literally hit us or exert a pull on us'[97]—back to Taussig's potato bag moment.

So, we arrive at the mode of address, how we tell. Often perceived as an aesthetic, it is instead a radical telling that starts from a broken middle. This narrative sits knowingly as that which is being 'made up' because there is no other way that these fragments can be told except through passion and speculation. Narrative thus is not *about* something but writing *to* or *from* a position removed from architecture. Writing through affect is dynamic and becoming, moving *from* somewhere or moving *to* architecture. Writing from the middle is thus, more-than argument, less-than authority; it is what Sara Ahmed calls 'writing as where', a writing conscious of its limits, biases, influences, reach, and one that is committed to that which unsettles.[98] Writing from the middle reorientates subjects, discourse and theories.

'Writing as where' is intimate and unsettling. Challenging the traditional structure of the historical academic text, the intimate narrative always needs to find its way; it needs to take detours around that which exerts a 'drawing-in-ness'[99] to first sketch out, and then make sense of, relationships unfamiliar to disciplinary hierarchies, which is to say that architecture is acknowledged after, or through the encounter, not prior to it. In this mode of narration, the narrator must negotiate between her extra-disciplinary evidence with the new knowledge it potentially constructs, and the gatekeeping tendencies of disciplinary methods and concepts. The taking of the detour sidesteps categorical and normative overviews embedded in the latter. Massumi calls this oblique mode of working 'a relation of mutual taking-account'.[100] Taking-account is not the same as accounting-for: 'It is transpositional, triggering in-between, roundabout and back and forth. It does not satisfy the criteria of any conventions of commentary-about, illustration, or analysis'.[101] The mutual relationship resists authority and overview (of architecture) in favour of the middle and the detail (of something else entangled with architecture). It compels looking from oblique positions.

'From', 'to', 'between', feminist architectural-art historian and theorist Jane Rendell argues are transformational prepositions in their ability to position, direct, connect and focus on relationships; 'they change everything around them'.[102] Rendell emphasises that knowledge is contingent on positionality: 'where I am makes a difference to what I can know and who I can be'.[103] 'Writing from' reappears in another form through Rendell's 'site-writing'—writing contingent upon site's constraints and potentials—where site and writing are made equivalent.[104] Site-as-situation/writing-as-site/situation-as-writing insist that contingent texts create architecture. Site-writing is both material and processual. Site and writing act as transitional objects to recreate the relation between writer and encounter. More recently, Rendell's work on auto-theory and fiction examines how modes of writing structure worlds, and how writing worlds knowledge.[105] Fiction, she argues, is exciting in academia not because of its creative content, but because its language accommodates subjects and voices that cannot exist in mainstream writing. Through these critical fictions where 'there is always something (of ourselves) at stake', theorisation of such omissions becomes possible.[106]

Examining blackness as visual, cerebral and bodily matter, architect and academic Lesley Lokko remarks that positionality is more than an 'organizational tool' for thinking through evidence. The precarity of the outsider lacking a permanent space, literally maps into how blackness can only be told through made-up fragments, '"Space" slips into the text via a network of connected peoples and places, some real, some not'.[107] Lokko's fragments combine photographs—close-ups of facial features, skin, surface textures at 10:1 and 1:10 scales—with text vignettes. The photographic details offer discrepant evidence on how blackness, something so materially visible, is silenced and displaced. Lokko's material response is to recount this obliterated history by mirroring its silence, 'How then to make out of material, form, space and light a response to the condition…?'[108]

Rendell and Lokko document and critique but they equally create and speculate on architecture's 'what-ifs'. In their texts, theory does not annihilate what it opposes, it seeds exchanges. Theory becomes a toolkit for thinking through and thinking around; it can be challenged, historicised and reconfigured by what is manifest and tangible. Thus, theory presents an 'immanent criteria', which 'bear(s) on how a practice (like architecture) speculates on what it can do', that is to say, theory enables a reaching for language, finding words which can re-body the relational encounter with that which pricks us.[109]

Immanent critique—what the following chapters seek to perform—is constructive. Coming from a point of unrest, it produces experimental, and often precarious, modes of knowing. Aligning immanent critique with activist philosophy, Massumi argues for the duality of such critique, in their two-fold dimensions of being both aesthetic and political, speculative and pragmatic—the latter referring not to practicality but an ability to recompose so as to remain relevant and operational.[110] Drawing from Alfred Whitehead's *A Function of Reason* (1929), Manning identifies a mode of immanent critique in speculative pragmatism, a critique which combines two kinds of reasoning. Whitehead shifts away from 'the godlike faculty which surveys, judges and understands' towards an instinct and appetite for life, which he compares to the reasoning of foxes.[111] In favour of the fullness of experience and without precluding possible anarchy or contradiction—what he calls 'appetition'—Whitehead criticises the persistence of 'methodology' as a way to dodge or to freeze knowledge. Speculative pragmatism embraces a 'process that remains open to the more-than' while being 'completely invested in its "something doing"'.[112] Thus, the act of getting to know, reveals its own reasoning and theory.

The speculative-pragmatic and aesthetico-political mileage is a feminist legacy pioneered by architectural historians and theorists such as Jennifer Bloomer, Katja Grillner and Karen Bermann.[113] Emphasising diversity and attentive to the political, recent work from Australia-based architecture academics Naomi Stead, Hélène Frichot and Anoma Pieris adopt affective histories, biographies, geographies, and philosophies to construct decolonised architectural subjects and their (lost) stories. [114] Combining fiction with criticism, Frichot and Stead advocate a world-making ficto-critical practice that performs 'the constructive, creative and critical situatedness of the thinking-designer in the midst of their problematic field'.[115] Materialised through fiction, critique is 'not the evaluative, judging, fault-finding, seeking-out of flaws kind—not the criticism of the pedant, but rather of the enthusiast, one which uses the scene or setting or object at hand as a point of departure, creatively, spiralling outward in arabesques'.[116] Critique-as-arabesque expresses the poetics and form of affective discourse, one that follows the contours of its subject with care, articulating how that subject has affected the writer, and vice versa.

Grounded by the feminist politics of creativity and ethics of care, affective architectural discourse ranges from interrogating interdisciplinary limits, to questioning identity, gender, sexuality, social justice and race. The writings experiment with aesthetic forms of radical telling—where the voice and shape of the intimate narrative is configured to embrace specific agencies, sites and issues. The outcome of hybrid texts—taking on the modalities and politics of allegory, metonymy, metaphor, indexicality, biography, fictions—defy bifurcation of thinking-feeling. Affective texts carry a documentary poetics that respond directly to disciplinary limits, loss of voice and archival omissions. They are activist texts both generous and generative: 'A text carrying affective intensity is a generative text … it inherits the surplus-value of always having more to say than it knew how to say…'[117]

Thus, theorising architecture through affect, is contingent upon its subject matter, with the subject matter being theory-producing, rather than being defined by theory. Stewart describes theory as something which rather than explicating the 'analytic object' needs to work hard to negotiate it, to come into some kind of alignment with the object of study which 'can only be approached awkwardly, described around'.[118] Theory *is* the practice of immanent critique, to find a language and a structure to linger in the moment and space of the subject, to be able to sense its potentials and to stay with what perturbs and pricks us. Theory comes from understanding how the fragment which grabs our attention, avails itself and unfolds to our knowledge.

Theorising architecture through affect entails theorising affect through architecture. Getting to know one in order to know the other. Getting-to-know involves risk—an elliptical and experimental exchange, marked by detours, errors, hesitance, failures. The precarious relationship 'between two' maps onto the curiosity that Whitehead calls appetition, the lure that Barthes calls advenience, the inchoate frame that Williams terms a structure of feeling, and the afterimage of encounter that haunts Taussig. Precarious because the terms of knowing are subject to change; getting-to-know is not mastering. 'Mastery of non-mastery', Taussig cites Barthes, '"[is] an ethic, a guide to life lived through twinklings of tact in an anecdotal discourse recruited to outsmart mastery"'.[119]

In *Architecture and Affect*, affective evidence which sparks the process of getting to know is theory-producing. Each chapter attempts to re-body the relational encounter with a piece of extra-disciplinary evidence, folding an outside into architectural knowledge. Evidence—manifested in stories, anecdotes, situations, fieldnotes—resists mastery. Figuring out how to make a place in architecture for this kind of oblique knowing and knowledge, a knowing which is intrinsically part of life, motivates this book.

IV. Precarious spaces: *Structure of the project*

> I can't help but dream of a kind of criticism that would not try to judge... It would multiply not judgements, but signs of existence; it would summon them, drag them from their sleep. Perhaps it would invent them—all the better.[120]

> ...or try to write theory through stories, or try, through descriptive detours, to pull academic attunements into tricky alignment ...[121]

Diverse corporealities including a tiger, a ruin, hoarded rubbish and stray cats, single women, ghosts and failed infrastructure, constitute the affective subjects and evidence of the forthcoming chapters. These varied, complex and porous bodies are affective intercessors to the environments they occupy and shape, which include two monuments (a hotel and a former train station), the public spaces and flat interiors of public housing blocks, and two landscapes (a cemetery and the sea). They are sited in Singapore, a hyper-capitalist, hyper-efficient and technocratic Southeast Asian city, where money and utility trump poetry and history. These subjects lead me to think otherwise about architecture and landscape. Through them, I encounter spurious evidence, the sort cast-off as 'no more than' anecdotes, situations and fieldnotes—flimsy, inefficient, unsanctioned, informal. But such discrepant evidence could not be ignored. Manifestly local and vivid, they impressed themselves firmly onto my architectural encounters. There was no other way but to write with and through the subjects and their anecdotes, situations, fieldnotes. What happens when these sticky entities are admitted into architectural discourse?

My encounters seed six precarious architectural accounts conveying aesthetics and experiences which question or contradict archival records and official narratives. I use the term 'precarious' intentionally. Conveying implications of risk, disenfranchisement, and vulnerability, 'precarious' firstly describes the intellectual and emotional labour expended to create a discursive space for these subjects. Precarity, Stewart writes, concerns itself with emergent forms of sensing, thinking and perceiving.[122] The affective evidence—existing loosely but compellingly as anecdotes, situations and fieldnotes—precludes the subjects' multi-faceted stories in extant architectural accounts. Indeed, to tell these other stories in academic discourse is to risk not being taken seriously, or to be architecturally irrelevant. Still, it is a risk worth taking if only because it links critique with the force of the encounter, with its density of narratives, and the cultural imaginings wrapped around the latter.

Challenging disciplinary conventions which prioritise the architect's intentionality and significance at the expense of all else,[123] I argue for the capacity to theorise and historicise architecture 'as one finds it', that is, to allow significance to emerge out of such occasions.[124] Literary theorist Jane Gallop emphasises the significance of lived activity even if it is 'occasional' because a particular moment may in fact present 'an opening—the occurrence of something new, something surprising'.[125] Elsewhere, feminist literary critic Nancy Miller reminds us that 'the fallout of the event' creates 'the chance for something to happen in the wedge of unpredictability not yet foreclosed by … discourse'.[126] What Gallop and Miller agree on is the importance of recognising situations which do not count as important but often mattering a great deal, of taking seriously serendipitous encounters which we cannot shake off because they point to something significant taking, or having taken, shape.

Secondly, precarity resides in the affective encounter. It denotes the ephemerality of situations/anecdotes/fieldnotes, and the fragility of their related subjects having to find a place in discourse. It refers to the omitted, and the feminist sensibility to acknowledge what is absent. The work pursues 'what it feels like to be carried by something on the move', of 'the dark wakefulness';[127] in effect, to not close down curiosity or instinct simply because it is not status quo. The marginalisation of evidence and subjects, and consequently, the side-lining of discourse enacted around such matters, are all forms of precarity. The chapters traverse architecture from positions of intimacy—often shifting between looking at architecture through the architectural historian's position and language, to that of looking at architecture from within, feeling (or groping!) around the nooks and crannies through the lenses of the occupant/subject, which in some cases include non-humans. Through this shift, a conscious decision has been made to leave behind some of the assumptions accompanying the conventions of architectural history and theory, and indeed to question what is seen or unseen, and why this is so.

Third and finally, the book is positioned at the still precarious disciplinary boundary between on the one hand, thinking through creative practice, and on the other hand, writing architecture through its histories and theories. The first works predominantly through creative leaps and gaps, the second moves cautiously through evidence and argument. As Bloomer writes half-jesting in *Hold It (Meditations upon a Gorgonzola Cheese)*, there are two camps fighting over what architecture is, or who may rightfully lay claim to it.[128] On one end, there is the '*architecture parlante*' of 'image, expression, and representation' held together by the primacy of form, geometry and symbols. On the other end, there is an *'architecture de l'ecriture'* concerned with the discourse of 'materials (physical or intellectual), connections and patterns of possibilities' which contests the former side for its 'often simplistic relationships (both literary and architectural)'. Bloomer leans more to the latter's grounding on 'inquiry and exploration' though she admits her fascination with the former's 'leap of faith' and 'mumbo-jumbo'.[129]

I share Bloomer's predilection. To this end, three image-essays are positioned between the six text-based chapters. *The Third Archive* exists as holdings for the making of unruly and minor subjects, with their attendant narratives and theories. These subjects and stories have no place otherwise in architectural discourse. Drawn obliquely (see 'dirty drawings') from the oppositional realm of the *'architecture parlante'*, they punctuate the *'architecture l'ecriture'*. The reader might see these archival intervals as ornaments or distractions, or such drawing discourse might slip into the main text as hauntings, becoming afterimages that will not go away. *The Third Archive* are what Bloomer calls 'dirty drawings'—instruments of incision, and equally, connections to the world.[130] The 'dirty drawing' is 'technically accurate but "improperly" ornamental', they are baroque, excessive and fetishistic.[131] They make homes for subjects that do not yet have a proper place in architecture. Conceptualised as dilations and openings, citations and critiques, they trace the moving of thought from one realm to the other. The drawings and models speculate around the gaps in the archival texts. The drawings thus accentuate the ambiguity of the archives while also embellishing these gaps with what can be imagined. Or what has been repressed. Additionally, each of the six text-based chapters opens with a situational/anecdotal/field drawing of a piece of affective evidence and closes with a worm's eye axonometric of the architecture. These drawings reiterate how the accounts move back and forth between affect and architecture, allowing one to inflect upon the other, reciprocally and, in equal turns.

Architecture and Affect is foregrounded in Singapore, a neoliberal city-island-nation state in Southeast Asia. Notable for its high standards of housing, health and education, a thriving economy, compliant citizenry and a stable government, the rich global city has its share of issues. These include widening income disparities, plunging birth rates, a large aging population, restrictive laws, geopolitical tensions which have driven its obsession with territorial expansion of land area, and with this, a domestic economy serviced by essential workers, domestic helpers and construction labour largely supplied through the influx of foreign workers from the neighbouring region and from South Asia.

Aside from my familiarity with, and access to, this city, the reading of its architecture and landscapes through an affective lens is itself a shift from existing architectural discourse. Extant work tends to be couched in a statist framework of national development, indigenous-to-modern perspective described through transnational narratives, and notably, postcolonial histories of its architecture. These modalities have largely focused on questions of architectural authorship, intentionality, design lineage and influence, or the politics, society and culture enveloped within the architectural production processes, and its commissioning structures. *Architecture and Affect* shifts the attention to sticky category leftovers entangled in architecture's afterimage. Too unradical, too ordinary, they lack the political profile expected of subjects featured in blockbuster theories like postcoloniality. Yet, as Berlant notes in her choice of 'silly objects', 'the waste materials of everyday communication in the national public sphere are pivotal documents in the construction, experience, and rhetoric of quotidian citizenship'.[132] In other words, such remaindered evidence is inevitably cultural. In this sense, other cultural artefacts—film, visual and installation art, and literature—are closely engaged as being critical to this architectural conversation.

The six text-based chapters in the book are organised in pairs within three sections that follow an ascending scale—Part One: Monument, Part Two: Block, and Part Three: Landscape. In each scalar category, a chapter presenting an architectural perspective, dialogues with an extra-architectural one looking at a similar space from visual art or filmic lenses. Mediating these paired chapters is a section from *The Third Archive*. Hence, each scalar category is conceived as a triadic composition featuring one architectural voice, one artistic or film-inflected perspective, and reciprocal drawing-based discourse from *The Third Archive*. The paired chapters are arranged as dialogues between architectural discourse perceived from the inside *and* from an outside. Three visual documents are closely engaged—an artist book and photographic series on the former railway station by Sydney-based Malaysian artist Simryn Gill, the architectural essay film *03-FLATS* by myself and Singapore director and filmmaker Lei Yuan Bin, and the ongoing work *SEA STATE* by Singapore artist Charles Lim. The visual documents function as interlocutors, alternative resources and archives holding evidence undervalued in extant architectural scholarship. In this schema, I have taken on three voices, or roles—as architectural theorist and historian writing about a historical present, as designer of *The Third Archives*, and as a collaborator in the essay film *03-FLATS*. The triadic composition of the book, and of my own roles, acknowledge the significance and influences of drawings, photographs and films, and the productive cross-disciplinary exchanges between architecture and these visual disciplines. These dialogical chapters question how one might understand and express architecture beyond what architecture knows of itself.

Part One: Monument looks at a hotel and a former railway station. Chapter 1 *The Ruled and the Unruly*, examines forms of evidence in the anecdotal alongside the literary, homing in on accounts of Singapore's famed Raffles Hotel and its unwanted guest, Stripes the tiger. The chapter posits how the masculine space of the Billiard Room is sustained by the imagination and humanity of a feminine form of storytelling, contrasting the latter to its colonial narrative. Chapter 3 *After the Last Train*, gives pause to the ruination of the former Tanjong Pagar

Railway Station, which ceased operations in 2011 now representing a vestige of Singapore and Malaysia's separation in 1965. The chapter examines the neglect of this building to the tropics through Sydney-based artist Simryn Gill's *Guide to the Murals at Tanjong Pagar Railway Station*, a book created for the Singapore Art Biennale in 2006. Gill's artist book is an appropriation of the architectural guidebook, implicating the passage of time and complicating how we constitute historical evidence and its narrative. Mediating between Chapters 1 and 3 is the first segment of *The Third Archive*. *Tracing the Last Tiger* is related to the famed tiger under the Billiard Room of the Raffles Hotel. It follows the archival evidence of the island's last tigers, reputedly sighted in the northern semi-forested regions of early twentieth-century Singapore, a time when the island was rapidly urbanising. Mirroring the anecdotes about Stripes, *The Last Tiger* is fashioned as a detective game, a historic *Cluedo*, featuring a trail of clues and a motley collection of objects standing in for the elusive and phantasmic beast.

Part Two: Block profiles Singapore's successful public housing. The architectural discourse of Singaporean public housing is focussed primarily on efficacy, principally framing the architecture as teleological, and subsequently traced and retraced through a progressive trajectory. Chapter 4 *Keeping Cats, Hoarding Things*, argues that standardised measures of performative and statistical methods used to achieve and justify efficacy fail when they are applied to two public spaces of the housing block—the void deck and the common corridor. Based on such domestic situations as keeping cats and hoarding, the chapter discusses how a discourse of efficacy represses the affective nature of these public spaces, reframing the limits of property and prohibitions, the sanctioned and the primal. Chapter 6 *03-FLATS*, discusses the titular film which tracks the domestic lives of three single women across Singapore's public housing 'heartlands'. The chapter contemplates how the architectural essay film opens up disciplinary questions of nuanced content and its intimate mode of representation, producing an alternative optics of home. Departing from film as an instrument of ethnographic and anthropological documentation to a mode of architectural representation, the chapter situates film as a creative, critical and projective research modality. *Anarchive of Public Housing*, the second segment of *The Third Archive*, is inspired by the infographic, a genre which compresses information for quick consumption. The drawing is an axonometric of a housing block, distorted in scale to contain information of extraordinary domestic routines, many testing the limits of compliance to public housing's rules and regulations. The situations created revisit questions of ownership, authorship, appropriation, or what architect and historian Jonathan Hill calls 'the creative user'.[133]

Part Three: Landscape negotiates the periphery of the island state, looking at a once abandoned cemetery and the ubiquitous sea which surrounds us. Chapter 7 *In The Midst Of*, tracks Bukit Brown Cemetery in 2011. Abandoned and dormant for almost 40 years prior, the cemetery was catapulted from obscurity to become a popular subject of research by both specialists and amateurs. The chapter frames the cemetery as 'field', from which fieldnotes produced by cemetery's visitors multiply evidence and stories through diagramming, mapping, narrativising, dramatising, and simply walking. It describes how an architectural ecology of care emerges in persistent acts of discovery, everyday occupation, and silent protest against the site's extinction. Chapter 9 *The Sea, and the Sea,* looks at the sea as the island's infrastructure. Surrounded by this sea and scarce in its terrestrial area, the Singaporean anxiety of diminutive scale yields an architectural commoditisation of its oceanic frontier as hinterland. Framed by Walter Benjamin's dialectical image, the chapter reconsiders the sea through artist and sailor Charles Lim's *SEA STATE* (2004-present)—a multimedia, ongoing visual art project—that makes such infrastructure accessible and vulnerable. It foregrounds the embedded narratives—mythological, historical, personal—in the sea's infrastructural objects. The final interpretative archival segment, *Holes in the Ground,* revolves around customs and taboos of tomb exhumation. It enacts the tensions evoked by clashes in urban systematisation and spectrality discussed in the chapter on Bukit Brown. *Holes in the Ground* constructs two-and-a-half dimensional drawing-models of two imagined tombs. The drawing-model combines the precision of the orthographic architectural plan, with the perceptual and material depth of the model. Piecing together documentary excerpts and grave diggers' anecdotes, the drawing-models embellish the exhumed tomb with superstitions and hearsay.

A final word to the reader. I have written each chapter so that you might have access to these sites and the affective evidence, in the immediacy of words, drawings and photographs. To this end, I have intentionally kept the philosophical complexities of 'affect theory' out of the chapters themselves. I have retained the discourse to be proximate with how each piece of architecture or landscape was encountered. Certainly, if a piece of affective evidence leads us to its own theories and histories, then to cast an overarching theory over every subject, even if this theory is about affect, is to lose the plot.

'With affect, a body is as much outside itself as in itself—webbed in its relations—until ultimately such firm distinctions cease to matter'.[134] Encountered in a state of distraction; something else brings us to architecture. This book explicates an understanding of architecture by following the sticky 'something else': it constructs a discourse by examining the scaffolds of encounter and architecture's webbed relations. Knowing otherwise, each chapter strives to record what we see and sense, to stall overarching conclusions, straining to find precise words and exact theories and histories, to capture and transport discourse to that place, and immediacy, of encounter.•

Part One: Monument

1 • The Ruled and the Unruly

Animality, anecdotes and storytelling

> In this way, a widening network of interconnected anecdotes was substituted for a linear history of leading events. ... In contrast to the official narrative—one in which the clearing of land and the harmonisation of different views proceeded with clockwork precision, fulfilling a predestined plan—these other historical phenomena obeyed a principle of ambiguity. They were not so much events as moments in historical consciousness when analogies were noticed between dissimilar things.[1]

THE TERM 'TROPICAL' HAS BEEN INSTITUTIONALLY PROFILED through periods of colonial and cultural imperialism. Since it first appeared as a defining concern of colonial governance, infrastructure and research in the twentieth century,[2] meteorological questions of climate slipped into matters of physical and moral degeneration, as well as 'environmental "othering"'.[3] In the early-twentieth century, the 'tropical' was a term annexed to a host of colonial pursuits for instance, tropical medicine, tropical geography, tropical ecology, tropical agriculture, tropical economics and tropical hydrology.[4] Uncertain and perilous, the tropics continued to be sustained as a hermeneutic construct, suited to scientific domestication and slippery enough to exist as 'other'.[5] Even today, the historical understanding of the tropical world as an environmental 'other' persists with the climatological vulnerability of the tropical belt—a zone which produces the bulk of the world's food, natural and labour resources—becoming a security threat to developed nations, who depend on the belt's continued economic productivity.

This chapter explores the affective confluence of architecture with the cacophonous tropics—a system connected not just to climate but to specific cultural, racial, gendered, social, and political affordances. It navigates through an Other architectural history that accounts for the ambiguity of experiences and meanings embedded in these insurgent spaces. Contesting the roles of colonial narratives and archives wherein architectural knowledge is principally derived and developed, it embraces the tropic's unruliness, animality, and fecundity as primary evidence, particularly as these are expressed in anecdotal and literary forms. How does the affective tropical world influence narratives, agencies and evidence when its architecture is looked at through a borrowed literary view? Beyond exoticising and aggrandising the tropics—temperaments of writing which are frequently employed to transpose the familiarity of the temperate climes into this Other milieu—does a literary imagination of the tropics that accounts for the embodied, vulnerable and doubting experiences of its subjects, offer something else to think around its architecture? The relationship between literature's text and architecture's 'texts', its discourse and modes of writing, might be configured around the image of grafting, offered by Jacques Derrida:

> … two texts … contaminate each other's content, tend at times to reject each other, or pass elliptically one into the other and become regenerated in the repetition, along the edges of an overcast seam. Each grafted text continues to radiate back toward the site of its removal, transforming that, too, as it affects the new territory.[6]

Literature articulates what is surplus to architecture, what might be considered peripheral to the latter discipline. This architectural surplus is central to the articulation of a literary atmosphere; it is what stirs the reader's imagination, it sets the scene. Thus, the sum effect of this alliance recalibrates architectural discourse such that it becomes aware that its theories are not exempt from fictions: 'a self-consciousness that points to the fictional strategies inherent in [its own theories]'.[7] Literature becomes architecture's conscience. It reveals that what is known and what gets told, are subjects that gain authority through the force of institutions, whether it is the weight of colonial architectural history, the promise of an urban plan, or the benefits of modern housing. Architecture, which can be traced to singular, often male identities,[8] and as the beneficiary of unremarked patriarchy, frequently describes and repeats its 'privileged topology'.[9] Literature corrects this myopia. It places architecture out-of-focus, and relocates it, dangerously perhaps, in the midst of Others.

In the opening epigraph, cultural theorist and artist Paul Carter argues that when anecdotes, 'other historical phenomena [which] obeyed a principle of ambiguity', punctuate a linear historical narrative, we begin to see these ambiguities as 'moments of historical consciousness'. Thus, the interconnected anecdotes do not merely assert an alternative historical perspective. Instead, the ambiguities they raise pose foundational epistemological questions about recorded history's premises, assumptions and lapses. The anecdote is a powerful mode of storytelling especially for groups without agency. The telling of tales conjoins architecture with stories told *around* it as opposed to about it, embellished as these are with unexpected and out-of-focus evidence. The architecture of the tropics subsequently emerges with new insights, protagonists and plots. Moving between colonial and contemporary milieu, three spaces are surveyed—an Indian cemetery and a Burmese billiard room—featured in the stories by Arundhati Roy and George Orwell respectively, and the billiard room at the Raffles Hotel in Singapore whose fame is tied to a tiger. While their historical periods and locations differ, these spaces—bound together by the atmospheres and biases of the tropics—offer possibilities for reimagining the shape of its architectural history, manoeuvred through an archive populated by tropical flora and fauna.

'She lived in the graveyard like a tree'[10]

Located next to the city's hospital and mortuary, the small but lush, and occasionally used cemetery is a ten-minute Tempo ride from Delhi. It is a residual place 'where the bodies of vagrants and the unclaimed dead were warehoused'.[11] Jointly occupied by stray dogs, addicts (who parked themselves by the northern edge of the hospital's dumpster in search of used syringes), and the homeless (who cooked their meals over an open fire on the southern end), it became the site of Jannat Guest House, whose landlord is one of Arundhati Roy's protagonists, Anjum. Anjum (previously Aftab) is a transgender woman who is a 'Hijra', a Hindu word for hermaphrodite. Alienated from her own middle-class family because of her sexuality, Anjum emerges the most celebrated resident of the Khwabgah, a sanctuary for Hijras until she witnesses and survives a massacre in Ahmadabad almost thirty years later. The traumatised Anjum moves out of the haveli into the Muslim graveyard where her own family and friends are buried. The cemetery evolves as the core space of Roy's epic second novel, *The Ministry of Utmost Happiness* (2017); it structures the patchwork of stories, tragedies and relationships which develop across six decades and three nations—India, Pakistan and Kashmir.

The Guest House originated from a tin shack, fashioned from Anjum's vitalist sensibility of living 'like a tree', merging and gathering with 'everybody and nobody, of everything and nothing',[12] an Other view of the 'Duniya' (world) in which relationality and difference, even trans-species, are privileged over a politics of sameness. Anjum's shack first accommodated a bed. It grew to include a kitchen, and subsequently a terrace where 'she surveyed the dominion of the dead'.[13] The Jannat Guest House flourished like a tree, spreading itself respectfully around tombstones. Its enlargement responded to need as more came to lodge. Its expansion was organic; each room enclosed a tombstone as homage to the dead who came first. That was how the guest house came to be: the living with the dead; humans, plants and animals accommodated in a symmetrical and reciprocal co-stewardship of the wasteland.

The cemetery is a witness to loss, not just of human but also animal life. In the epigraph to the first chapter, Roy draws our attention to the other denizens of the tropical graveyard:

> At magic hour, when the sun has gone but the light has not, armies of flying foxes unhinge themselves from the Banyan trees in the old graveyard and drift across the city like smoke. When the bats leave, the crows come home. Not all the din of their homecoming fills the silence left by the sparrows that have gone missing, and the old white-backed vultures, custodians of the dead for more than a hundred million years, that have been wiped out. The vultures died of diclofenac poisoning.[14]

Traditionally, vultures were part of the sacred Hindu ecosystem. Symbols of the dead, they scavenge for dead animals in the wild. Vultures were also part of the natural burial 'pyre' for Parsi Indians who practised the ancient tradition of 'sky burials' whose bodies left at the tower of death the vultures consumed. Following the birds' extinction from the effects of Diclofenac, the Parsi Indian's century-old practice is likely facing the same fate.

The missing birds are urban collateral damage. They are casualties of a city which has changed not just its streets and buildings but also its desires, traditions and culture. This city cannot accommodate the poor, '"People who can't afford to live in the cities should not come", a Supreme Court judge said, and ordered the immediate eviction of the city's poor'.[15] Delhi has climate-controlled cars and apartments; garden cities where there are cars for dogs, and gardens for cars; and while thousands of 'surplus people' die on the streets unnamed, in the wealthiest suburbs, the pavements are unpeopled, lit brightly by tall street lights which look 'encashable—columns of liquid gold'.[16] At the fringes of the city where forests used to be, productivity is enhanced in 'skyscrapers and steel factories' but the air is polluted and the water poisonous.[17] Instead, 'rivers were bottled and sold in supermarkets, fish were tinned, mountains mined and turned into shining missiles'.[18] In this urban inventory, Roy lists each natural resource as eventually enclosed and redefined by architecture, industry or infrastructure. Unlike the organic evolution of the rooms at the Jannat Guest House, these enclosures are unforgiving of what came before.

At the cemetery, there is hardly a problem with power cuts—a perennial urban scourge even in the wealthiest properties—because the Guest House's electricity is siphoned from the mortuary. In a sardonic twist, the city's paupers ultimately 'lay in [its] air-conditioned splendour', enjoying round-the-clock refrigeration.[19] The airconditioned spaces of the city numb the occupant from the sweltering heat outside. At the mortuary, air conditioning delays decomposition and decay. One of the residents of the Guest House, Saddam Hussein, recounts the story of his father being murdered when the stench of a dead cow (an animal sacred to the Hindus), which the latter had cleared from a farm, becomes offensive to the Hindus celebrating the festival of Dussehra on the streets of Old Delhi. It could be argued that the tragedy was hastened by putrefaction in the tropics—the foul smell of rotting cow flesh instigated a Hindu mob to beat the Muslim man to death.

Significantly, air conditioning is a 'selfish technology'[20] one that works through the maintenance of an unequal temperature gradient between the interior and the exterior of a building. Its net effect is an increase in heat; it contributes significantly to global warming. An air-conditioned city also maintains a gradient of class and wealth; the poor and lower castes are kept outside. As a technology of climatic and social boundary-marking, the air-conditioned city still fails to rid itself of the surplus people that it works hard to keep out. The old city of haphazard streets littered with traders, pilgrims, cripples, beggars, the homeless, livestock and stray animals, is much more hospitable and forgiving. When Roy tells of the cemetery, the *haveli* and these narrow streets, her stories are situated in-the-midst of these spaces. They are entangled with the roots of the banyan trees, the flock of migrating birds, the adopted stray animals, and the heat, sounds and smells of Shahjahanabad, located near to Shah Jahan's historical Red Fort.

The porosity of the old city is mapped out through its relationality with time, exemplified in the bonds between its inhabitants and a network of lived architecture with increasingly ambiguous boundaries. Although surrounded by a high wall, the *haveli* is easily accessible from the street through a blue door that opens into a calm and empty courtyard populated by a single pomegranate tree and a handpump. With the exception of the Ustad who stayed in a room (the second room was already a ruin and was occupied by a family of cats), its other occupants lived on the verandah, 'their bedding rolled up in the day like giant bolsters'.[21] The Jannat Guest House was unruly in structure, occupancy and composition—rooms arranged around tombstones instead of a grid; occupants (humans and nonhumans) bound by kinship rather than family, and an inn which incorporated a funeral parlour, a pool, a school and a zoo. In its expansion, the Guest House was located in-the-midst-of the cemetery and the tomb stones, nestling between trees, wildlife and the incongruous inhabitants of the graveyard. It adjusted, made do, grew with, and amongst, the unruly tropical space which it did not attempt to change or make into its own. How can we call this, what disobeys the defining principles of architecture, architecture? How do we discern architecture from within haphazard occupation and visceral experience when the discipline emphasises architecture as a distinct and authoritative specialist art of spatial ordering?

'They share it with tigers, hamadryads, bootlace-snakes, leeches, *pelandoks* and the rest of the bewildering fauna...'[22]

The most enduring image when one speaks of the tropics is the jungle, wherein all manners of decorum, reason, distance, focus and scale begin to shift:

> He saw it, saw a landscape, a tropical swampland under a cloud-swollen sky, moist and lush and monstrous, a kind of primeval wilderness of islands, morasses and muddy alluvial channels; far and wide around him he saw hairy palm-trunks thrusting upward from rank jungles of fern, from among thick fleshy plants in exuberant flower; saw strangely misshapen trees with roots that arched through the air before sinking into the ground or into stagnant shadowy-green glassy waters where milk-white blossoms floated as big as plates, and among them exotic birds with grotesque beaks stood hunched in the shallows, their heads tilted motionlessly sideways; saw between the knotted stems of the bamboo thicket the glinting eyes of a crouching tiger; and his heart throbbed with terror and mysterious longing.[23]

In Thomas Mann's *Death in Venice* (1912), the protagonist Aschenbach pictures the tropics whilst wandering along the edges of a cemetery, just at the outskirts of Venice. His vision of the tropical jungle was evoked by the sudden presence of a man whose clothes, looks and mannerisms exuded an 'exotic air'.[24] It did not matter where the stranger hailed from. The mere speculation of an 'exotic air' elicits 'a desire to travel'[25] to a 'colourful external world'—an outside—where privileged European society, specifically the well-heeled traveller, had escaped for 'health precautions', under prescriptions of 'leisure and pleasure'.[26] Aschenbach's tropics was a double space. On the one hand, it captured the charm of a faraway paradise—passive, rich, resource-laden, open to intervention, and ready for the taking. The tropical jungle was 'moist', 'lush', fertile in its 'muddy alluvial channels', giving growth to 'fleshy plants' and 'exuberant flower[s]', refreshed with 'glassy waters where milk-white blossoms' were 'as big as plates'. On the other hand, this imagination leaned towards an excessive otherness that was 'monstrous', a 'primeval wilderness', 'misshapen(ed)' and 'stagnant', 'grotesque' and teeming with hidden 'terror and mysterious longing'. This wild and ugly Other was in need of the enlightened touch to groom, subdue, put into order, and domesticate. Aschenbach's tropics stood as distinct and distant from the familiarities of a civilised society. It was simultaneously better than the modern European city thus warranting a desire to travel, but also primitive and backward.

In Roy and Mann are polarised modes of engagement with the unruly tropics—one entangled and implicated, the other distanced and exoticised. Each experience is, of course, historicised and particular to its own time and place. The social inclusion and political emancipation of Roy's cemetery—a tropical utopia—is coincident with the novel's relational articulation of space and belonging. The cemetery is an undesirable and unconducive place to live. It is taboo, derelict, and only occupied by a dehumanised public: the un-homed and unhomely, the transgendered 'different', illegal stayers and non-citizens of the state, the outcast, the grotesque, the impoverished, the downtrodden in mainstream society. In the book's epigraph, Roy tells her reader that the extinction of the cemetery's fauna, its oldest inhabitants, forewarn the intricate balance between the city and nature. The solid boundaries of architecture cannot protect against the ills of chemical foods, hormonal concoctions and poisonous pesticides that seep into ice creams, milkshakes, the soil and the air. Instead, life in the cemetery is made possible, later even desirable, with tacit knowledge of the unknown, recognised through signs in the 'presence and proximity of nonhuman interlocutors'.[27] Writer Amitav Ghosh calls this an 'environmental unconscious' which necessarily constitutes knowledge weaving together fact with myth, fable and imagination, and the intertwining of human with non-human forces: 'In an overlapping of the pragmatic and the poetic, a broad acknowledgement of mutual dependence, in which rights, mutual obligations and a sense of wonder are seamlessly merged'.[28] Like Roy's cemetery, Ghosh's 'environmental unconscious' acknowledges what is there but cannot be seen nor understood. The tropical jungle implicitly demands a heightened environmental awareness, respect and mindfulness for the jungle demon's wrath, 'not [to] urinate, defecate or spit while collecting honey or firewood'.[29]

Yet for Mann, who was writing just before the First World War, the tropics was mostly conjured. In that sense, the vision, aesthetics and vocabulary of the tropics were more tangible and immediate than the distant place itself. Literary and environmental theorist Monique Allewaert argues that the tropics was cast in colonial writing as an alternative form of the sublime, an Enlightenment aesthetics which equates distance with control. The 'tropical sublime' is aesthetically beautiful and other-earthly, yet dangerous and life-threatening to the subject. Confronting that which was *both* better-than-life *and* life-threatening, the '[human] subject stand[s] apart from the object world that he or she would master'.[30] The unknowable tropical sublime, in turn, became a reason for imperial conquest, or as enlightened logic would have it, reason to domesticate an otherwise dangerous landscape for the good of humankind.

Thus, an imperious global tropical consciousness, which precedes our contemporary understanding of ecological precarity already came into play as the tropical parts of the globe were being 'civilized'. The tropical world, ostensibly lacking in intellectual, moral, social, political and cultural attributes, needed correction and had to be salvaged. In imperial travel writing, there was a morally inflected notion of weakness and lack, which implicitly embedded an imperial hierarchy of the coloniser towards the colonised tropical world. This superiority evolved into a 'power to transform a vertiginous instantiation of Otherness into a source of self-edification'.[31] The weakness of its tropical Other guaranteed the strength of the coloniser.

The tropical landscape—particularly its flora and fauna—reinforced the 'system of social relationships' between the empire and its natives, persistently keeping separate and distinct those belonging to 'the ruler, the ruled, and the unruly'.[32] The domains of the unruly extended from plant and animal life, to 'the people—the natives—the childish and half-civilised: "half-demons and half child," "lesser breeds without a law"'.[33] In Anjum—'a woman trapped in a man's body; … she who had lived for years like a tree in an old graveyard; … she, who never knew which box to tick, which queue to stand in, which public toilet to enter; … she, who knew she was all wrong, always wrong; … she, augmented by her ambiguity'[34]—one comes face to face with the sense of tropical risqué, and discrimination.

Yet, was there a moment when subjects who stood at the threshold of the tropical world were armed neither with the certainty of entanglement with, or control over, their environment? Was there a moment of ambivalence which perhaps seeped into literature, which in turn might destabilise the magisterial historical narrative of colonial architecture?

In his essay 'Seker Ahmet and the Forest', cultural critic John Berger observes that in painting the forest (or the jungle in our case), artist Seker Ahmet intervenes with the perspectival and physical placement of the trees and its lone human subject, a woodcutter. The effect is one of 'double vision'—the viewer glances at the woodcutter vanishing amongst trees which appear instead to lurch towards her (as though she, the viewer, is in the forest herself); she simultaneously sees the trees moving away into the background (as though she is looking into the forest, from its outside). The densely wooded space creates an experience which 'depends upon your seeing yourself in double vision. You make your way through the forest and, simultaneously you see yourself, as from the outside, swallowed by the forest'.[35] In Ahmet's painting, the idea of figure versus ground—a schema of active subject versus passive background—is contested. The painting depicts a doubling of vision, one in which the viewing subject's control over a vantage point surrenders to multiple perspectives, resulting in a kind of blindness, or a way of seeing differently.

The final sections dwell on this doubling of vision, which entails being 'inside' and 'outside' of the tropics, at the same time. Moving from the unusual domestic arrangements of the Jannat Guest House in Old Delhi, we journey to two billiard rooms—one fictionally set within a colonial club, and the other, a piece of architecture in a colonial hotel in modern day Singapore. Entering the first billiard room within George Orwell's *Burmese Days* (1934), I argue for how Orwell's literary space not only intersects, but also enables and critiques the architectural one at the Raffles Hotel.

'Was there really a tiger under the billiard table?'[36]

In the early waking hours of 12 August 1902, a tiger by the name of Stripes was spotted by a nervous hotel attendant who quickly raised the alarm. In no time, Mr. C. M. Philips, the head teacher of the Raffles Institution, a premier boys' school located down the road from the hotel, was summoned from his bed. Philips had been to a fancy-dress ball the previous night at the Government House. He was still nursing a hangover. Dressed in his pyjamas, the head teacher aimed, fired, and missed the tiger several times, hitting instead the brick pillars of the Billiard Room. When he finally managed to kill the beast, the tiger's body was dragged out from under the raised building by nervous bystanders. The episode was reported blow-by-blow in a leading local newspaper but there were neither photographs of the ill-fated tiger nor of the head teacher, who reputedly tore his coat in the event (Fig. 1.1).[37] The hotel has been tethered to the story of Stripes ever since.

Despite the fanfare around Stripes, the accounts describing the nearly botched kill at the Raffles Hotel remained largely anecdotal. From the newspaper report, it seemed that the reporter and the crowd found Philips's performance lacking lustre. The reporter clearly sympathised with poor Stripes, a runaway circus tiger which was probably terrified by the noise of the gathering crowd and had kept himself out of sight in the dark undercroft of the Billiard Room, until he was rudely shot. The hotel's archive, an expectantly weighty one considering its status as a gazetted colonial monument, does not hold evidence of the tiger in its Billiard Room although it was through the notoriety of the 1902 shooting that the Billiard Room became significant. Instead, fact slipped into fiction. In the friction/fiction of information, gaps widened. The historical anecdotes of the

AUGUST 13, 1902

A TIGER IN TOWN
SHOT AT RAFFLES HOTEL

Under The Billiard Room

A TIGER was shot under the Billiard room of Raffles Hotel early this morning. Lest any one be inclined be doubt the veracity of the foregoing statement, a representative of this paper who saw the dead body of Stripes soon after he was shot, is prepared to bet a new hat that a live, loose tiger slept under the billiard room of Raffles Hotel last night, that Mr C. M. Phillips of Raffles Institution was informed of the fact early this morning and that the said gentleman dispatched the tiger in the presence of an excited band of the hotel residents and others.

The details of this unique incident are as follows:- A day or two ago, a tiger belonging to a native show performing on the Beach Road reclamation broke from captivity and to all intents and purposes disappeared. Soon afterwards he is alleged to have made his appearance near the Drill Hall where some means or other he got on to a tongkang moored near the reclamation work now going on just there. The watchman alleges that the tiger interviewed him and after giving him.

A Few Friendly Scratches

made off, presumably swimming gaily up the Singapore river. From this point he was again missed until the closing hour of the Raffles Billiard rooms last night when, lo and behold ! His Majesty of the jungle gave the bar "boy" a stiff shock by staring through the low verandah railing on the hotel side of the Billiard room. This was rather

Too Much for the "Boy"

who promptly secluded himself- under the counter or in some other handy place - and awaited developments. Finding the tiger did not seek a personal interview with him, the "boy" stealthily emerged, hurriedly shut up the billiard room, and "scooted". his departure being considerable hurried- so it is alleged- by sundry scratchings and other noises under the floor of the billiard room, which it may here be explained is built some feet from the ground, on brickwork pillars. The news of the tiger's presence was communicated to the management of the hotel, and early this morning, it having been ascertained that Stripes was still reposing under the billiard room floor, a messenger was sent across to Mr Phillips at the Raffles Institution, asking him to deal with the situation. Mr Phillips was in bed at the time, and having been at the Government House ball a few hours previously naturally did not feel in particularly excellent trim for being "jerked"-so to speak-within a few feet of a loose tiger.

However, Mr Phillips hurriedly tumbled out of bed, and securing his Lee-Enfield rifle and a few hollow-nosed (Mark IV) cartridges, proceeded to the scene of action.

In His Pyjamas

The tiger was still under the billiard room floor. There was no doubt about that; yet no amount of peering into the gloom of the under-floor region could discover his presence. Suddenly Mr Phillips saw the tiger or thought he did. He promptly put three shots in on the spot, and there was no apparent reason why the tiger should not be as

Dead as a wall

But unfortunately it was a bit of a brick pillar, or something similarly indifferent to rifle bullets, that Mr Phillips saw; so the stalking operations were resumed, and continued for sometime. The noise of the three rifle shots had created some consternation, and the hotel residents soon trooped out to see what was the matter. There was no difficulty in keeping them at a reasonable distance; so Mr Phillip's movements were not impeded ! At last the Shikari got a sight of the tiger; that is to say he

Saw his Eyes Gleaming

in the gloom. There was no doubt about it this time, and Mr Phillips soon put one of those nasty hollow-nosed bullets right between the pair of eyes. The eyes disappeared, and then after some further walking round the tiger was seen lying at length on the ground, but not dead; for he raised his head, into which another bullet was at once placed, and Stripes laid him down and died.

It required a good deal of persuasion to get several onlookers to crawl beneath the floor and drag out the body; but at length one, bolder than the rest got hold of the dead tiger's hind legs, and with assistance he was dragged forth to the light of day, to the great relief of all those persons in the hotel whose rooms adjoined the spot. The tiger was found to measure from tip to tip 7 feet 8 inches, and 3 feet 4 inches in height. The effect of the hollow-nosed bullets was tremendous. The point of entry was small, but where the bullets came out behind a big portion of the tiger's head was blown away. After being extensively "viewed", the body was carted away.

Mr Phillips

Spoilt His Coat

and we may perhaps suggest that he might be provided with a new one at somebody else's expense. There is no doubt those residents of the hotel who came home from the Coronation ball at an early hour today feel, grateful that they did not make a closer acquaintance with "Stripes". One gentleman is said to have slept last night in a long chair on the billiard room verandah. No doubt he feels particularly grateful, especially as the tiger is said by those who saw him alive to have looked a very hungry specimen.

1.1 *The Straits Times* newspaper reporting on Stripes' appearance at the Raffles Hotel, 1902.

tiger incident, recounted by witnesses, chart a frenzied and non-corroborated route of the tiger's movement through the hotel. The tiger's body retrieved from the undercroft of the billiard room, migrated to several spaces. In turn, the animal was reputedly discovered in various states: dead *in* the billiard room; asleep *on* the billiard table and jabbed awake with a billiard cue,[38] then finally, dead *under* the billiard table.[39] Although the flimsy fact of the tiger anecdote is still in question today, its significance as a founding moment in the history of the Billiard Room is unshakeable.

I am reluctant to dismiss the unruly evidence of the tiger for two reasons. First, anecdotes have been claimed by feminists as a critical mode of knowledge, circulated primarily by word-of-mouth. These stories empower the teller and her audience, enabling both to gain imaginative access to an otherwise masculine space. Second, the newspaper report and its later spin-offs portray the colonial hunter (Fig. 1.2), a masculine figure identifiable with the gendered Billiard Room, as flawed or emasculated. Cultural theorist Meaghan Morris suggests that anecdotes map 'the contours of a social terrain', modelling the way a world is working.[40] Furthermore, the anecdote, literary theorist Jane Gallop argues, is not only insistently gendered in how it is informally and intimately disseminated, it is also a form of resistance, being associated with 'exteriority … exits, departure, attempts to get out, and in particular the attempt to get out of a rut'.[41] The anecdote leads us to the frame of the encounter, and to express what cannot be said otherwise. Consequently, both method (the anecdotal storytelling) and content (the botched-up hunt) poke at the insularity of the Billiard Room as a sacrosanct male domain.

1.2 Tiger hunting in Singapore, 1890s.

Two pieces of writing by English writer George Orwell provide critical entry points into the problematic human-animal relations under colonial rule. Orwell's novel *Burmese Days* (1934) and essay *Shooting an Elephant* (1936) were written based on recollections of Orwell's service with Burma's Indian Imperial Police between 1922 and 1927.[42] Both writings, as with the anecdotal fragments that recount the tiger under the hotel's Billiard Room, present architecture and the city as interwoven with the animal-as-detail. It is through the animal that the relationships between coloniser and colonised, ruler and ruled, are simultaneously produced and destabilised. By no coincidence, Orwell's two pieces constitute accounts of a waning British empire in the tropics.

> It was perfectly clear to me what I ought to do. I ought to walk up within, say, twenty-five yards of the elephant and test his behaviour. If he charged I could shoot, if he took no notice of me it would be safe to do no such thing. I was a poor shot with a rifle and the ground was soft mud into which one would sink at every step. If the elephant charged and I missed him, I should have about as much chance as a toad under a steam-roller. But even then I was not thinking particularly of my own skin, only the watchful yellow faces behind. For at that moment, with the crowd watching me, I was not afraid in the ordinary sense, as I would have been if I had been left alone. A white man mustn't be frightened in front of 'natives'; and so, in general, he isn't frightened.[43]

> For it is the condition of his rule that he shall spend his life in trying to impress the 'natives' and so in every crisis he has got to do what the 'natives' expect of him.[44]

Orwell wrote *Shooting the Elephant* in 1936, almost a decade after he left colonial Burma for England. Disillusioned by the declining British Imperial system, Orwell resigned as a civil servant and became a writer upon his return to England.[45] His time in Burma arguably influenced his choice of subjects, particularly his decision to 'rebel against his own class, … its view of the world and man's duty within it'.[46] In the passage above, Orwell is the lone (colonial) figure sandwiched between two forces of nature. In front of him, some eighty yards away, stood the monolith of an Indian elephant; an unpredictable force of the wild. Earlier on, the beast crushed an Indian to death. Behind him, another force had to be placated: the entire population of native villagers, at his heels, excited at the prospect of witnessing a kill. When at last, Orwell released some unconfident and imprecise shots, he botched 'the job, but finally kill[ed] the beast'.[47]

Orwell recalled himself that in that moment, he was unsure which to agitate and which to placate. Despite the shotgun in his hand, he was rendered all but helpless and powerless. Orwell's tropics, unlike Mann's, pointed to a destabilising 'other'; marked by ambivalence of one's authority. In Orwell's *Shooting an Elephant*, the faceless and nameless Burmese people transform into an animated and unruly body; a fearless mob. This unsettling animal imagery extends across much of his writing on Southeast Asia wherein the unexpected intelligence of the animal/native close inwards to further debilitate the weakening of empire. For Orwell, Burmese animosity was the outcome of the empire's own hostility towards the locals, who were mistreated and dehumanised. In *A Hanging* (1931), he observes that 'condemned men [were] kept in animal cages, [and] the warders handle[d] the prisoner as if he were a fish'.[48] The refusal to relate also resulted in misapprehension. In *Burmese Days*, Elizabeth Lackersteen—an English lady who recently arrived in Burma—watches in horror as a traditional dancer's writhing, flexible and agile body twisted and contorted to snake-like and grotesque effect.[49] She is as repulsed by the smell and noise of animals, birds, and crowds thronging through the market at Kyauktada as she is reviled by the strange movements of the traditional dancer. Eventually, Elizabeth seeks the sanctuary of the Colonial Club, an unimpressive place by Orwell's accounts:

> … when one looked at the Club—a dumpy one-storey wooden building—one looked at the real centre of town. … the European Club is the spiritual citadel, the real seat of British power, the Nirvana for which native officials and millionaires pine in vain. It was doubly so in this case, for it was the proud boast of Kyauktada Club that, almost alone of Clubs in Burma, it had never admitted an Oriental to membership.[50]

Symbolic of the 'customary ways of imperial life, with its rigid patterns', the Club spatialised the sacrosanct divisions of race, gender and class privilege.[51] Orwell pairs the declining colonial power with a 'dumpy one-storey wooden building' where European masculinity was reinforced through single sex membership, and an exaggerated indulgence in the masculine game of billiards or sport hunting. Clubs were associated with the 'the consolidation of bonds between public men through the physical and ideological exclusion of women' through patriarchy, 'as well as the contestation of rivalries between certain male groups' through fratriarchy.[52] Although satellite clubs such as Singapore's Raffles Hotel and Burma's Kyauktada Club were housed in modest settings, the ideological sentiments and key symbolic spaces remained intact. As the last 'all-white' colonial bastion in the tropics, the Club replicated the social order and lifestyle of the metropole. More importantly, it spatialised 'the imperial caste system, the manifestation of the separateness of the ruling elite from the subject race', ensuring and preserving an incorruptible barrier between the ruler and the ruled.[53]

> Inside, the Club was a teak-walled place smelling of earth-oil, and consisting of only four or five rooms, one of which contained a forlorn 'library' of five hundred mildewed novels, and another an old and mangy billiard table—this, however, seldom used, for during most of the year hordes of flying beetles came buzzing around the lamps and littered themselves all over the cloth… For ornaments there was a number of 'Bonzo' [a cartoon featuring a dog] pictures, and the dusty skulls of *sambhur* [a kind of tropical deer]. A punkah, lazily flapping, shook dust into the tepid air.[54]

The colonial epicentre in Kyauktada revolved around this interior—an earth-oil scented Club in the Burmese colonial outpost with two key spaces—the library and the billiard room. Popularised as a middle-class pastime in Edwardian England, the billiard room had become an essential feature in the homes of the rich,[55] and in gentlemen's clubs.[56] The game represented a vital aspect of club life, both in London and overseas. In his chronicles of mid-nineteenth-century expatriate life in Singapore, John Cameron, the editor for Singapore's broadsheet *The Straits Times*, noted that billiard rooms were attached to respectable residences, serving as sanctuaries for the host and his male friends to retire after dinner to discuss serious business or to boast about a recent hunt.[57] Conversely, Orwell's Billiard Room was overcome by animal life of another kind. It was forlorn and atrophied. Emasculated by the heat and humidity of the tropics, decay was imminent everywhere. The library was overwhelmed by mould and mildew; the building smelt of earth-oil used to desist termites, and the billiard table was rendered useless as its smooth surface was infested with dead beetles. The animal trophies—the sambhur deer skull hung in the Billiard Room and a leopard skin—would eventually be devoured by damp and rot. The beetles gravitating towards the brightly lit club interior, migrated from the jungle into the semi-enclosed Billiard Room. Burrowing into all kinds of upholstery and fabric, including the green baize cloth of the billiard tables, the unruly tropical mob resized itself into diminutive tropical insects responsible for irreparable damage.

Drawing from Orwell's writings, the architectural history of the Raffles Hotel's Billiard Room cannot be evaluated on the basis of its design pedigree alone, even when this is corroborated by a colonial archive. The power of anecdotes, the persistence of storytelling as the primary mode of dissemination, and the centrality of the Orwellian animal imaginary as its founding moment all reaffirm the critical importance of fiction and narrative to the Billiard Room's architectural history.

Confronting the tiger, and its architecture

There are two billiard tables in the middle of the room. Though rarely used, the tables are scrupulously maintained and occupy a significant position within a building which is known today as the 'Bar and Billiard Room'—or simply the Billiard Room (Fig. 1.3). The Billiard Room is an unassuming, low-lying, single-storey building with a deeply pitched roof, large windows, and an airy verandah encircled by white pilasters. Slightly raised off the ground, its typology is consistent with the Anglo-Indian architectural tradition, popularised

1.3 A billiard table in the Bar and Billiard Room, 2019.

during British colonial rule in late nineteenth-century Singapore (Fig. 1.4). Yet, more beguiling than its architectural lineage is how this space is persistently touted for its alleged association to the tiger. In the present-day hotel's private museum, a series of framed archival newspaper articles together with historical and more recent cartoons recount in some detail how wild animals besieged the esteemed premises. In this museum, there are neither photographs of the ill-fated tiger nor the uncoordinated head teacher. There is no taxidermic specimen of the prize trophy. All that remains of this incident is the anecdotal newspaper report, of a tiger shot by a Mr. Philips. Thus, the reputation of the Billiard Room is implicitly wedded to its animal anecdote.

Despite the head teacher's dismal performance at the hotel, game hunting was fashionable in Singapore at the turn of the twentieth century. In Charles Burton Buckley's *Anecdotal History of Old Times in Singapore*,[58] a key historical chronicle recording Singapore's early social and cultural history, tiger stories are liberally sprinkled throughout the book. Yet, even as the animal was seen as lucrative bounty (large rewards were offered to those who successfully killed a tiger) and persisted as a threat in the expanding business of plantations, the tiger was also a potent figure in Southeast Asian mythical cultures, believed to be a creature which straddled the dual boundaries of the human and animal worlds. Beyond Buckley, the tiger story has also inspired contemporary cartoons and two illustrated children's stories—Kathy Creamer's *The Tiger Who Came to Tea* and Kelly Chopard's *Terry's Raffles Adventures*.[59] During the hotel's centenary celebrations in 1986, a white female Bengal tiger from a visiting circus act was

1.4 Advertisement for the Raffles Hotel showing the Billiard Room on the extreme left, c.1902.

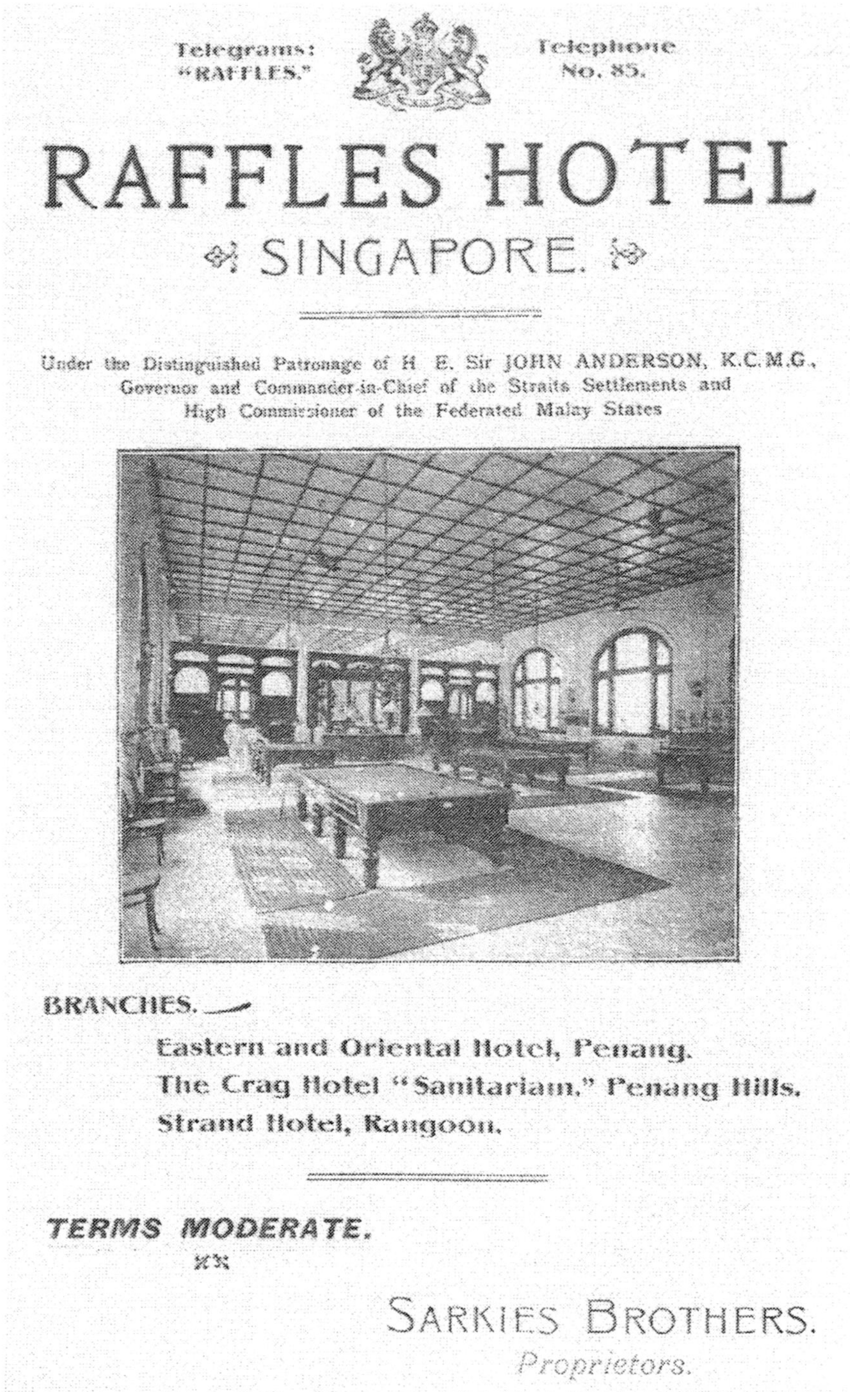

1.5 Advertisement for the Raffles Hotel showing the Billiard Room after its 1907 renovation, c.1909.

roped in by the hotel's manager Roberto Pregarz, who dressed himself up as a colonial hunter to re-enact the primal scene from a hundred years before.[60]

Stories about Stripes getting shot became so widespread that its whereabouts shifted from under the Billiard Room to under the billiard table, a point which I shall elaborate in some detail later. Given that the Billiard Room was operational between 1896 and 1917, and then only reinstalled in 1989, the circulation of these and other remarkable stories in the intervening 70 years when this space was physically non-existent suggests that the Billiard Room effectively existed as an accessory to its tiger stories—very much how I have described the incident vis-à-vis the hotel's architecture itself.

While these episodes may be read as part of the hotel's aggressive commercial propaganda, their effectiveness is fuelled by the renewed existence of the Billiard Room and its table, these acting as physical anchors for the controversial bodies—the tiger's and the white colonial male's—that are no longer present (Fig. 1.5). Indeed, the aura behind such a space implies a masculine presence, not inconsistent with the influence of colonial patriarchal power during the Billiard Room's heyday in the early 1900s. Billiards were popular in the Victorian and Edwardian eras, coinciding with the period when the Billiard Room was fully operational (1890–1917). When Tigran Sarkies, the founder of Raffles Hotel, took over the property formerly called Beach House in 1887, he commissioned Regent Alfred Bidwell, the hotel's architect, to design and construct a purpose-built billiard room (Fig. 1.6).[61]

1.6 The Bar and Billiard Room, c.1905, prior to its 1907 renovation.

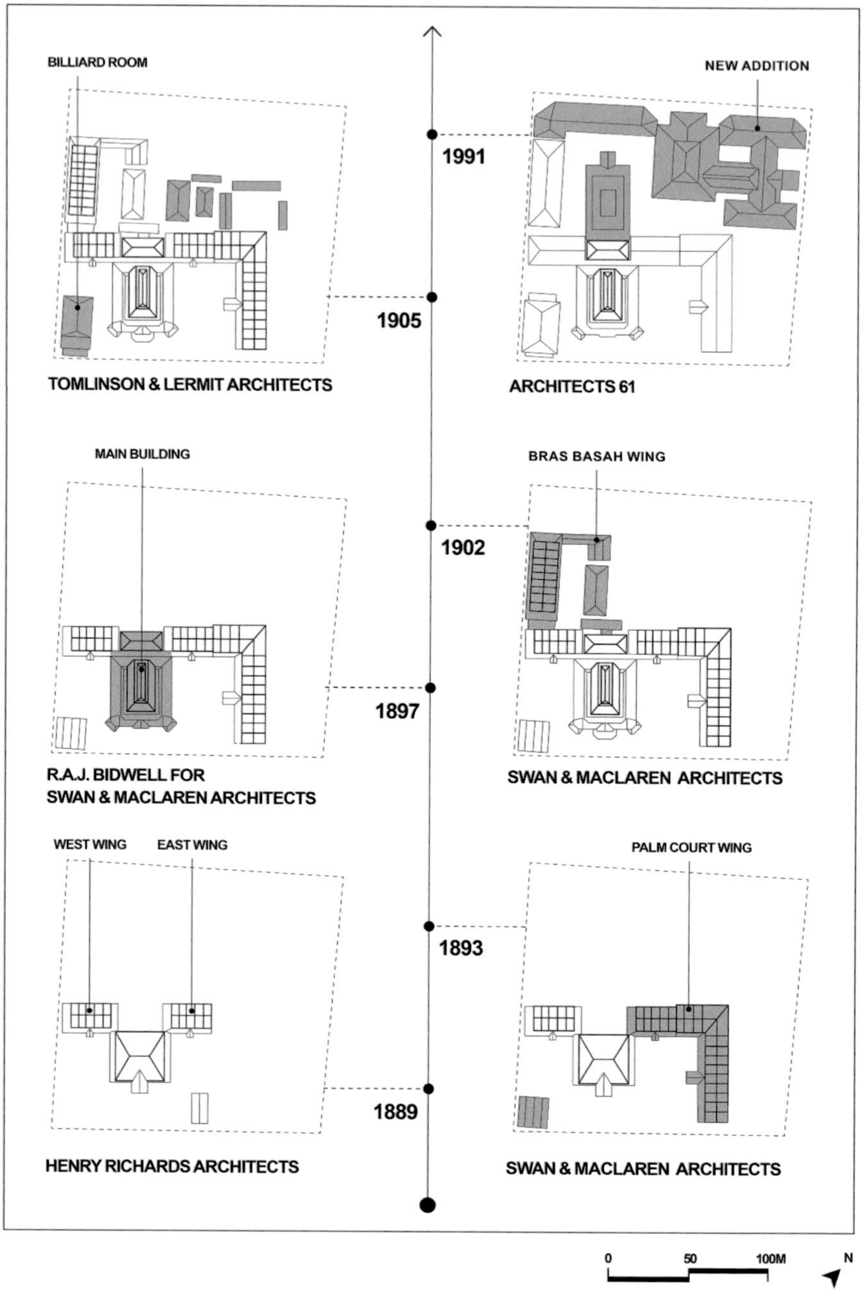

[Above] **1.7** Additions and alterations at the Raffles Hotel between 1889 to 1991.

[Opposite] **1.8** Southwest elevation of the Raffles Hotel and elevational detail of the Bar and Billiard Room facing Beach Road in 1902; longitudinal section of the Billiard Room in 1905.

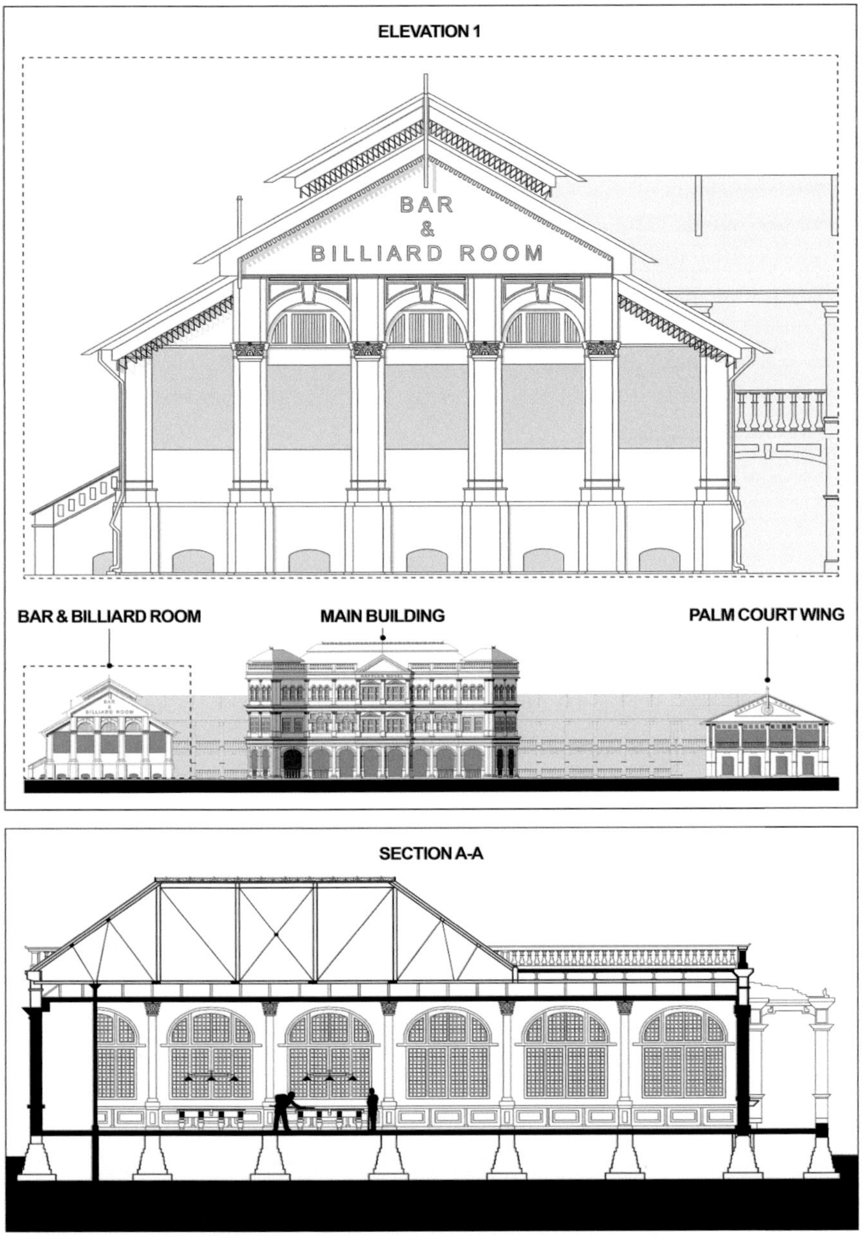
ELEVATION 1
BAR
&
BILLIARD ROOM
BAR & BILLIARD ROOM
MAIN BUILDING
PALM COURT WING
SECTION A-A

NOT TO SCALE

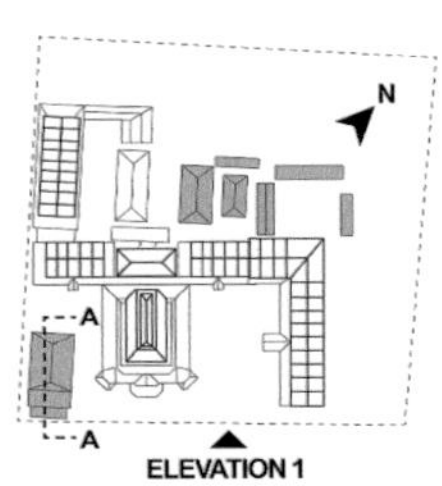
N
A
A
ELEVATION 1

Bidwell's original building was ready in 1896 and featured a raised structure on brick pillars to ensure a level floor for the game. By 1906, the popularity of this first Billiard Room (Figs. 1.7-1.8) necessitated larger accommodations.[62] It was likely that the Billiard Room at the Raffles Hotel had also doubled up as a club to the expatriate British community,[63] since, in 1911, male residents outnumbered women residents by eight to one, and the majority of those staying at the hotel were bachelors.[64] The new addition, by architects Tomlinson and Lermit, was modelled after the Anglo-Indian detached bungalow, and the undercroft space where the escaped tiger hid in 1902 was eventually sealed off. By 1917, demand for rooms trumped the game and the Billiard Room was converted into guest suites. It was only during the restoration of 1989 that a decision was taken to reinstate the Billiard Room as a key space for the hotel, then newly gazetted as a national monument.

Natural history and local myth

> It is not surprising that in the presence of animals in general (metaphorical, mythical, or otherwise), we are forced to consider the mechanisms of control, and simultaneously, waywardness, and thus to consider morality, rationality, order, civilization, cities, and architecture.[65]

Panthera tigris has haunted the Singapore psyche long before the 1902 incident at the Raffles' Billiard Room. In fact, the persistent return of the tiger, a species that has existed in the Riau-Lingga archipelago, which includes this island since its geological beginnings a million years ago, indicates traces of unresolved conflict between humans and beast, colonial and native, culture and nature, the natural and the supernatural, fact and myth. The tiger has accumulated a contradictory reputation. It is simultaneously seen as 'treacherous and the source of death and fear'[66] as much as it is lauded for its stunning beauty, 'with his velvety step and undulating movements, the firm muscles working beneath his loose glossy skin, and the cruel yellow eyes blinking under the sun over a row of ivory teeth'.[67] The tiger in the Billiard Room marks the repetition of a primal repressed scene, one which needs to be taken apart in the local context in order to understand its political significance.

In the colonial tropics, the perceived threat of the beast was strategically symbolic since it was commonplace for colonial governments to assert their supremacy over territory that was deemed 'dangerously animal, wild and unfree'(Fig. 1.9).[68] According to historical accounts, the tiger menace in Singapore (and contiguous southern Johor) escalated in the 1840s after large tracts of forests were cleared to make way for lucrative black pepper and gambier plantations, which formed almost 75 per cent of Singapore's cultivated area in 1851. Buckley records that the majority of bourgeois colonial occupants supported a more liberalised land leasing system under the hand of Chinese planters as this simultaneously modernised 'the jungle and marsh' as well as encouraged the island to grow as 'a highly productive settlement'.[69] However, enthusiastic efforts to clear the interior reaches inevitably destroyed the tigers' natural habitat. The rise in importance of such economic crops coincided uncannily with the increased threat of tigers on the island. Taming the 'tiger-infested' jungle, a description synonymous with sparsely occupied and risky areas outside the city limits, was one way of keeping the beast at bay and expanding the island's productive boundaries. One could speculate that the tiger's appearance in the city was a calculated attempt by the colonial government to justify impending land clearances.

Subsequent to the colonising of arable land, livestock and plantation workers became prey to tigers. The exact number of casualties remained unverified, being often recklessly over-inflated to illustrate the risk of the oriental natural world through western perception, or guardedly underreported by plantation owners cautious about scaring off the local Chinese workforce.[70] For example, mid-nineteenth century colonial accounts reported that '365 men per annum have their lives dashed out by the crushing stroke of the tiger's paw',[71] and further in support of this estimate, that 'five men in eight days, as early as 1840 … was not improbable'.[72] Indeed, apart from India, 'there are no historic records of any country suffering a mortality from man-eaters comparable to that experienced for so long and so widely' in Singapore.[73] At the same time, rewards to trap and kill tigers were sponsored by the colonial government. These handsome

1.9 Heinrich Leutemann, *Interrupted Road Surveying in Singapore*, c.1865.

bounties, which fluctuated in value between 100 Straits Dollars (in the 1840s and 1860s, the price for a tiger either captured dead or alive) and 50 Straits Dollars (in the 1850s), successfully lured both local and European game hunters.[74]

Beyond historical statistics, the wild beast and the threat, or promise, posed by such animality is an especially enigmatic part of Singapore's (still contentious) founding story. In the popularised version of the *Malay Annals*, a compilation of oral narratives of archipelago Malay history, a Sumatran prince, Sang Nila Utama, also known as Sri Tri Buana, arrived at an island called Temasek, the ancient name for Singapore, in the thirteenth century and spotted an ungainly beast:

> And they all beheld a strange animal. It seemed to move with great speed; it had a red body and a black head; its breast was white; it was strong and active in build, and in size was rather bigger than a he-goat. When it saw the party, it moved away and then it disappeared. And Sri Tri Buana inquired of all those who were with him, 'What beast is that?' But no one knew. Then said Demang Lebar Daun, 'Your Highness, I have heard it said that in ancient times it was a lion that had that appearance. I think that what we saw must have been a lion.[75]

Sang Nila Utama thus established his kingdom *Singa Pura* (Sanskrit for 'Lion City') possibly based on what he had witnessed. But as the Asian lion has not been known to exist east of India, it was likely the Sumatran prince and his entourage had encountered a tiger. Here, faced with the wild beast for the first time and 'working on the flimsiest of heresay', Sang Nila Utama's aide Demang Lebar Daun declares it as the revered lion or *singa,* henceforth putting the unknown animal in 'allegorical disguise for the establishment of Singgahasana or "lion throne"' symbolising Malay royalty.[76] Following this popular myth, it is significant that this city was established not only on the uncertain sighting of a beast, but also that its foundations may have been laid based on a fictional identification of this animal, two points which lend comparisons with the founding moment of the Billiard Room at the Raffles Hotel, in the wake of its own animal encounter and its early stories.

In the Southeast Asian context, the tiger is believed to be an animal endowed with supernatural powers. As historian Peter Boomgaard writes, tigers are 'protagonists' in all manner of stories, be they 'myths, legends, fairy tales, and fables', even visually subjects in 'paintings, carvings and sculpture';[77] there were superstitions about tigers. Hence, prior to the nineteenth century, Southeast Asian cultures, particularly those of Java, Sumatra and the Malay Peninsula, loathed slaying tigers. More than a control for crop-damaging and foraging animals such as the wild boar, tigers were believed to be an embodiment of one's ancestors and could provide protection to humans; they were said to be 'receptacles for the souls of departed human beings',[78] and thus, stood at the boundary 'between human and animal, and between the living and the dead'.[79] It was also believed that the tiger had shamanic powers and—through a process of *jadi jadian* (the changing of form)—morph into human by day, and back into beast by night[80]—'weretigers'.[81]

Thus, tigers 'as a mythic reality, played a powerful role in boundary maintenance. They were transitional creatures, spiritual in-betweens who variously cautioned, cajoled and assisted humans in relating to the life-world'.[82] Significantly, as Boomgaard and other historians like Robert Wessing and Kevin Chua unanimously argue, the tiger did not come into existence apart from, or outside of, human civilisation but more as a consequence of humanity, 'not [a] distant beast of nature, [it is] instead [a] permanent spectral embodiment of urban civilization, conjured up by and for us' (Fig. 1.10).[83] This point reinforces the tiger's cultural and political roles, and its need to be 'tamed' in deference to colonial territorialisation.

Thus, it is unsurprising that there is an enduring relationship between the animal narrative and the founding of space in this region. However, the animal narrative also establishes spatial and political boundaries through a 'logic of ambiguity', that is, through both sanctioned storylines as well as versions which misrepresent the story's content. In this way, the tiger stories create what Michel de Certeau calls 'a frontier'.[84] Referring to a third element or a space in-between, this frontier bridges what constitutes status quo spatial knowledge, such as the architectural histories established by sanctioned producers including architects and planners or legislated according to construction and planning laws, with what we viscerally experience, see, and understand. The latter category exists as fragmented, particularised and polyvalent sets of knowledge, not easily co-opted into 'authorizing references'. At the Billiard Room, the tiger story sets up frontiers of knowledge that filter, mediate and transform the predetermined history of its architecture. These stories challenge conventional wisdom, altering the contours of this space from that of austere masculinity to a frontier marked by tropical surplus and ambiguity.

Telling the tale from under the table

In one of his many memoirs recalling past adventures in Singapore, wildlife expert, hunter, actor, circus performer, world traveller and wild animal purveyor Frank Buck—an American who maintained a Singapore compound with the largest private collection of live animals in one place, and a frequent patron of the Raffles Hotel during the interwar years—tells how the hotel's tiger story was so popular that different variations kept cropping up. Grounding his

ENTRANCE HALL.

1.10 Taxidermy specimen of tiger in a vitrine at the Main Entrance of Singapore's Raffles Museum, 1908.

own reports on the factual solidity of the architectural space, Buck reiterates, 'There is no questioning that a tiger was found in the old billiard room at the Raffles Hotel', and adds that, 'it would have been a simple matter for a tiger … to stray into the Raffles billiard room, which was then a big open space, with no doors or windows, on the ground floor'.[85] Buck then cites three different accounts of this story:

> One of the stories they tell you with a straight face is that the Chinese boy who came to clean up the room at six in the morning had found the tiger asleep on a billiard table and had prodded it with a cue. I thought that the native waiter who told me this story was joshing, and my response to it was a hearty laugh. I soon saw I had offended him and pretended to believe the story. [C]an you imagine anyone crazy enough to jab a tiger with a billiard cue? [A]nother version (the story of one of those eye-witnesses, who, it later developed, wasn't even in Singapore at the time) is to the effect that the tiger, in his rush to get out of the room, knocked over the Chinese boy and made a dash for the kitchen. There, so the yarn goes, the animal helped himself to half of the beef which he dragged out on the porch, where he parked himself and ate his meal…
>
> Mr Aratoon, the manager of the hotel, with whom I discussed this Singapore legend, offers the only sensible explanation. Unquestionably, he tells me, a tiger had entered the billiard room. Three reliable people had seen the animal make its escape. The big cat, he believes, reached the island shortly before dawn. Bewildered by its surroundings, it had dashed from street to street in an effort to find a way out. Seeing the open space on the ground floor of the hotel, it had entered, hoping that perhaps this would lead to open country. [W]hen the boy entered to clean the room, the animal was probably pacing up and down wondering what to do next. Seeing the boy, the beast's first though would be to escape doubtless then it made its dash to safety.[86]

In his account, Buck begins with what he takes to be the immutable facts of the Billiard Room—a simple open space on the ground floor, which momentarily housed the tiger. But in the first story which Buck tells us, the tiger is no longer 'factually' placed under the Billiard Room as documented in the 1902 newspaper report, but found *inside* the room itself, on the billiard table. In the second story, the tiger is relocated to the kitchen, before landing up on the hotel porch. The structure of the old Billiard Room is confirmed in architectural drawings and in the first newspaper report, as a standalone pavilion-type building, raised on brick pillars. Thus, it is difficult to imagine how the tiger could have rushed into what is essentially a space raised off the ground, mistaking a roofed structure for 'open country'.

Even as he admonishes the fabrication of stories, Buck is himself indulgent. He anticipates the 'beast's first thought', tells his readers what the tiger would have experienced and projects how it would have acted. Strikingly in Buck's versions, the tiger appears not to have been shot, nor is there mention of Mr Philips' involvement. In these accounts, the tiger stories ground the experiential foundation of the Billiard Room. Yet, these stories tend to be embellished. They change according to the time of telling and fluctuate in relation to the audience and the storyteller. The story is always anecdotal, which makes it immediately problematic to structures that require certainty and fixity, such as scripted history. The anecdote, as architectural historian Barbara Penner points out, is regarded with suspicion and often alienated in academic discourse.[87] Dictionary definitions link the anecdote to 'a tendency to tell too many stories', and oppose it to 'corroborated evidence or proof'.[88] Nevertheless, the founding moment of the Billiard Room in the public's imagination is implacably structured on the anecdotal, as the *The Straits Times* report awkwardly prefaces: 'A tiger was shot under the Billiard Room of the Raffles Hotel early this morning. Lest any one be inclined *to doubt the veracity* of this foregoing statement'.[89] But even more importantly, the power of the story is its degree of emancipation. The storyteller internalises the story, either from first-hand experience or through another's telling, and in retelling it anew, explores and constructs 'the rhythm of the work (that) has seized her'.[90] The audience also becomes instrumental to the construction of the story as the teller surrenders the narrative to them, going beyond 'information'.[91]

At the Billiard Room, the animal story transforms how one perceives this space. The tiger was originally reported to have hidden in the underbelly of this space, but later mistakenly believed to have strayed under one of its billiard tables. While conventional architectural meanings linked to billiard rooms are based on enforced patriarchal hierarchies, in the Raffles' Billiard Room, the anecdote exposes the space to factual irregularities, excessive speculation and subjective claims. It is significantly entered from the 'lower' edges—from under the floorboards, between an animal's hind legs, framed by a pair of pyjamas, and surrounded by a cowering group of male hotel boarders, who reputedly lacked the courage to pull out the dead tiger's body from under the building after it was shot.

Also, rampant word-of-mouth fuelled by intense public curiosity about the incident coupled with the physical porosity of the Billiard Room were reasons enough for the tiger's whereabouts to be severely distorted. Like the stories told by Buck, erroneous reports began to circulate about the tiger being spotted under the billiard table. Notably, a later newspaper account from the *The Straits Times* published a while after the first and only tiger sighting at the hotel similarly misreported that in 'the recent tiger incident ... one of these "terrors of the jungle" was shot beneath one of the billiard tables at the Raffles Hotel'.[92] These reports were all the more confusing given that some of them emanated from the hotel itself. For example, the hotel's cultural historian Raymond Flower wrote in its 1986 centenary commemorative volume that the head teacher Philips saw the tiger under the billiard table, 'and fired three rapid shots, only to demolish one of the legs, and to bring down the table with a crash'.[93] On another occasion, also in 1986, Pregarz, the hotel's manager, placed a rented circus tiger (brought in to celebrate the lunar 'year of the tiger') on top of the billiard table, thus embellishing the mistake made by Flower and further confounding the puzzle of where the tiger hid.[94] The aim of reiterating these different anecdotes is not to locate an 'authentic' version of events but rather to show how the event was amplified and repeatedly misrepresented.

Today, in the absence of the undercroft space where the tiger was shot, the billiard *table* itself restructures the architectural experience of the Billiard Room. With the disappearance of the building's undercroft space following renovations undertaken in the first quarter of the twentieth century, the billiard table began to assume the status of the imaginative object onto which the animal anecdote hinges. Significantly, the technical specifications for a billiard table reiterate the importance of its uprightness, solidity, firmness, elegant design and the smoothness of its surface to ensure the precision, speed and accuracy of the game.[95] It is striking that there is a resemblance between the qualities of a reputable billiard table and the Vitruvian principles of architecture, the latter similarly exalting beauty, commodity, uprightness, firmness and strength as architectural virtues.

To look under the table is to look in an unauthorised place. Metaphorically, a perspective of the room from 'under the table' implies an upside-down view. It reflects a child's perspective, which may explain why the hotel's animal story survives most vividly today as part of a collection of children's stories. It challenges what is upright, normal and orderly. Something that is 'under the table' can also be something hidden or repressed, while to turn the tables on someone or something suggests a subversive reversal of positions. This conceptual recasting is rehearsed in the grotesque presence of the animal.

Animality and grotesque architecture

In the humid tropics buildings are susceptible to the forces of nature, corrupted eventually by damp and wood rot, for example. As architecture, matter and time are evanescent, so too is history especially in its oral forms—becoming more fluid, more story-like, and more susceptible to hearsay and erratic narration. Anchored by stories woven around, and under, the billiard table, the formation of its architectural knowledge shifts from the customary emphasis on immutable and controlled facts, be they about architectural style, history or masculine meanings associated with the Billiard Room, towards the unregulated potentials of the animal anecdote which was transmitted and received by different groups of people who encountered this space. The ill-fated tiger in the Billiard Room was manifested as a sensual, feminine and grotesque body overlaid on a heterosexual, masculine, colonial stomping ground. The grotesque body, according to literary critics Peter Stallybrass and Allon White, is marked by 'impurity (both in the sense of dirt and mixed categories), heterogeneity, masking, protuberant distension, disproportion, exorbitance, clamour', they go on.[96] It is Other, unhomely, a gross mélange.

This grotesque body operates by inversion, that is, by turning something inside out, or upside down. Thus, the tiger anecdote effectively produces the Billiard Room's 'new' architectural body from bottom up as it were: The hunter could not see into the dark of the undercroft and so was forced to go down on all fours. This unusual posture robbed him of accuracy given by upright vision, and thus, he surrendered to his gut instinct. In his 'overturned' condition, he shot wildly and erratically. The space drawn out from this encounter is not one commandeered from an authoritative one-point perspective, a vantage point customary for a standard architectural account. Instead, we are given a fragmented, animalistic, out of control and lowly view—a simulation of a quadruped's perspective, or an unknowing child's perception of the world. This concept aligns with the distorted tiger lore, which displaces the animal from the undercroft of the building to under one of its tables. It articulates a similar inversion, by shifting the experiential construction of this space from architect-centred to storyteller-invented.

What I find compelling is how the tiger story destabilises the Billiard Room as an eccentric space by punctuating the account with irregularities, misinterpretation and individual embellishment, all of these working hard to disrupt scripted architectural accounts. The tiger, a trope that represents the untamed jungle, is also a tropical symbol par excellence. Overlaid onto the Billiard Room, this tiger, Stripes, presses forward the incommensurable differences between the monumentality of empire and the unruliness of the tropical world and its denizens. 'Architecture', as architectural theorist Catherine Ingraham argues, is conversely the instrument of empire since it acts to control that which it confines:

> Architecture is, in some sense that which acts in opposition to the animate. It is the embodied principle of the inert. It would be interesting to say that architecture negotiates the divide between the animate and inanimate, but I think a more accurate description is that architecture controls or attempts to control this

> divide in order to maintain the propriety, the seemliness, of the body in space.[97]

Thus, just as architecture may be associated with progress and empire, the question of animality is linked to empire's 'other'. Drawing on Levi-Strauss's observation in *Tristes Tropiques* that '"architecture" seems to arrive in cultural history at the same time as the idea of "empire"' and empires are significantly 'created through a massive infusion of political and hegemonic force into the material world', Ingraham emphasises that 'architecture is routinely used as one of the symbolic markers of this empirical dominance, since it can concentrate material wealth, natural energy, and political process into one act, one site'.[98]

The 'uprightness' of architecture is key to such symbolism, alluding not only to 'structural and morphological analogy but also architectural associations with civic appropriateness and human propriety'.[99] In other words, colonial architecture's uprightness is both structural and metaphorical—its claims to be evidentiary and empirical, thus also serving as 'the model for and about reality', is predicated on a clear separation between the coloniser (who is here credited with the active production of 'architecture') and the colonised (who is assumed to have passively inhabited this space). Animality is conceptually regarded as 'atavistic' and 'primitive', the 'other' to civilisation, and thus opposed to the empirical ideals of architecture, which in this case is equated with the colonial era hotel's Billiard Room.[100] Writer David Quammen reminds us that colonisation is not just the conquest of land and the disciplining of its subjects but entails a weeding out of its archaic superstitions: 'until you've exterminated their monsters'.[101] There is more at stake in the architectural archives' omission of the Billiard Room's tiger.

The tiger anecdote is a subversive instrument, antithetical to the aspirations of colonisation and its monument, even when it appears in its most benign form, as a children's story. In the late 1990s, following major restoration work at the hotel, the Billiard Room became the subject in two works of children's fiction—Kelly Chopard and Patrick Yee's *Terry's Raffles Adventures* (Fig. 1.11), and Kathy Creamer's *The Tiger Who Came to Tea* (Fig. 1.12).[102] In these accounts, the grotesque body is replayed through the recurrence and recasting of the animal story whose 'invitation to pleasure is invariably an invitation to subversive pleasure'.[103] The stories perform the historical scenes within the Billiard Room using humour and caricature to parody the overblown sense of self-importance and power invested in colonial identities. As cultural theorist Steve Baker and feminist writer Ursula Le Guin underscore, to identify with the animal is ultimately to consort with an inexplicable 'other'—specifically with 'women, children and animals'.[104] The Billiard Room cannot be divorced from its animal, and thus, is ultimately also implicated in the feminised foundations of these Others. As these new 'animal stories' are circulated to another generation of readers, it is important to consider how a historical masculine hunting tale is

transformed, and how the Billiard Room is inevitably encountered in gendered terms, bearing in mind that these stories were written for and read by an Other—that is, they were written to be read to children by their mothers. It also raises important questions as to how the architectural knowledge of a masculine space such as the Billiard Room is ultimately sustained by a fundamentally feminine practice with a strong oral tradition. De Certeau argues that storytelling is the 'delinquent' double of history while Jane Rendell sees it as a political activity that enables a reclamation of a history dominated by 'men's anecdotes'.[105]

Today the tiger story has taken yet another turn in that it has been commodified by the hotel; tiger lore still thrives in the hotel, only commodified. In its shop, there are cuddly tigers, tiger print t-shirts and tiger posters on sale. The Billiard Room continues to honour the game by keeping two tables, one marked as an 'original' dating back to the time of Stripes' visit. A tiger-themed billiard exhibition match was hosted here in the 1980s. At the bar, guests drink out of mugs emblazoned with a tiger motif and are given keepsake matchboxes adorned with a tiger image—its image is everywhere. If the tiger is emblematic of tropicality and its attendant constructions, then this problematic transformation

[Opposite] **1.11** Recounting Stripes' misadventures in Kelly Chopard and Patrick Yee's *Terry's Raffles Adventures* (1996).

[Above] **1.12** The tiger chase from Kathy Creamer's *The Tiger Who Came to Tea* (1996).

also reveals something about how the tropics is contextualised today—as manipulatable and consumable. The Billiard Room encountered through its tiger story—through the embodied acts of looking from the underside, hunting down uncertain facts, and the oral traditions of small talk and storytelling—is not a passive space. The architecture of the Billiard Room with which Stripes abets, is reproduced through this animal. If the tropics has been linked to the feminine, the wild, and the degenerate, then the Billiard Room at the Raffles Hotel is evidence of a distinct tropical colonial space which destabilises its masculine foundations. Encountering the Billiard Room through its tiger lore tells of the tropics' 'subversive pleasure'.•

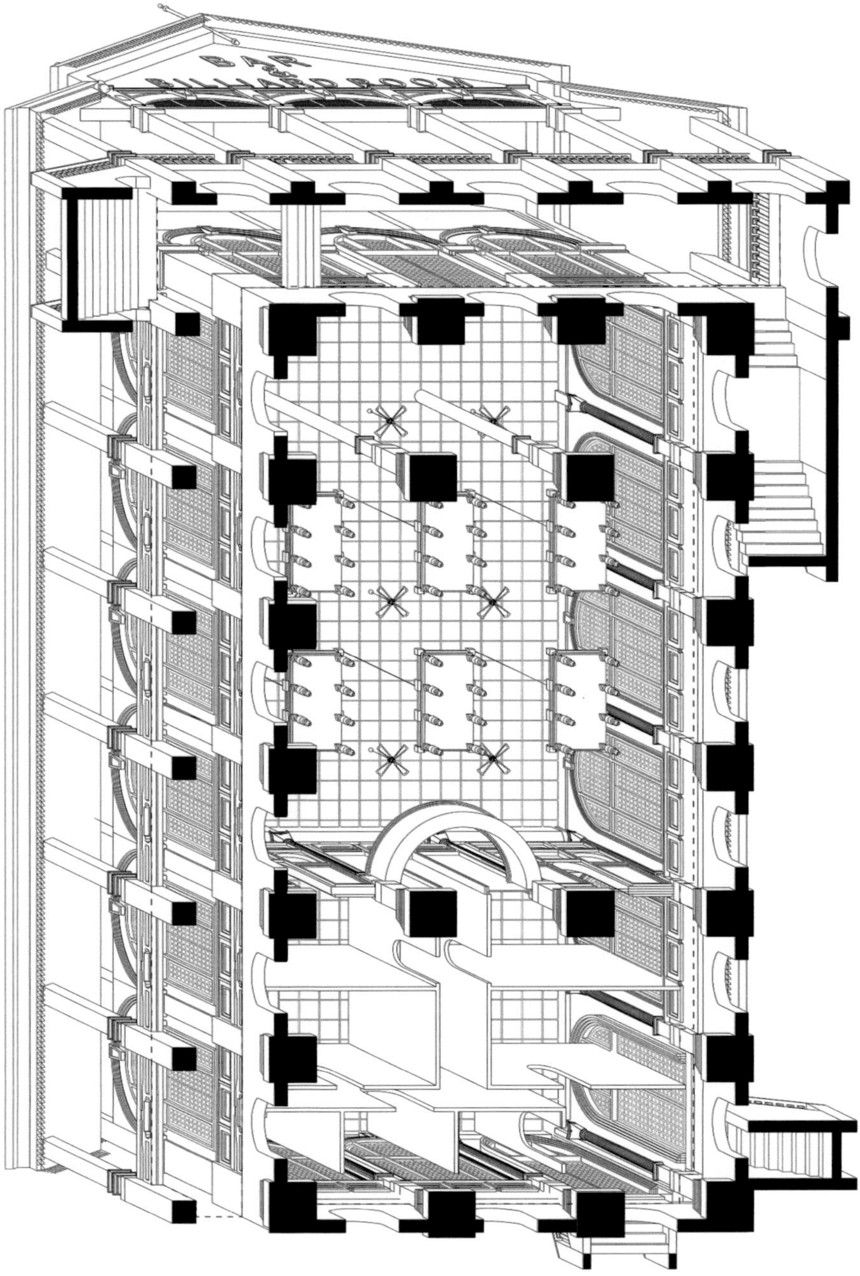
BAR
BILLIARD ROOM

THE THIRD ARCHIVE

2 • Tracing the Last Tiger

Lilian Chee and Toby Fong, 2020.

Tracing the Last Tiger revisits the anecdotal encounters with free-roaming tigers at forested locations in a still undeveloped Singapore between 1923 and 1930. Three reports featuring The Cashin Tiger (1923-1926), The Bukit Timah Tiger (1928), and The Choa Chu Kang Tiger (1929-1930) were recorded. Scientific evidence postulates that Singapore's wild fringes could only have supported a maximum of three tigers. Corroborating this evidence, archival records affirm only three or four free-roaming tigers in Singapore then. However, no less than twenty tiger incidents were officially reported in the decade leading to the last tiger's death in 1930. These reports suggest a sizable local tiger population in the rapidly depleting jungles of the island. Yet, with only three tiger killings officially registered in the same time period—on the West Coast Road (September 1928), in Sungei Tengah near Jurong (February 1929), and—the last known free-roaming tiger—on the Choa Chu Kang Road (February 1930), questions surrounding the missing tigers have emerged.

Why was there such a large discrepancy between the actual and imagined tiger population?

Located in the small but intense tropical secondary jungles of early twentieth-century Singapore, *Tracing the Last Tiger* offers a speculative spatial reconstruction of the contradictions surrounding the official, and unofficial, tiger sightings. Conceptualised as a long landscape scroll with a bidirectional timeline linking together the entangled jungle in three sites and their accompanying events, the drawing performs the experience of these tiger sightings, stringing together repetitions, coincidences and disjunctions which resulted in a tiger lore of disparate and conflicting anecdotes. The drawing is inspired by the workings of a detective game, wherein a series of clues encourage suspicion and speculation.

The language of the drawing is a twist on the popular nineteenth-century colonial tiger paintings found today in major art galleries. In particular, it takes after a lithograph made in the 1880s titled *Road Surveying Interrupted in Singapore*.[1] This print shows architecture's proxy in the figure of George Drumgoole Coleman, Singapore's Superintendent of Public Works, being ambushed by a tiger in the midst of surveying the jungle with his theodolite.[2]

The last tiger was metaphorical in its significance. It was depicted as a protean, shapeshifting creature which mirrored the anxiety precipitating change as Singapore's rural fringes were progressively transformed by urbanisation. The drawing depicts perception in movement. Giving obsessive attention to archival 'witnesses'—often attributed to specific objects scattered in the jungle—and in attempting to create a precise reconstruction of the encounter, the scale of the scroll drawing frequently changes from the very small to the very large. When moving from left to right, the sequence of official news reports unfolds chronologically and reads factually. However, read in the reverse direction, the anecdotally informed 'facts' quickly unravel. All that was once precise becomes approximate and circumstantial. The veracity of the archive is questioned, as is historical fact since these are ultimately installed and given their status through privilege to an archive.

Tracing the Last Tiger visualises the outcomes of a research constructed through the intermediaries of the tiger encounters, the erratic information and partial perspectives of anecdotes, and their disputable material traces. These anecdotes, however, continue to 'stick'. They are a crucial part of material thinking, and they continue to demand attention. They affect how we see, relate to, and understand a space—a jungle—which no longer exists. They are speculative; they are historical.•

1. Heinrich Leutemann, *Unterbrochene Strassenmessung auf Singapore (Road Surveying Interrupted in Singapore)*, 1865-85, wood engraving, Collection of National Museum of Singapore, National Heritage Board, Singapore.

2. Ho Tzu Nyen's *One or Several Tigers* tells the story of Singapore through the tale of the Malayan tiger – a very powerful symbol that dominates traditions, lore and religions throughout Southeast Asia. The two key characters in Leutemann's print, *Unterbrochene Strassenmessung auf Singapore* – the tiger and the surveyor – are referenced in *One or Several Tigers*, where the former embodies nature's untameable powers and the latter represents the conqueror's ruthless reasonings. See: Ho Tzu Nyen, *One or Several Tigers*, 2017, shadow puppets, LED lights, smoke machine, show control system, and video (2-channel, 10-channel sound, 16:9 format, colour), 33 min, National Gallery Singapore, Singapore, https://www.nationalgallery.sg/see-do/programme-detail/740/light-to-night--one-or-several-tigers-by-ho-tzu-nyen.

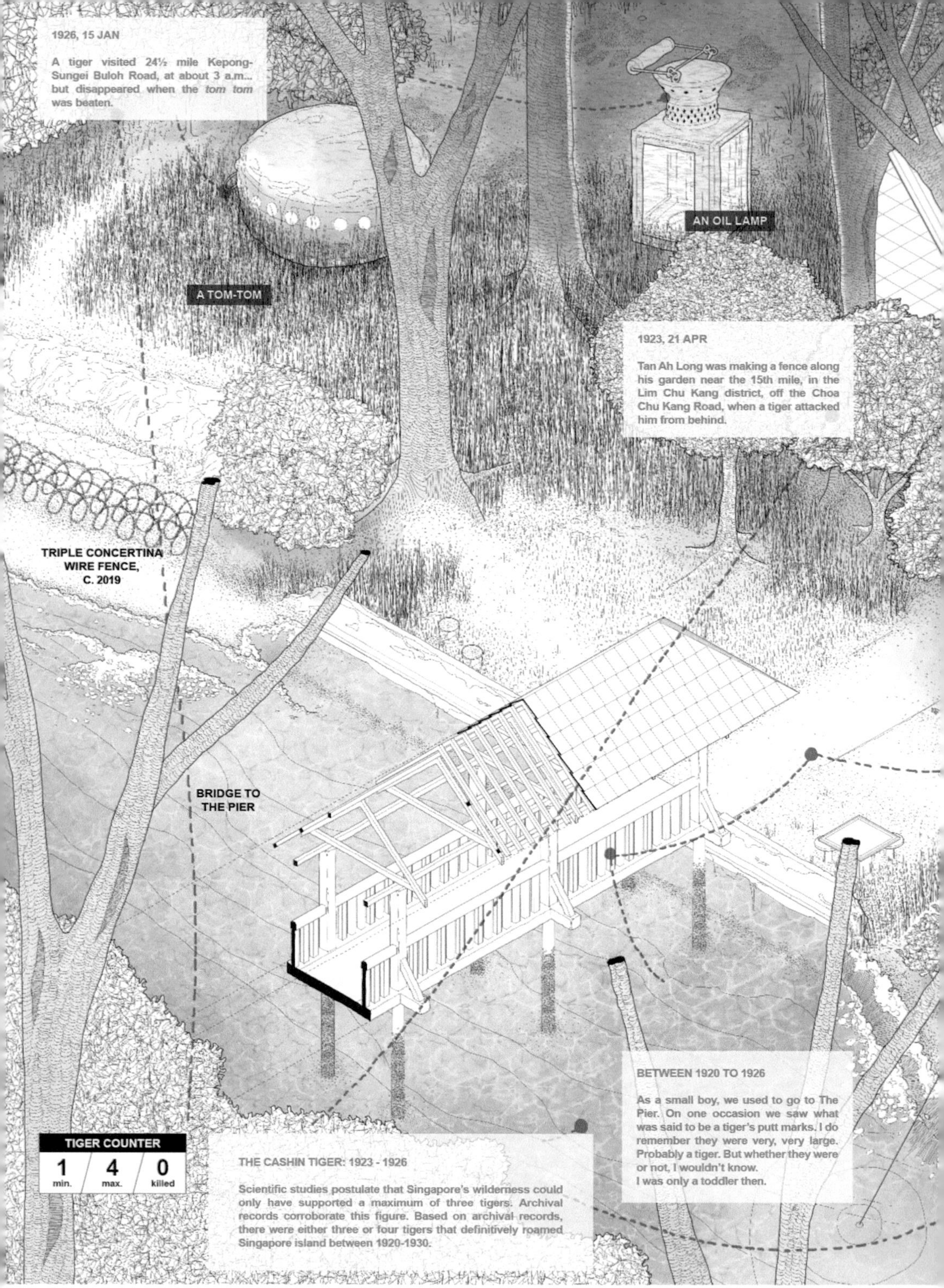

ARCHIVAL NEWSPAPER REFERENCES:

A collection of newspaper articles between 1923 to 1930: *Straits Times*, 'Tiger on Singapore Island', April 21, 1923, 8; *Straits Times*, 'A Fastidious Tiger: Remarkable Incident At Sungei Buloh', October 30, 1924, 9; *Straits Times*,

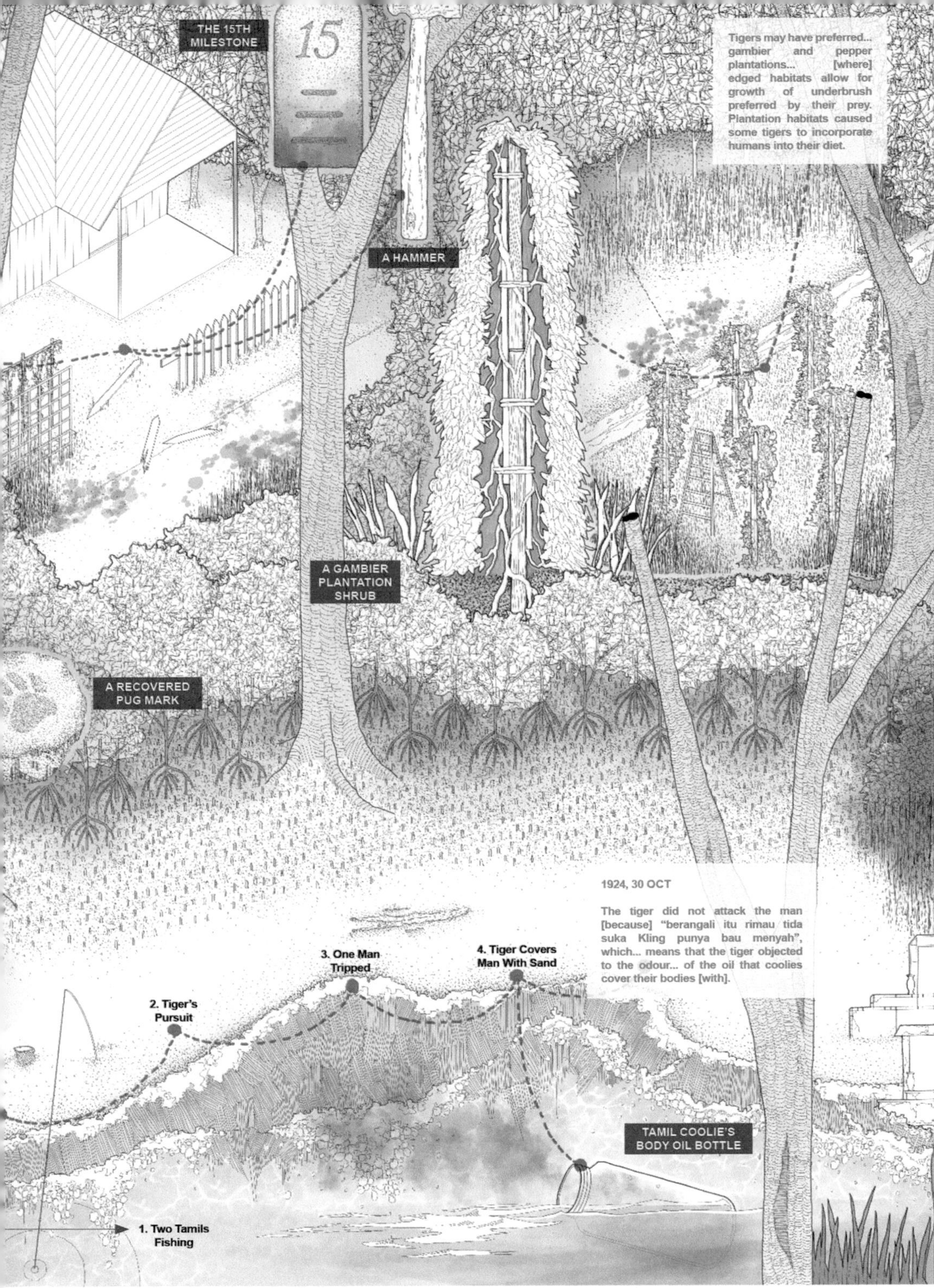

untitled, November 19, 1925, 8; *Straits Times*, 'Coroner's Heavy List: Clawed by Tiger: Alleged Murder and Suicide', November 20, 1925, 5; *The Singapore Free Press and Mercantile Advertiser*, 'Tiger At Seletar', November 25, 1925, 9; *Straits Times*, *untitled*, January 22, 1926, 8; *The Singapore Free Press and Mercantile Advertiser*, 'Tiger At Large, February 6, 1928, 9; *Malaya Tribune*, *untitled*, February 6, 1928, 8; *Straits Times*, *untitled*, July 19, 1928,

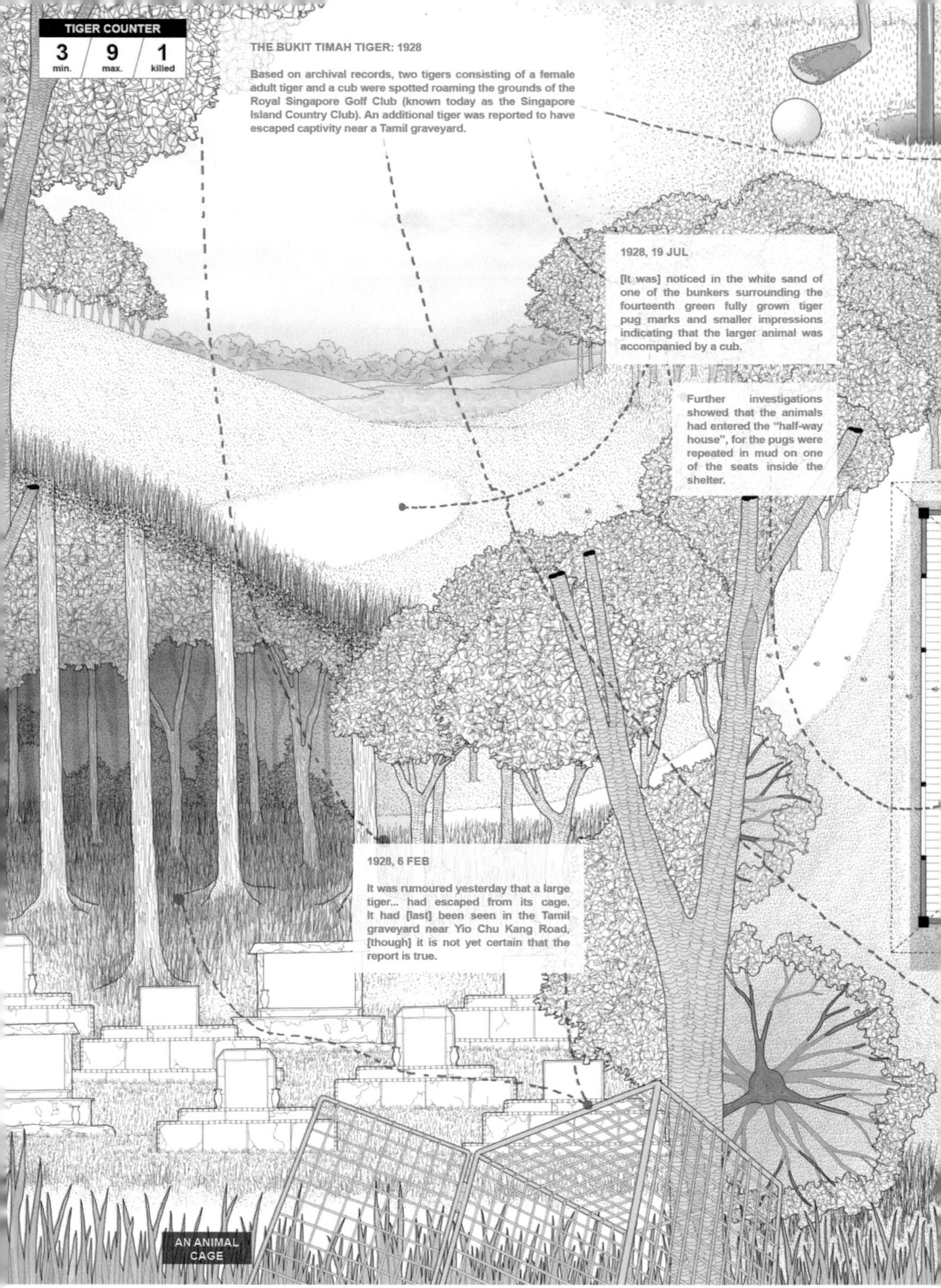

10; *The Singapore Free Press and Mercantile Advertiser*, 'Brighter Golf: Tigress Roaming Bukit Timah, Pugs in the Bunker', July 19, 1928, 16; *Malaya Tribune*, 'Tiger Shot in Singapore', August 13, 1928, 10; *Straits Times*, 'Tiger Shot on Island: Hunter's Success on West Coast', August 13, 1928, 8; *Malaya Tribune*, *untitled*, August 17, 1928, 10; *Malayan Sunday Post*, 'Tiger Shot in Singapore', August 18, 1928, 5; *Straits Times*, 'Another Tiger?: The Terrier,

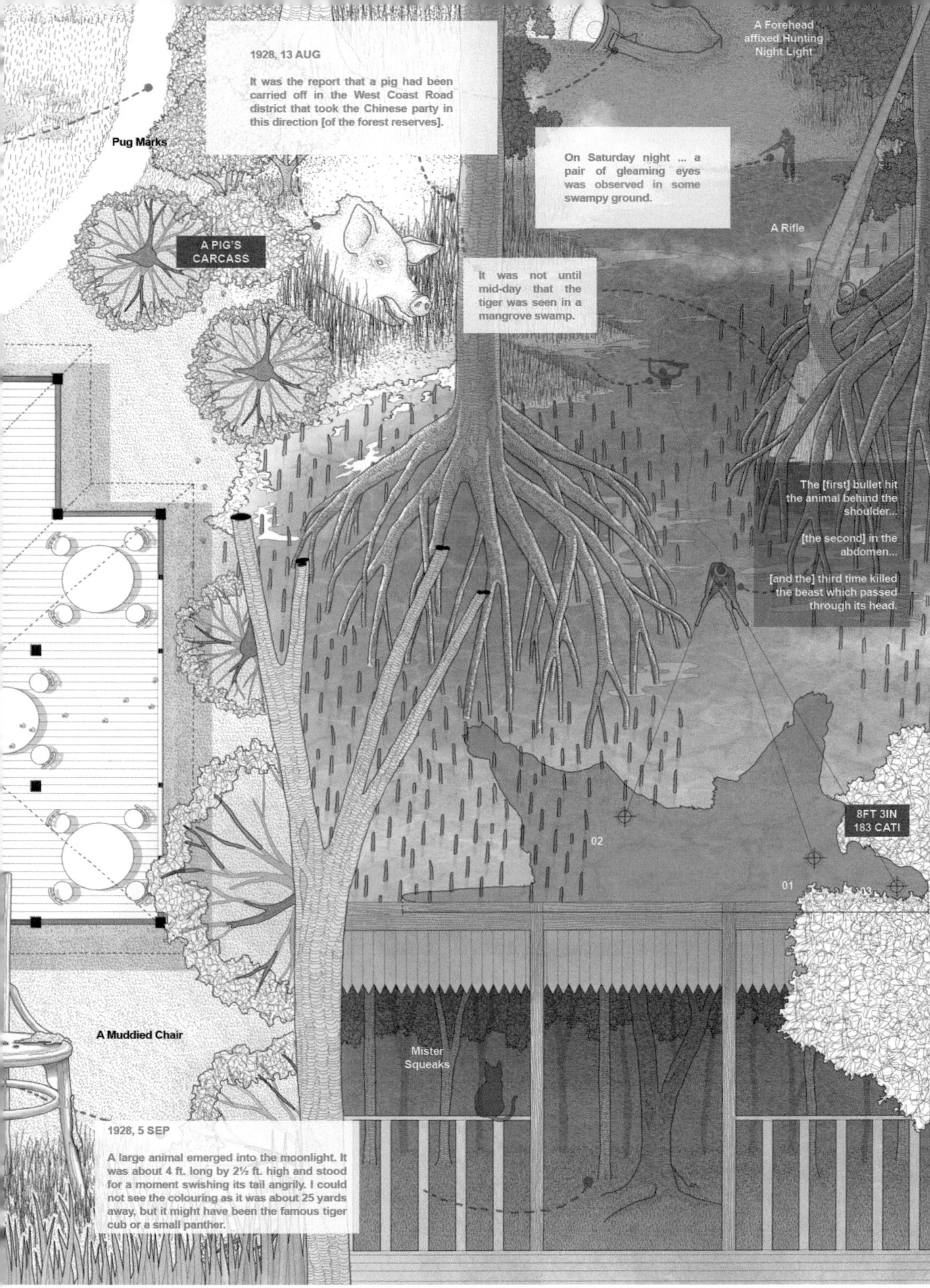

The Cat, The Soldier and…?', September 5, 1928, 8; *Straits Times*, 'Tiger At Seletar: Singapore's Motorist's Thrilling Experience', October 30, 1929, 12; *Straits Times*, 'Tiger in Suburb of Singapore', December 16, 1929, 11; *The Singapore Free Press and Mercantile Advertiser*, 'The Boxer and the Tiger: Unpleasant Encounter at Serangoon', December 17, 1929, 20; *Straits Times*, 'Singapore's Tiger: Seen Near Naval Base on Tuesday', December 20, 1929,

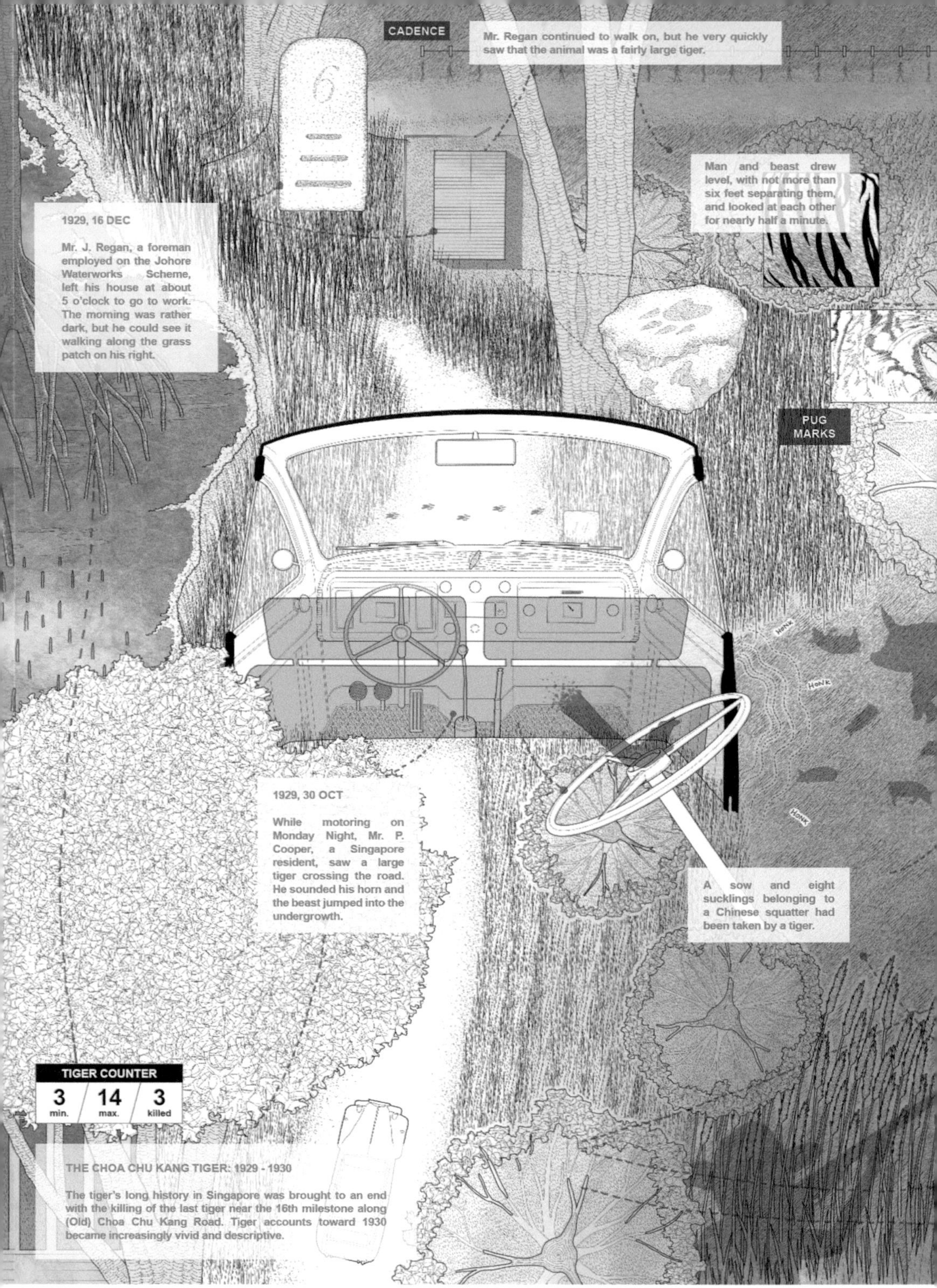

14; *Straits Times*, 'An Elusive Tiger', December 30, 1929, 17; *Malaya Tribune*, 'Tiger At Changi: Partly Eaten Pig Confirms Tapper's Store', February 13, 1930, 10; *Malaya Tribune*, 'Bukit Timah Tiger Shot: Devoured Sow and Eight Sucklings', February 20, 1930, 8; *Straits Times*, 'A Menace Removed: Fully-Grown Tiger Shot At Jurong', February 21, 1930, 14; *Malayan Saturday Post*, 'Tiger Shot in Singapore', March 1, 1930, 13;

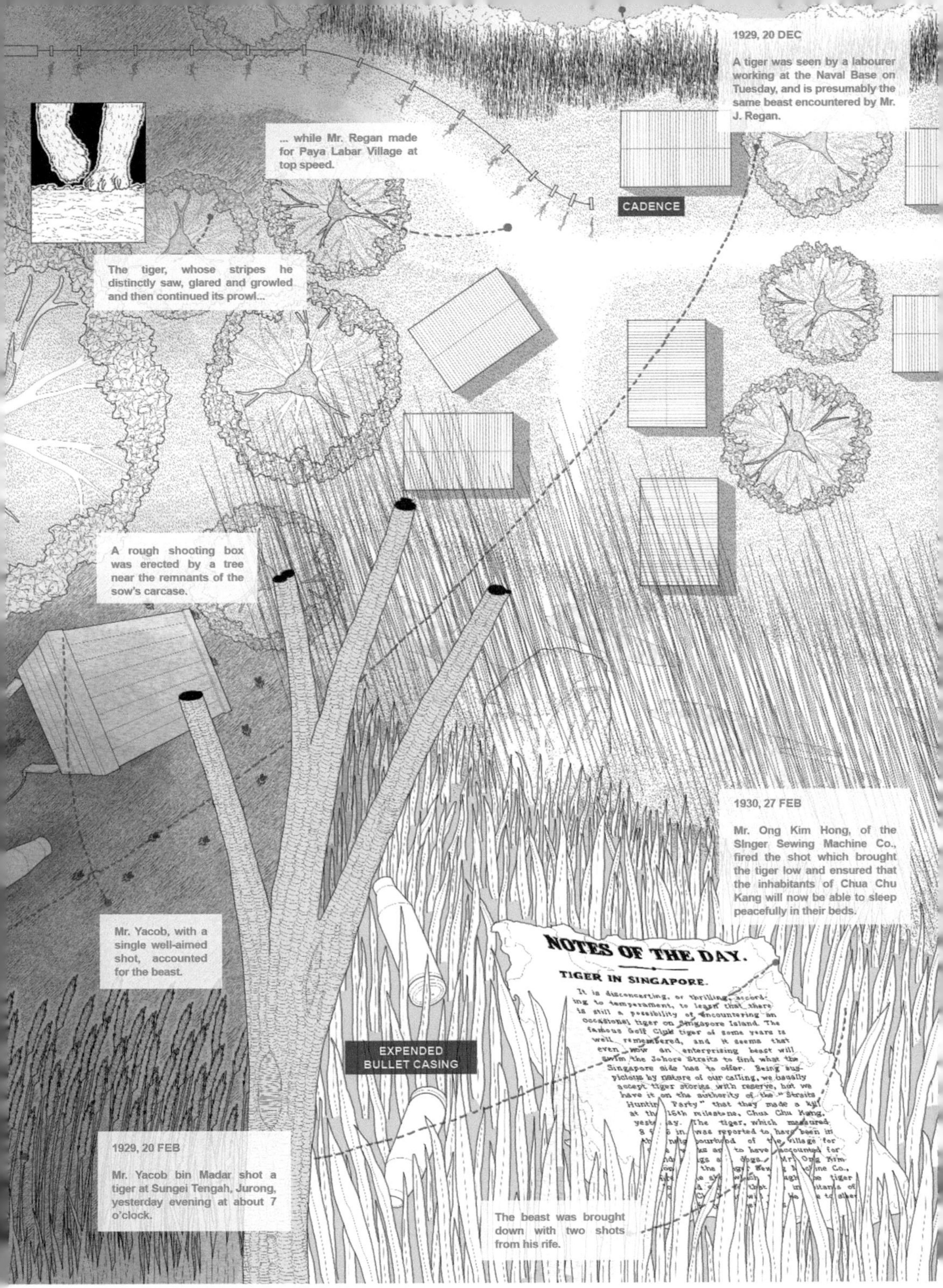

Straits Times, 'Notes of the Day: Tiger In Singapore', October 27, 1930, 10; *Malaya Tribune*, 'Tiger Shot At Chua Chu Kang', October 27, 1930, 8.

3 • After the Last Train

Remainders at the Tanjong Pagar Station

> In the wide space of architecture, that which is not the building is of no consequence. Ideas, descriptions, critiques, theories, even ideology—all abstractions—are, in the end, passive and inert, the ether of architectural space. The object—separate and privileged—is the sole subject of an enclosed and centripetal order. Architecture is a collection of ruins that closes at six o'clock.[1]

As the last train drew out of the Tanjong Pagar Railway Station on 30 June 2011—driven ceremonially by Sultan Ibrahim Iskandar of Johor,[2] whose grandfather had opened the causeway linking Malaya and Singapore in 1923—the architectural narrative of this station found itself again at the brink of abrupt change. After years of territorial dispute, a large section of the originally Malaysian-run and owned Keretapi Tanah Melayu railway line from the Woodlands Checkpoint to the Singapore-based city terminus ending at Tanjong Pagar, was decommissioned (Fig. 3.1). The architectural presence of the terminus—a once imposing Art Deco building completed in 1932—was overshadowed for many years before, on the one hand by political wrangles

3.1 D. S. Petrovich, exterior of the Tanjong Pagar Railway Station, completed 1932.

INDUSTRY

over territory, and on the other by the building's slow and silent descent into oblivion as the two neighbouring nations parted ways.

Prior to 2011, the station was an anomaly by Singaporean standards: a sign on its platform read 'Welcome to Malaysia'; in the concourse people sat idle reading newspapers or drinking cheap coffee in the middle of the day; commuters shuffled along the platforms where announcements were sporadic and unreliable; pigeons, cats and bats occupied the same space. This was perhaps the only location on the island where litter-strewn floors were not obsessively cleared. The station was suspended in a space and time which jarred against an over-regulated and sanitised Singaporean sensibility. Its architecture lived with, and was misshapened by dirt, dereliction and decay.

When Singapore separated from Malaysia in 1965, the station constituted one of several cross-border infrastructural assets holding joint Singapore-Malaysia interests. Managed originally by the Federated Malay States Railway (FMSR) on a 999-year leasehold term, the land on which the station and the remaining railway line were located constituted 200ha given over to the Malayan state by the British colonial government in 1918. Upon separation from Malaysia, it was agreed that Keretapi Tanah Melayu (KTM) which took over from FMSR would be allowed to retain control and ownership of this Singapore-based asset. This meant that the station was legally in Malaysian sovereign territory. Border irregularities were cagily tolerated. Passengers embarking in Tanjong Pagar were cleared for entry to Malaysia even before they had reached the Singapore immigration checkpoint in Woodlands, a good half-an-hour train ride northwards where passengers were then cleared for exit from Singapore. The station manifested an unsettled border region between the two nations.

Yet its architectural lineage suggested something more enduring. Designed by Serbian architect D. S. Petrovich who practiced with the British firm Swan and Maclaren, the three-storey railway terminus was destined to be more than a large train shed. In his speech at the grand opening of the station in 1932, Singapore's governor Sir Cecil Clementi declared the significance of Tanjong Pagar as a 'terminus of world importance', the 'future junction of transcontinental lines'.[3] Two main lines would converge here—one 'reaching across Asia and Europe to the English Channel, and the other through Siam, French Indochina and China to Manchuria, connecting up with the Trans-Siberian railway'.[4]

Petrovich's eclectic Art Deco-Neoclassical design was inspired by Eliel Saarinen's Helsinki Station (1909-1919). The scale of the barrel-vaulted building reaching up to 22-metres is immense, demonstrating the kind of aspirations the architecture once vaunted (Fig. 3.2a). Together with the City Hall and Supreme Court buildings erected around the same time, the station was intended as a monument with colonial gravitas. Evidence of its privileged standing are still manifest in its exterior façade where four vertical reliefs of once white marble, now irrevocably discoloured by dirt, showing the allegorical figures of Agriculture, Commerce, Transport and Industry, pillars of Malaya's colonial economy—assert colonial dominance and order (Fig. 3.2b). The pillars were originally thought to have been the work of Italian sculptor and architect C. Rudolfo Nolli who was also responsible for Singapore's City Hall façade but recent evidence suggests they were carved by Angelo Vannetti who was Nolli's contemporary. The marble engravings form the triple-arched portico at the station's entrance. The portico also features distinctly green ceramic roof tiles, similar to the ones used in Chinese temple architecture. A triple-volume barrel-vaulted roof which is both imposing in stature and pragmatic in function, graces the station's main ticketing hall. Two of its facing walls have tall louvered windows which enable cross ventilation and efficient daylighting. On the other two walls are hung six vividly coloured murals made of coloured rubber tiles produced by the Singapore Rubber Works using a patented process (Fig. 3.3).

Initially, the station was a hardworking commuter and industrial space. Located opposite the busy Keppel shipyard and port, it complemented the land-to-sea routes across the Malay Peninsula for export and trading of goods and raw materials. Its platforms were designed to accommodate the longest mail trains of that time, measuring around 950 feet.[5] And before air travel became popular and affordable in the late 1970s, the train journey to Malaysia was a customary one taken by individuals and families for business and holidays. Some commuted daily, others weekly. Travellers were treated to a host of modest, small-scale amenities which contrasted sharply with the station's purported grandeur—two popular 24-hour Indian-Muslim eateries operated since 1958 by the migrant Hasan brothers from India, the Habib Railway Book Store, a convenience store and adjacent money changer established in 1936, and a 34-room hotel, pub and restaurant run solely for over 60 years until 1993 by one Lim Jit Chong and his family.

Yet as border disputes between Malaysia and Singapore intensified on this site and subsequent bilateral agreements were reneged, the station slowly diminished in physical presence (Fig. 3.4). It was dwarfed by public housing blocks. The station endured in the shadow of the eight-lane Ayer Rajah Expressway built in 1988 to connect the eastern end of Singapore to its western fringes. The massive flyover obscured the station's façade and bypassed the long disregarded marble reliefs. Perhaps because of such impasse, the station continued, even when it was still operational, to be sustained by its own peculiar ecology of small-scaled businesses run by small-time proprietors, who were in turn, subscribed by a dwindling pool of commuters. Anecdotal evidence from newspaper interviews in the last year of the station's life showed that lasting friendships were struck, eccentric habits accommodated, and in all these, the outdated train shed was ultimately spared from large-scale redevelopment. The station fell out of step with the rest of the globalised city which abounded with shiny surfaces and constant improvements. In contrast, the former was rendered anachronistic

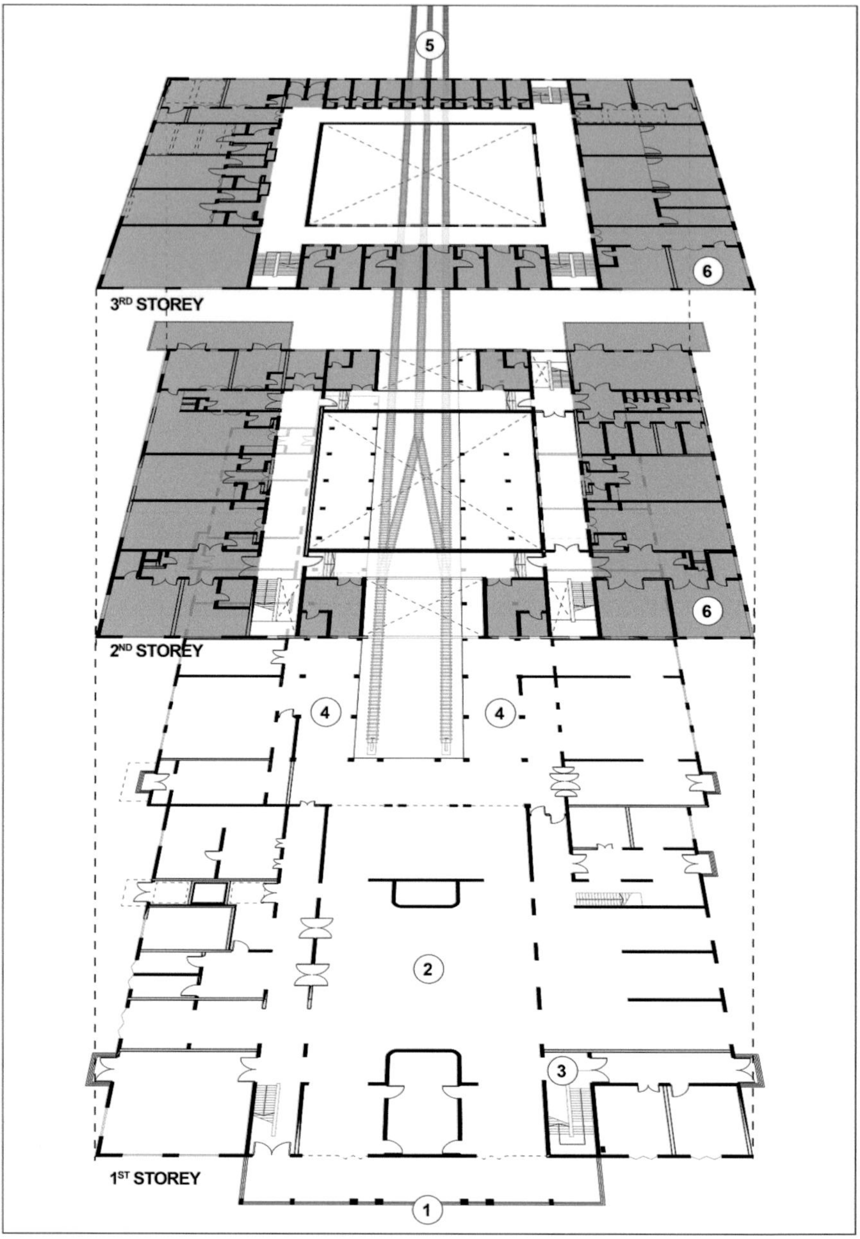

LEGEND

NOT TO SCALE

1 - Driveway/ Drop-off
2 - Central Hall
3 - Hotel Access
4 - Platform
5 - Runway Tracks
6 - Hotel Rooms

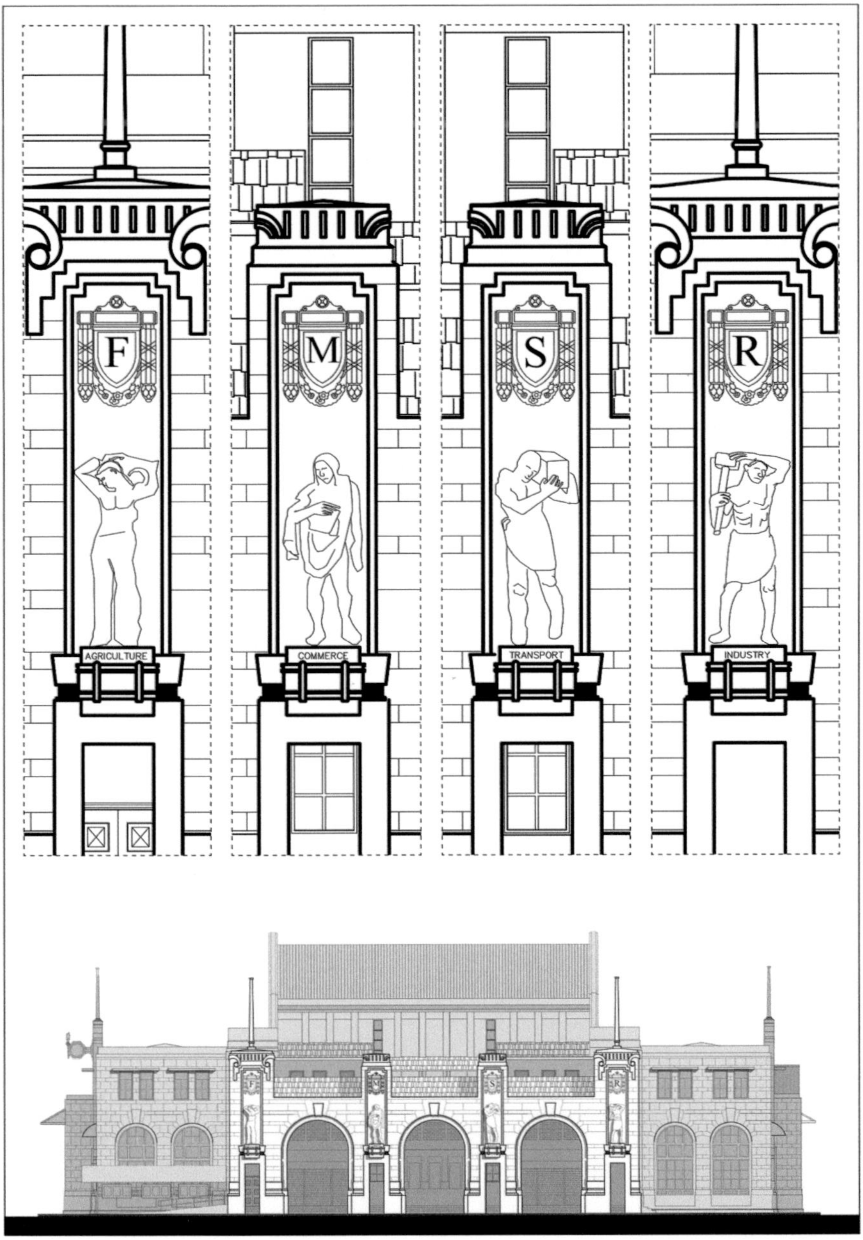

NOT TO SCALE

[Opposite] **3.2a** Architectural plans.

[Above] **3.2b** On the east elevation, four vertical marble reliefs depict the bedrock of British Malaya's colonial economy: Agriculture, Commerce, Transport and Industry.

3.3 Station ticketing hall.

3.4 The station is located between the eight-lane Ayer Rajah Expressway (left) and public housing tower blocks (right).

110
109

years before it became functionally obsolete (Fig. 3.5). Two months before its closure in 2011, the station, then under Singapore ownership, was declared a national monument.[6]

It is within this uncertain landscape that Sydney-based Singapore-born Malaysian artist Simryn Gill produced a site-specific artist book titled *Guide to the Murals at Tanjong Pagar Railway Station*, her contribution to the Singapore Biennale held in 2006 (Fig. 3.6). Gill's book documents two forms of architecture at the station—'Architecture' in the sense of the stylised Art Deco monument as the enduring *objet d'art*, and 'architecture' as an ephemeral map of space undergirded by commuter experience and atmosphere. A textual description of the station's architecture and its murals are counterposed by photographs of commuters. These black-and-white photographs depict commuters in motion, reinforced by an emphasis on their movement and their luggage. The photographs do not glamourise train travel. They do the opposite. They gather forgettable scenes of flux that gave the station its life and reason of being. And although few details of the station are intentionally picked up in these photographs, fragments of the building which do appear further reinforce the banal and the disregarded: run-of-the-mill ceramic floor tiles, plastic chairs, untidy exposed electrical wiring, stained walls, uncoordinated signage, cheap plasterboard grid ceilings and glaring fluorescent lights. *Guide to the Murals* treats the station as a tentative structure in flux, a reluctant monument-to-be which embodies the evidence of power—the fickleness of politics and capital—and its imminent decay brought on by the attrition of the very same power.

As if nodding to the latter, a second work by Gill comprises a set of photographs of the now defunct station hotel interiors. She chanced upon the abandoned hotel while she was working on *Guide to the Murals*. Gill remembers being a guest in one of these rooms when she was a child visiting Singapore with her parents in the 1960s. The Station Hotel was then a grand hotel connected to an imposing colonial railway station. This place was second only to the famed colonial Raffles Hotel.[7] Shut down prematurely by the Malaysian government in the 1990s, the abandoned hotel was rumoured to be haunted. Gill remembers going up alone to photograph every single room in 2006. The stench of decaying inorganic building fabric mixed with organic animal waste from bats, birds and stray cats is something she vividly recollects.

An 'unfinished' piece, Gill calls these photographs a 'bottom-of-the-drawer' project.[8] She has expressed the importance of not too quickly foreclosing what the photographs might become—that these images in their unfinished and inconclusive state perhaps mirror the real time decay and uncertainty of the abandoned hotel rooms themselves. Thus, in both the guidebook and the untitled photographs of the hotel, the monument in perpetuity ('Architecture') is cast alongside another transient space ('architecture') created incidentally amidst bilateral tensions and ongoing border crossings. The second 'architecture' is what Gill tries to capture and hold in place. It is 'a mode of remembrance that is erratic and ephemeral', or as geographer Caitlin DeSilvey so eloquently poses, 'the forgetting brought about by decay allows for a different form of recollection'.[9] Here, the work forges another kind of memory and monument, which Gill is asking us to consider, or at least to question.

I use the term 'monument' in this chapter intentionally, particularly since Gill's work predates the formal designation of the building as a national monument by the Preservation of Monuments Board in 2011. By 'monument', I refer to how Gill's work emotes the intensity of conflicted political tensions impossibly restraint within a single piece of architecture for almost half a century. Her affective portraits are consistent with the station's steadfastness, its obduracy, a quiet, reserved, but disobedient building. The station represented personally for me, a Malaysian who left her country for Singapore over 30 years ago, an unsettling piece of architecture. It was a colonial building given late national recognition because it belonged to another country. The station was most poignant and powerful in its last days when it existed as a memorial, singular and enduring but already in ruins, thus tentative, and precarious. It was a memorial that held the rite of passage between two nations. Thus, the term 'monument' here holds this duality—of its capacity to manifest history (a memorial) but also to inscribe what was unfolding (its changing significance), not omitting, or adulating the past.

Significantly, Gill had chosen to work on the station because she was familiar with it. She had used it herself with her own parents when she was a child. Later in the 1990s, when Gill herself became a mother, she used the rail service again to shuttle her two young children between Singapore and the Gills' ancestral home in Port Dickson, a small Malaysian seaside town. Gill's relationship with this site is not nostalgic. The politics of this location—an uneasy boundary between two neighbouring countries for which she was part of—mark her own identity. Born in Singapore in 1959 before the republic's painful separation from Malaysia in 1965, her family moved back to Port Dickson when she was a child. Gill has spoken of the vulnerability she felt possessing a British-Malayan birth certificate that seemed to alter its alliances too quickly—first, the British colonial government withdrew from Malaya and subsequently, Singapore separated from Malaysia.[10] Although the artist refuses to cast her work against her personal histories, the readings and strategies informing *Guide to the Murals* are inseparable from the question of where the artist might be positioned in this contested territory. The nation-to-nation contestation was not played out openly. Strategies and counter-strategies were enacted through diplomacy and government-to-government agreements while the effects of these agreements were frequently neither clearly articulated nor properly debated. The animosity between the two countries vacillated wildly between the furtive and the blatant. To what extent did such strained bilateral relations affect their publics? Apart from an argument fought almost entirely through third party premises—the

3.5 Empty tracks and platforms.

3.6 Simryn Gill, *Guide to the Murals at Tanjong Pagar Railway Station*, 2006. Published initially for the Singapore Biennale, this guidebook was stocked exclusively at the station's newsagent.

media and the international justice arbitration mechanisms—were there physical traces which remain embedded in the urban milieu?

Gill's work perhaps re-envisions the Tanjong Pagar Railway Station as one such affective repository; a space which held oblique traces of the repressed hostilities between the two nations. The work highlights the transient and often forgotten aftereffects of such animosity—paying attention to the erosion of power as it slowly percolated into the architecture, and into the daily experience of a place that was literally coming apart, unmaking itself through use, altered by the punishing tropical weather and ultimately ravaged by neglect. In this way, the tensions can be seen, felt and at some points, even aestheticised. Resonant with the artist's enduring interests of how location informs who one is or is not, her project raises difficult questions about the provenance of this monument's memories and its values. To understand the import of such tension to an architectural reading of this station, some background of the Singapore-Malaysia relationship and the contestation for both station and railway land follow.

A station between two nations

The separation of Singapore from Malaysia in August 1965 reputedly devastated the usually energetic and often combative first Prime Minister of Singapore, Lee Kuan Yew. Lee was taken ill after months of strenuous but unproductive negotiations with Kuala Lumpur. There were too many irreconcilable differences on economic, political and ideological fronts between the two governments. The disputes over taxations supposedly levied on Singapore, the potentially problematic issue of a large Singaporean Chinese majority versus the majority of Malaysian Malays, Lee's outright demands for egalitarian policies based on meritocracy instead of race, and not least, his openly confrontational politics, all unnerved the Malay politicians.[11] When the merger collapsed, Lee publicly burst into tears as he announced to Singaporeans on national television that his hopes for a prosperous, united and equal '"Malaysian Malaysia" based on meritocracy rather than ethnic—Malay—preferences' had been irrevocably shattered.[12] Anthropologist Yao Souchou writes that the statesman who was normally 'given to emotional restraint and tough-mindedness' was by then 'personally and politically' traumatised.[13] Lee himself recalled how he was 'emotionally overstretched'[14] by the occasion. Debilitated by physical and nervous exhaustion and surviving the long days only by will of prescription sedatives and pep pills, the resilient leader eventually collapsed, succumbing to a 'minor breakdown'.[15] He retreated to the Changi government barracks for six weeks to recuperate.[16]

This affective outburst was precipitated two years before, in 1963, when Singapore signed a merger to join Malaya, Sarawak and North Borneo (Sabah) as part of the Federation of Malaysia. Speaking from London on 9 July 1963 where the merger was signed, Lee said that he had endured an 'arduous and gruelling week of detailed arguments and negotiations', where 'harsh words' were exchanged, and compromises were in the end inevitable.[17] The merger was vital for both countries, 'all of us knew that Malaysia with Singapore is so important that a fair solution must be found and it was found'.[18] For Singapore it meant the opportunity to access a common market crucial to the entrepot trade and Malaysia's vast hinterland to compensate for its smallness; for Malaysia it promised secure financial capital and expertise since Singapore could become 'the New York of Malaysia'.[19] Both countries were committed to an anti-Communist stance.

Despite all efforts, the union rapidly unravelled. The common market did not materialise as quickly as Lee had hoped. He was also impatient to obtain pioneer status for his government from Kuala Lumpur, who seemingly prevaricated on the matter. At the same time, a gnawing mistrust had developed between the major political parties—Singapore's People Action Party (PAP) led by Lee and the United Malays National Organisation (UMNO) headed by Tunku Abdul Rahman. Such ill feelings were fuelled and sustained by the deeply imbalanced racial composition of Malays and Chinese in both countries. Subsequently, two separate racial riots broke out in Singapore on 21 July and 2 September 1964. The incidents claimed forty-six lives and injured almost six hundred people. Despite more bilateral efforts at diplomacy, the relationship continued to deteriorate. The crux of the problem was Lee's insistence on decreasing Malay dominance and achieving 'a less communal Malaysian society'.[20] This ideology coupled with his constant questioning of Malay power in the Malaysian political system ultimately became 'a challenge that UMNO needed to remove'.[21]

When the moment came, the separation of Singapore from the federation of Malaysia was effected swiftly and in secrecy by both sides. Former Malaysian diplomat Kadir Mohamad points out how the separation has been described by various writers, researchers and journalists as an 'expulsion, extrusion, exclusion, severance, ejection, secession, disengagement, hive-off, break-up, cut-out and kicked out'.[22] There is a sense of the visceral and the bodily in these descriptions. They alluded to an inextricable relationship between two countries whose peoples share similar cultural and social norms. Yet the racial question had become too immense for Malaysia's Malay majority who were made, by Lee's interrogations, to feel increasingly vulnerable. The Malaysian Prime Minister Tunku Abdul Rahman writing to his deputy Abdul Razak Hussein from London on 1 July 1965 said, 'I fear that we will have no choice but to cut out Singapore from Malaysia in order to save the rest of the body from gangrene'.[23] Conversely, Lee himself saw the separation as one of Singapore's inevitable choices. He appropriated the Muslim men's irrevocable pronouncement of divorce to end a failed marriage, 'Talak, talak, talak'.[24] Curiously, both men again evoked affective bodily metaphors to describe the failed union—one of a painfully diseased body; the other of the wedded body bitterly torn asunder. The separation was necessitated by a body in trauma.

In the years following the 1965 separation, the clean amputation or irreversible divorce which Tunku and Lee intended was continually foiled. Ironically, the two countries found themselves bound not just by proximity but increasingly by necessity and pragmatism. There was, and continues to be, a shared sense that the communal stability between the two countries must not be ruffled. Interdependencies on bilateral trade, limited joint army defence exercises, common air defence assets and shared foreign policies against communism and insurgency bind Singapore and Malaysia. Yet as political historian Tim Huxley reiterates, 'beneath this veneer of cooperation ... severe stresses and strains have persisted'.[25] Several issues continued to entangle, if not also jeopardise the prickly relationship. The pre–1965 disagreements over racial supremacy mutated to other areas where each country tries to assert dominance over the other. These included the tariffs of raw water (supplied by Malaysia to Singapore who has limited water resources) and treated water (which Singapore resells to Malaysia), dispute over airspace for military exercises, contestation over environmental and sustainability issues resulting from Singapore's land reclamation works at Tuas, Malaysia's on-and-off plans to remove the causeway linking Johor Bahru to Woodlands, displeasure over Singapore's ban on the wearing of the *tudung* in its schools, economic competition over port trading specifically the rise of Johor's Tanjong Pelapas port which has taken business away from Singapore's lucrative shipping trade, and two high profile territorial disputes.

In the last instance, two significant concessions were made in Singapore's favour—a win in the internationally arbitrated judicial case of Pulau Batu Putih or Pedra Branca in 2008, and more recently in 2011, the bilateral resolution of land rights to the Tanjong Pagar Railway station on the southern tip of the island and the old KTM tracks running from the station to the Immigration, Customs and Quarantine checkpoint at Woodlands in the northern part of Singapore. For Pedra Branca, one of the islets located at the eastern entrance to the Straits of Singapore, the dispute started in 1979 when Malaysia published a map claiming ownership of it. Singapore argued Pedra Branca was in fact *terra nullius* or no one's land, and that Singapore had kept the island running in various ways including building and maintaining a lighthouse since colonial rule, installing a military asset there, studying the feasibility of land reclamation in the vicinity, and requiring Malaysian officials visiting the island to obtain permits from the republic. These actions had not been contested by Malaysia. Significantly, the 2008 verdict was swung over by the maintenance of the colonial Horsburgh

Lighthouse. In this sense, territory was claimed on the basis of an architecture—the lighthouse ultimately defined the shape of island and the legitimacy of its claimant.[26]

In the case of the Tanjong Pagar Railway Station, the situation was much more complex. The land was legally and indisputably under Malaysian sovereignty. Furthermore, the leasehold of 999 years bequeathed by the British to the Federal Malayan States (FMS) in 1913 was nowhere near its expiry. As a consequence of the 999-year leasehold, the station and the railway line connected to it were the only Malaysian stock physically embedded in independent Singapore. The station and the line constituted a country uneasily accommodated within another country. The land stocks under Malaysian ownership and leasehold were substantial—81.74 acres were in perpetual ownership and 353.52 acres on lease for 999 years. Besides opportunity costs sacrificed in the face of foiled urban development, the Malaysian track and station was, according to Kadir, 'like a long sword piercing right through the heart of Singapore ... It was a pain to the national psyche and an affront to Singapore's sovereignty. It hurt their national pride to have a foreign country owning and running a railway service in their own backyard'.[27]

Lee personally embarked in rectifying this anomaly before his tenure as Prime Minister ended. He had pragmatically argued that the foreign-run railway constituted a weak security link which penetrated the whole length of the island and compromised its sanctity, cultivating maleficence by Singaporean standards at least, in acts like drug trafficking.[28] Meeting with the Malaysian Finance Minister Daim Zainuddin in 1990, a 'Points of Agreement on Malayan Railway Land in Singapore between the Government of Malaysia and the Government of Singapore' (POA) was signed to encompass seven points pertaining to the removal of the Malaysian Customs, Immigration and Quarantine checkpoint from the centrally located Tanjong Pagar to Woodlands in the northern reaches of the island, and the subsequent conversion and development of all Malayan Railway land for other purposes within agreed compensatory measures.[29] In this agreement, Malaysia had agreed to vacate the Tanjong Pagar station in return for three other land parcels in Singapore.

Nevertheless, the 1990 agreement fell through. Both governments persisted with different interpretations of this agreement. Malaysia argued that the compensation was inadequate as only fifty-eight percent of Malaysian land stocks would be compensated. It refused to relocate the immigration checkpoint to Woodlands. Singapore conversely saw the agreement as clear and binding, and proceeded to move its own checkpoint northwards following the timeline set out in the 1990 POA. A flurry of tense and futile negotiations ensued. In the light of this impasse, an irregular arrangement was made in 1998 to have passengers board the train at Tanjong Pagar without having their passports stamped by the Malaysian immigration there. The passengers then travelled for half an hour until the Singapore checkpoint at Woodlands where their passports were stamped as having exited Singapore. The Malaysian immigration then proceeded to stamp the passports on board the train as it passed the Woodlands checkpoint. This practice carried on for some years until a new agreement was drawn up between the two countries in 2010.[30] The Malaysian railway premises were moved to Woodlands in 2011, and the railway line dismantled. A new cross-border Rapid Transit System (RTS) Link connecting Woodlands to Johor Bahru is planned for 2026 when the Malaysian Railway service to Singapore will eventually be terminated.[31]

For over four decades, the intractable Singapore-Malaysia diplomatic relationship with its dead ends and cul-de-sacs converged uneasily at the Tanjong Pagar station. No matter how awkward, this remains the station's undeniable past. In a perverse twist of political impasse, the station became an anachronistic structure in the face of bullish urban development (Fig. 3.7). Its dysfunction marked the split between two nations, and the continuing physical consequences arising from this bitter separation. Yet a monument in the conventional architectural sense of the term cannot be dysfunctional, let alone be in ruins. Can the 'monument' following architectural logic accommodate other forms of existence?

Relic or remainder

> A building is flow, not form; it is creative, not merely a creation.[32]

The announcement of the green railway corridor development on the former railway track site in 2011 promised better public amenities.[33] The developmental potential of these unlocked parcels of land is undeniably important especially in increasing rare public open space through careful conservation and design. However, there is also the possibility that in time, with diminishing physical traces, the significance of the historical relationship between these two nations might also fade. Can such history still meaningfully resonate amidst change?

These relations had concentrated particularly at the Tanjong Pagar station which when stripped of its purpose in 2011, had become almost instantly an object of curiosity. As part of Singapore's Urban Redevelopment Authority's (URA) new masterplan forged through public consultation and open architectural competition, the station was slated for adaptive reuse. A civic landscaped space will be placed at the forecourt overlooked by the four marble pillars, and a mass rapid transit exit for the Cantonment MRT Circle Line will be housed in a lightweight steel structure addition scheduled for completion in 2025.[34] More importantly, the station will

3.7 Corroded gates and weathered walls.

become the new terminus to a twenty-four kilometre stretch of former railway land which will traverse 'diverse landscapes, linking high-density public housing estates with pockets of private housing, business parks, industrial areas, nature reserve, parks and land that have yet to be developed'.[35] Already here, there is a sense of what is expected of the soon-to-be-conserved railway station. Its future architectural role is akin to a modern relic. It will be a setting for civic and public activities; a pretty backdrop for special occasions; a spectacular locale for key national events. The conserved station will recede into the background. Its politics eventually neutered and neutralised.

It is at this point—when there is a will to gradually but persistently evacuate and smoothen the kinks of the past by beautifying, developing and utilising the station, arguably brushing out its still troubling past with an equally problematic conception of 'heritage'—that Gill's two works matter. Their insertions into this new urban utopia may be made on some of these non-political registers: as civic documents, artistic statements, anecdotal records and/or personal memoirs. Yet, none of these registers can be read meaningfully without reference or allusion to the railway station's historical and affective past, and its inevitable stakes in the troubled politics between two nations.

Geographers Brenda Yeoh and Lily Kong have been circumspect in their discussions of Singapore's attitudes towards heritage and redevelopment.[36] However, they are critical of the impulse to make heritage a 'repository of spectacle' and the instinctual move to reduce 'history to architecture' alone can leave out much of what is crucial to its construction.[37] Heritage is not free from ideology. Thus, every negotiation with heritage is a transaction of power. Subsequently, the 'naturalising' of heritage as part of the 'everyday, visible world' can potentially transform heritage into an ideological tool which 'masks the artifice and ideological nature of its form and context. Its history as a social construction is unexamined. It is, therefore, as unwittingly read as it is unwittingly written'.[38] When placed in contact with something as permanent as a monument, heritage transplants itself into a singular narrative unquestionably perceived as growing out from the monument. The narrative and meanings of the monument consequently become *fait accompli*, implicated by the ideologies that undergird heritage in that particular context.

Furthermore, the link between a monument and heritage might be argued to be an alliance pragmatically forged on the grounds of 'value'. The expedient alignment of monument to heritage, the latter possessing both cultural and commodified value, indeed stems from architecture's legacy of needing to legitimise its worth. Architecture achieves this through its own valuation (notable design pedigree, rare historical property, lucrative siting) and in its ability to generate value (through intensified use, inventive purpose, historical resonance). Architectural theorist Stephen Cairns and geographer Jane M. Jacobs jointly argue that architecture's propensity to produce purposeful value originated from Enlightenment principles of rationality. As a 'technology of enclosure and improvement' and tasked with producing myriad values which embellished the Vitruvian precepts of utility, beauty and permanence, the purpose of architecture was to provide a link between 'reason, value, and order'.[39] However, in raising questions about the obverse value of an architecture that may be derelict, obsolete or decaying, Cairns and Jacobs highlight how the often narrow conceptualisation of value is only ever more strongly identified with architecture's utility and fiscal potential. This is in contrast with what Barbara Herrnstein-Smith has called a 'double discourse of value' wherein architecture's subjective and often transcendental values of taste, creativity, influence, inspiration and experience are pitted squarely against a more finite 'economic sphere of value (calculation, references, costs, benefits, prices, utility)'.[40]

Drawing on philosopher John Scanlan's reading of the Enlightenment's emphasis on a 'cosmetics of order' which alienates 'the life and death of matter', Cairns and Jacobs suggest that architecture appropriates such an emphasis to substantiate its absolutist claims for order, life and purity, while at the same time distancing itself from disorder, and denying architecture's 'future states of decay'.[41] Just as the Enlightenment aesthetics of light and transparency chimed with modern architecture's penchant for the clean and the new, the Enlightenment project required a kind of radical mental 'clear-out and clean-up' in order to access clear-minded knowledge.[42] Not coincidentally, the procurement of knowledge was cast as being central to an inhabited space, or an architecture opposed to darkness and shadow. These two interchangeably physical and metaphorical conditions—darkness and shadow—were thought to obscure full visibility, prevent clear thinking and deter truth. They were believed to make men prone to superstitions and plots; ignorance and deceit; ambiguity and uncertainty. Thus, the Enlightenment project instilled in architectural practice and discourse a notion of value expressed not just in terms of functional purpose but significantly one responsible for upholding moral rectitude and mental clarity. The mental and moral clear-up was manifested in architecture's subsequent ideological role of 'expunging waste and wasting'.[43] Waste was that which was useless, but equally that whose use was uncertain or ambiguous.

Cairns and Jacobs bring up the contrasting definitions of an aging building by comparing the opposed views expressed in art and in architecture.[44] Art historian Alois Riegl recognised the aging building's blemishes as an affective condition, as architecture's irreplaceable 'age value' wherein architecture's beauty was simultaneously expressed and felt through its 'patina'. Yet, thirty years later in 1935 when Le Corbusier visited New York City for the first time, he was enamoured by the city's shiny, ordered and geometric skyscrapers which he contrasted to medieval European cathedrals 'blackened by soot and eaten away by wear-and-tear' and cultivated through what he saw as a culture of 'dirt and filth'. Where Riegl saw beauty in patina, Le Corbusier was repulsed by dirt. In this disjunction between filth and patina, architecture's originary impulse accompanied by its 'natalist imaginary' of the single architect-creator,[45] continue to valorise the originary and creative act of design over the co-production of architecture beyond design, particularly the transformation of architecture through use and through occupation. Not only is architectural history 'written around the monuments that were built to last', the frequently singular architectural narrative defining a building also disregards the 'noisy, dirty entrepot of multitudinous architectures in the process of constant change'.[46]

Notwithstanding architecture's complicated value system, the tagging of heritage onto the monument further confounds. Given that architecture legitimises itself through both its calculated values (for example, its cost and its marketability) as well as its transcendental ones (for example, its design pedigree and its distinctive atmosphere), the interchangeability of the monument for heritage means that heritage inadvertently becomes entangled with architecture's complex value systems. The already difficult questions confronting heritage—what, whose, why and how—are conveniently elided when heritage is simply collapsed into the architectural monument. As architecture's and heritage's value systems intersect, the reciprocal simplification of each value system to the most finite and calculable form—money—intensifies. Architecture as spectacle and heritage as commodity, and vice versa, become common purpose and reason.

In their provocative article on the preservation of monuments, preservation historians Aron Vinegar and Jorge Otero-Pailos argue for the importance of granting a monument its diverse responses and meanings because 'haunted by an irrevocable absence, the loss of a "world" or "culture" surrounding those material remains', we are already separated by 'inevitable distance and alienation from these worlds and works'.[47] Vinegar and Otero-Pailos emphasise a more oblique and imaginative narrative to drive preservation because the monument is never 'inertly there' but must be continually 'exuberant;' it must be made and remade. For them, the monument is only bodied and storied forth through lived experience and 'prodigious events'[48] which could be collectively or individually significant, but should embed 'multiple meanings, fantasies and desires'.[49]

How can we talk meaningfully about an architecture that for forty-six years failed to be a monument, but is now bestowed that role? If the monument is a cipher for a memory or a past, whose past is being preserved and narrated? What can be said about such a space without risking oversimplification or crude commodification? Is it possible to talk about a monument with a past that continues to evolve, without directly evoking, and foreclosing, architecture's value system?

In Simryn Gill's two works, the headlines of heritage and monumentality—soon to be the defining terms of the former railway station—are fastidiously thwarted. Her artist book takes the form of an unassuming 33-page architecture guidebook measuring 15 x 21 cm (Fig. 3.8). The guide was stocked exclusively at the Habib Railway Book Store for the duration of the Biennale, and continued to be sold at the same bookshop for several months after the festival ended. Its understated distribution network meant that the book remained for the most part, under the radar of the Biennale crowd. Describing the 'quietitude' of the guidebook which distinguishes it from the headlining work that most projects strive for at an art biennale, artist Philip Brophy emphasises the alternate positioning of the document as one which persuades the reader to inhabit the station; to become invested in giving the space personal meaning and relevance as the station itself was changing in its reference and relevance. In effect, the work leads the reader to enact Gill's still evolving relationship with an architectural space that cannot be easily surmised through the familiar headings of 'colonial', 'national', 'heritage' or 'historical':

> (It is) not as an act or statement, but as an intersecting point between artist and place. Presented completely outside of the visible scaffolding of the Biennale (no 'big concept' stencilling here), it refuses to stand as an act of intervention in the space … it maintains itself at the level of conversation … (it) is that brief but affecting chat by chance you have with someone waiting for the train.[50]

The guide features a bilingual text of English and Malay. It is accompanied by a series of black-and-white photographs on the inside. The cover, which is rendered in full colour, unfolds to reveal the titular murals found in the main hall of the station. The mode of delivery used in the text is officious and factual. It tells about the building as a colonial edifice with design pedigree provided by 'D.S. Petrovich, a Serbian architect of British training … in the employ of … Swan and Maclaren in Singapore'. It boasts about the building's 'Modern' style, the kinds of materials used ('reinforced concrete'), its architectural influences ('Eliel Saarinen's Helsinki Railway Station'), its ambitions 'to impress the natives' through 'grandiose and impressive structures' manifested no less in its 'cavernous central hall' where six triptych murals are placed 'up high' between 'cathedral-like windows'.[51]

The murals, we are told, were designed in the London studios of world-renowned ceramists Doulton. 'They depict six scenes of Malayans engaged in "heroic labour"'. There is a clear division of labour according to race—Malay women in a rice field, Indian labourers working in coconut and rubber plantations, Chinese workers beavering away in a tin mine. Although they are made to look like stained glass, the guide tells us that these are in fact painted rubber sections, 9000 pieces in all. It also highlights that there was once a hotel

3.8 Gill's photographs show the transient atmosphere of commuters waiting.

of thirty-four rooms on the two upper floors of the railway station which was of the same standing as the prestigious Raffles Hotel.

Accompanying the written history, the guide offers a set of images that forefront the commuters' experience of the railway station. We see fragments of raced bodies—Malays, Chinese, Indians—waiting, sitting, walking, queuing, pulling their luggage, pushing prams, pushing luggage trolleys, lugging overnight carryalls and plastic bags (Fig. 3.9). The racial mix of commuters milling about in the hall is a doubling of the racial composition depicted up high on the titular murals. Hints of racial difference are deciphered from the commuters' attire. We see the train passengers mainly as bodies in motion or temporarily at rest; phantom limbs waiting restlessly to leave as soon as the next train pulls into the station. In contrast with the described formality and scale of the cavernous hall, the photographs reveal an embodied space energised by use, wear-and-tear, a mundane and fragmented aesthetic that eventually becomes inconsequential or repressed in the architectural narrative because these subjects are cast outside the privileged objecthood of architecture.

A Guide to the Murals unsettles and reanimates the architectural monument. It focuses on the station not just as a space of transit but one that is itself altering, uncertain, ambiguous in its objecthood. It shifts our perception of the building from a static and immobile object propped up by an architectural style and lineage, into a space of flows, populated by people, things (trains, chairs, luggage, trolleys, prams, food, newspaper kiosks, magazines, leaflets, money), animals (cats, bats, birds, ants, termites, spiders) and intangible forces (political tensions, racial tolerance, financial capital, social capital, neocolonialist prejudices, wind, rain, damp, sunlight). It reminds us that the ways we order and structure our experiences of the station do not coincide with the ways architecture is ordered and preserved. There are shifting materialities, bodies and agencies entangled in the monument that cannot be accounted for in its preservation narrative.

Gill mimics the format of the architectural guidebook, even adopting its language of objective facticity. The monument is only ever felt in the writing voice which she consciously chose for the guidebook. The authoritative voice is de facto in numerous architectural descriptions, particularly historical descriptions of monuments. It is one which appears whole, unaffected and unassailable. Yet it might be riddled with omissions and contradictions. Fascinated with the import of words we unconsciously adopt, Gill's works have highlighted how words are never neutral. She questions particularly how meaning is generated through words which are 'produced, circulated and continue to colonise our thoughts'.[52]

For instance, in her earlier but ongoing work, *Pearls* (1999-), Gill transformed and exchanged the books given to her, cannibalising the book pages into beads. These book-beads were strung into necklaces to be worn on the body. Book to necklace; mind to body; intellect to viscera; words to things; reading to looking,

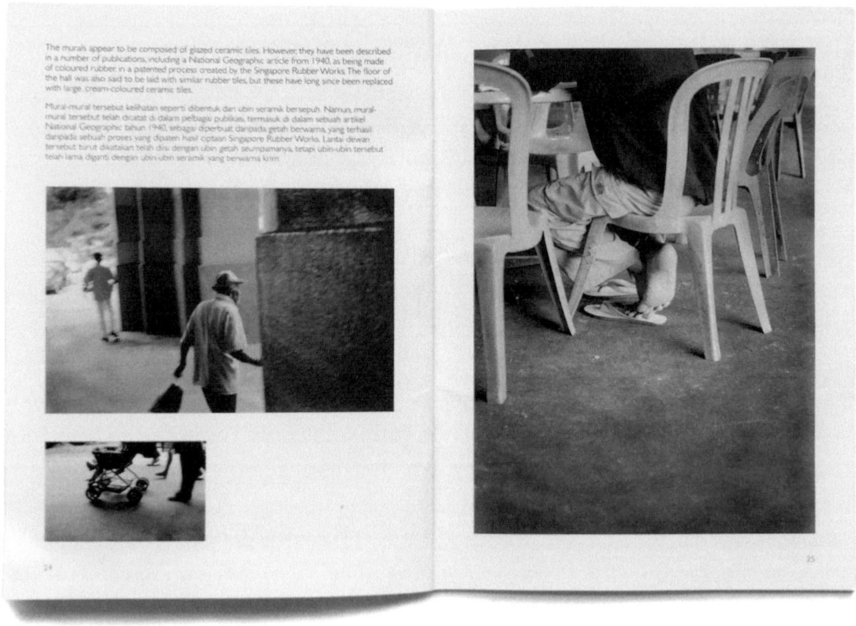
The murals appear to be composed of glazed ceramic tiles. However, they have been described in a number of publications, including a National Geographic article from 1940, as being made of coloured rubber, in a patented process created by the Singapore Rubber Works. The floor of the hall was also said to be laid with similar rubber tiles, but these have long since been replaced with large, cream-coloured ceramic tiles.

Mural-mural tersebut kelihatan seperti dibentuk dari ubin seramik bersepuh. Namun, mural-mural tersebut telah dicatat di dalam pelbagai publikasi, termasuk di dalam sebuah artikel National Geographic tahun 1940, sebagai diperbuat daripada getah berwarna, yang terhasil daripada sebuah proses yang dipaten hasil ciptaan Singapore Rubber Works. Lantai dewan tersebut turut dikatakan telah diisi dengan ubin getah seumpamanya, tetapi ubin-ubin tersebut telah lama diganti dengan ubin-ubin seramik yang berwarna krim.

24

25

3.9 Bodies in motion or temporarily at rest; phantom limbs restlessly anticipating the next train.

holding, wearing. In between one state and another, something inevitably slips. The gravity of words, once abstract on the page become tactile and heavy as beads. Yet as beads, outside the page of the book and thus also, outside the guarantee of the book's disciplinary limits, the same words lose their power. They are only as valuable as the painstaking labour it takes to transform a sheet of paper into a tightly coiled bead. In *Untitled* (2006), installed at London's Tate Modern, over 100 books were placed on tables. Many of the books were left open exposing the detail that small sections of text had been cut out from their pages. In see-through pouches found beside the books, pieces of identical words that had been removed from the books were found. Amongst the words in the pouches were 'because', 'always', 'then'. Unsettled from their place on a page, and now collected as objects, these words become oddly 'sinister',[53] almost as though they were already prejudiced with particular meanings and distinct contexts. As Gill insists:

> Everything has its specificity, like those words that I tore out from those books. In its own place each 'because,' for instance, sits in a very particular constellation of words, but when you take it out and it's all by itself, what is it? ... There's no such thing as blankness. And that very idea comes from power. It comes from a kind of white centralisation of the world.[54]

Similarly on closer scrutiny of Gill's guide, image and text contradict each other. For one, we never see the six murals inside the book until we start to fiddle with the book jacket which unfolds into a full-colour double-sided poster picturing the two sides of the grand hall with its six murals (Fig. 3.10). The images seem to bear no relationship to the text, which in turn, talks only about the architectural features of the building. Through the monochromatic photographs of commuters within the guidebook, only parts of the station emerge. Nothing is seen of its 'impressive structure'. From the guide we encounter distinctly contradictory views of the station—the text reads of magnificence, remark and grandeur; the photographs of ordinariness, neglect and impending decay. If we are compelled to read again in search of some kind of correlation between image and text, we must read in-between image and text. Herein, we detect insinuations of colonial superiority, racism, border rivalry, nationalist tensions which are surprisingly locked into the architecture. Not the architecture in the text, but the one we are forced to make up ourselves between what we read and what we see.

In *Guide to the Murals*, the architecture outlined (in the text) is disturbed (in the images) by a disorder projected through ungainly fragments of nonexemplary ordinariness: the commuters' casual footwear, their plastic carrier bags and worn-out luggage, cheap plastic chairs left in empty halls and platforms, unremarkable gridded plasterboard ceilings, jarring grid fluorescent lighting, unadorned walls, generic ceramic-tiled floors. As the text of the guidebook demonstrates, architecture remains inhospitable to such inconspicuous

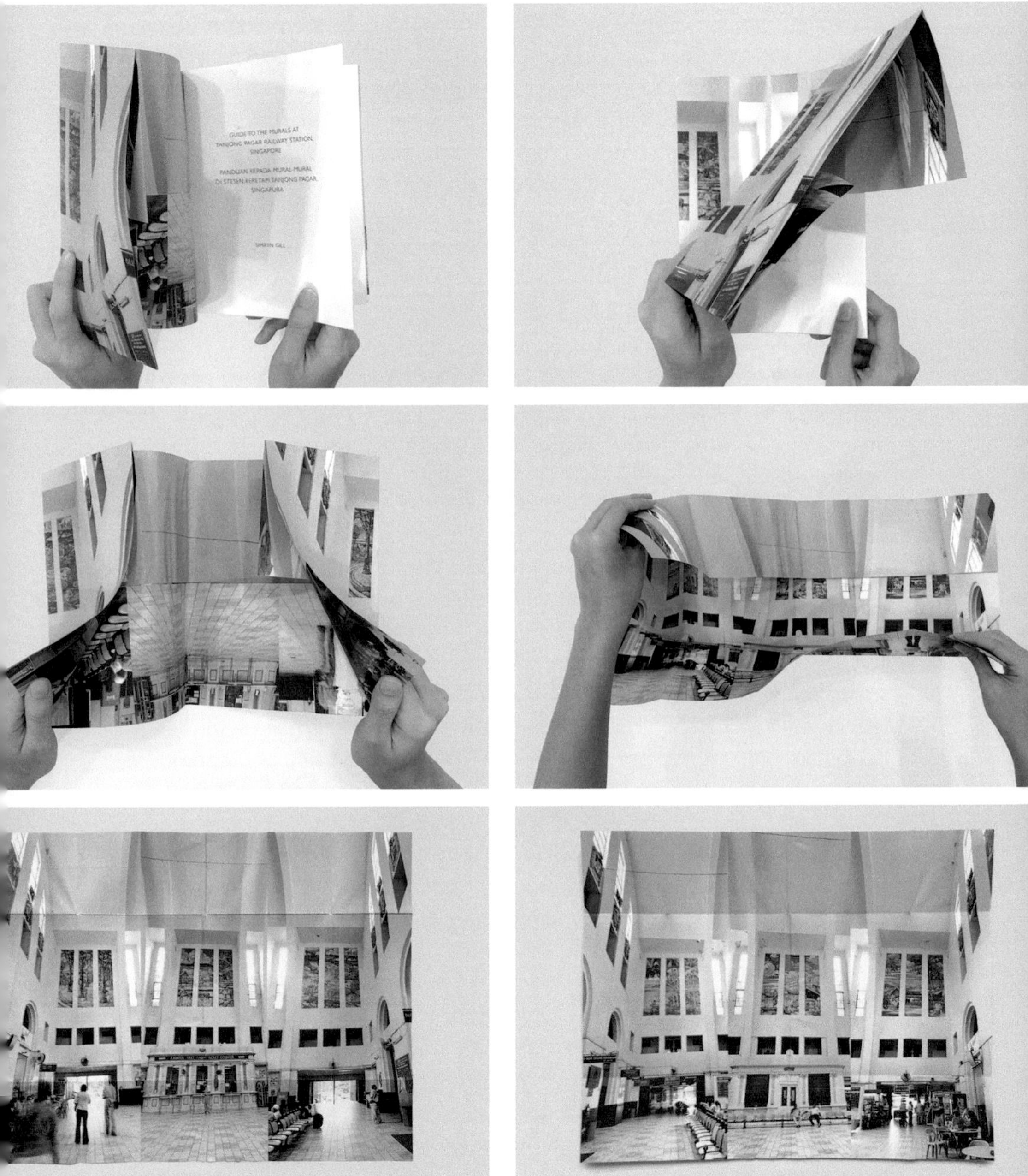

3.10 Unfolding the jacket of Gill's *Guide to the Murals* to reveal six murals in the Station's ticketing hall.

elements. And yet, this is the cartography of the station; this is what and where commuters negotiate their bodies around. The images defy and frustrate the authoritative voice of the text. There is no narrative propelled here so we can hardly inhabit these images. The subject of the images—passengers coming and going—are completely banal. They are ordinary to the point of becoming unremarkable. Yet, the commuters cannot be denied. They depict a space which is changing and moving; a perishable and fragile architecture constructed through 'fragments of trajectories and alterations of spaces … (which) in relation to (its architectural) representations, … remains daily and indefinitely other'.[55] Unlike the precise building details articulated in the text, the architecture appears only accidentally in the images, perceivable almost as a footnote at the periphery of each photographic frame. This architecture is rudimentary, makeshift and indifferent. It is a transient environment which only exists to facilitate the prosaic flow of people and things, in and out, of the station.

Gill's modest volume is critical in how it deals with these things together—both architecture and its users—as a problem of interpretation, wherein a dichotomy of one versus the other cannot be so easily insisted. The structure of text (the architecture) and image (architecture's 'outside') are here dialectically positioned—the text is teleological and finite; the images are open and associative. The guide's strategic rehearsal of text and image also highlights the disjunction between architecture and affect. The text follows in the tradition of uncompromising canonic architectural historical narrative. Its authoritative and factual voice—even exaggerated in two languages—unequivocally instructs and informs. The images on the other hand, are witnesses to an unfolding scene, acting in some cases in the capacity of what Roland Barthes calls the 'punctum',[56] or that which pricks our conscience and carries us into a space beyond the image. We are taken out of the frame of these photographs into a prepersonal embodied itinerary—a childhood journey, the anticipation of a loved one arriving, the missed train, the lovers' tryst: 'The most powerful space of the photograph resides in its peripheral space and the blank space, the glow, extending beyond and around the frame. This is the space of accidents, "failures", social movement, contemplation. It is in the peripheral space that images turn into language'.[57]

The ultimate achievement of *A Guide to the Murals* is Gill's ability to make ambivalent what is normally perceived as unshakeable and solid—a colonial monument and its factual history—not by forceful opposition but by mirroring and doubling the geometries and strategies of power that have dictated the rise, fall, and rise again, of this station. In the guide, the two agencies of power—the magisterial influence of colonial architecture and the seething tension in Singaporean-Malaysian relations—are drawn out as absences. Rather than an in-your-face confrontation, the influences and traces of such powers are delineated as blanks. Their influences may be inferred through what is missing, through gaps and lapses in text and image, and grasped only ever fleetingly, such as through casual conversation on the station platform.

The guide exposes two sides of a monument—its official representation as an immutable designed object, and its inevitable transformation through occupation and use. But beyond emphasising this habitual divide, Gill's original contribution—both to the architectural discourse of the station as well as to the idea of monumentality—is how perverse the guide to the murals and photographs of the hotel appear in terms of their affiliation with an architecture that is literally coming apart. It is the abject sense of decay which distances Gill's work from the customary narratives of nationalism or heritage. The perishable monument is subtly but persistently evoked at the periphery of every commuter image. This unspoken state of dereliction, which is already beyond normal wear-and-tear, ultimately alerts the reader to something 'indefinitely other' unfolding in their midst. The unkempt station and the ruined rooms of the station hotel are evidently scarcities in an unforgiving hyper-urbanised city like Singapore where renewal, erasure and redevelopment are norms. And given the convoluted political and historical contexts of the Tanjong Pagar railway station, the dialectical relationship between image and text in Gill's two works raise questions, perhaps even doubts, about any attempt to smoothen out the architectural representations as the station now shifts from contested territory to national monument. Once the Enlightenment remits of value and purpose rush to fill the vacuum that has thus far surrounded the station, it is likely that the blunt definitions of 'heritage' and 'national belonging' will also bear down on the station, consolidating it as a national asset, and co-opting its meanings into something less mutable.

In this sense, Gill's insistence on speaking through and around the edges of the monument is key. If one looks properly, a kind of minutiae overwhelms her two works. It becomes impossible to gloss over anything. There is plenty of detail, but the kind which is rejected in architecture—unsanctioned alterations to the building fabric, decaying parts of the building, filth and ruination, disuse and waste. Instead of a theory that projects architecture as exemplar and canon, architect and academic Robert Segrest suggests an alternative 'method and theme' for thinking about an 'out of order' architecture—one that violates 'the shape of the field' which is predicated completely on an object without an outside; one must think through the edge, the perimeter, the border and the itinerary.[58] Segrest cites Michel Serres, who proposes an architectural cartography of fragments, '... a mosaic of knowledge made up of borrowings, detours, codes and messages that cross each other, creating unforeseeable connections and nodes'.[59] In Gill's guide and hotel series, it is at the edges of the monument, where breaches into the unwieldy topics of territory, politics, nationalism, history and heritage begin to occur. These subjects are knotted into the architectural residue of the monument. Such residue is at risk of becoming that which architectural discourse inevitably distances itself from, having neither language to articulate nor will to engage. Yet in Gill's work, the questions concerning politics, nationalism and heritage become necessarily drawn through such awkward leftovers. The residue cannot be disregarded.

Rust and ruin

> This reality is mobility. Not things made, but things in the making, not self-maintaining states, but only changing states, exist.[60]

For her exhibition at Philip Cox's now dismantled Australian pavilion at the Venice Biennale between June to November 2013, Gill applied for the pavilion's roof to be partially removed so that the space could take in the changing light and weather as the warm and long summer days began to shorten. Curiously enough, Cox's steel-frame kit-of-parts pavilion—two rectangular volumes of different heights with varying floor levels and a wavy roof—was initially meant to be a temporary structure when it was first erected in 1988. The structure eventually lasted 26 years.[61] As the last exhibition to be held in that pavilion, Gill's approach and pieces were perhaps pensive about what might come after the show. There was a sense that she brooded over the kind of ending that would be meaningful for such an enduring structure. Placing *Half Moon Shine* (2013), a large cast-metal bowl custom made in Mumbai for the artist, under the now exposed canopy of a tree, Gill encouraged the influx of autumnal rainwater and falling leaf-litter into her installation and into the once sacrosanct interiors of the pavilion. She painted India ink along the pavilion walls, intending for rain to slowly wash away the ink, leaving only an indecipherable trace in its wake. She also coated nails which would be used to hold up some of her wall pieces with water and molasses, creating a patina that would readily degrade with the passing of days.[62]

The fading monument is a rare event because ideologically, architecture works hard to repress any kind of decline. Rather than waste away, a building simply ceases to be—it is swiftly decommissioned, immediately changed in purpose, hastily refurbished or unceremoniously razed to the ground. 'Obsolescence', architectural historian Antoine Picon argues 'is not exactly the same as death'.[63] The obsolete is an outcome of something that stops being useful. Instead of a state where aging buildings 'surrendered themselves progressively to nature in the form of the ruin', in the contemporary urban milieu, if buildings are not completely and instantly demolished, they are 'as if by magic ... relegated to obsolescence': 'We have gone from ruin to rust, from trace to waste'.[64]

What may be argued here is that Gill has attempted to give a different value to 'rust' by transforming what is usually repressed or hidden away into a process that can be viscerally contemplated. And although Gill produced art, which should neither be confused nor conflated with architecture, she was also working with an architecture as it approached the end of its lifespan in the Giardini. If the pavilion was symbolic of a nation—Australia—and had operated efficiently as a shelter, in opening its roof and thus undoing its basic architectural function, Gill pushed forth the difficult question of what we should make of an architecture when it is about to lose its original purpose and meaning(s). As in the case of Cox's pavilion, architecture's mission is to uphold permanence, utility and strength. This mission rehearses and reinforces architecture's Enlightenment values. Yet rather than posing a lack, is it possible to see a decaying monument as architecture which necessarily elicits a more speculative and indefinite response from its future audiences and architects? This is the state of 'in-betweenness' that Gill prefers to occupy, of 'place as verb rather than noun which exists in our doings'.[65] The same thinking undergirds her understanding of Cox's pavilion in its dying light, not as a terminal condition but a process that one must work through and engage with. The pavilion at the end of its lifespan was a place which would also only 'exist in our doings'.

The in-between states of failed building sites have fascinated Gill enough that her representations of such spaces are becoming a genre of their own within her vast body of work which take the subject of 'place' as its core. In her previous work with abandoned pieces of architecture in the tropics—*Standing Still* (2000-2003) a photographic series of abandoned buildings and failed projects across Peninsular Malaysia; *Power Station* (2004) a photographic essay revolving around her family bungalow and the obsolete power station next to that property, and *My Own Private Angkor* (2007-2009) a set of photographs focusing on an abandoned and unfinished housing estate in Port Dickson—the emphasis has been to think through the building's recalcitrance to entropy as part of the architectural problematic. The photographs catch the buildings or fragments of buildings as part of the natural evolution of space. They probe our relationship with such spaces as these spaces dismantle and fall apart, returning reluctantly but inevitably back into nature.

In these three works, architecture is neither a static object nor an isolated occurrence. The works narrate an open circuit of actions and inactions initiated and/or abetted by human will, which for architecture, usually involves money, legislation, design and technology. They are also telling of architecture's vulnerability to nonhuman forces like the weather, the animal world and simply building matter—concrete, plaster, glass, metal—falling apart. In the lifespan of the architectural monument, the nonhuman elements coalesce with the fabric of the building until they become indistinguishable from one another. Yet Gill is suggesting it is not just nonhuman forces which hasten architectural entropy. The buildings in disrepair here are equally the result of shifts in power and capital: for instance, the station and its decrepit hotel are evidently remaindered objects of specific political manoeuvres. Their imminent restoration into an impossibly pristine condition is also part of the same shifting political narrative.

Further, Gill's photographs of such 'doomed' spaces operate apart from standard architectural photography. Focusing in-between, around, beneath and beyond, the photographs perceive architecture as a kind of fog: the building is broken up into fragments which are sometimes recognisable and other times not. Architecture is glimpsed and made sense of, as Walter Benjamin proposes, in this state of distraction.[66] It emerges and is inflected through the transient processes of occupation, use and decay. Drawing the viewer into moments of longing, despair, affection, poetry, resignation and desire for this architecture, and into pondering about one's relationship to this architecture, to its past and its future that no one has yet properly talked about, Gill's photographs are affective rather than instructive. 'Affect arises in the midst of in-between-ness … (these are) visceral forces beneath, alongside, or generally *other than* conscious knowing, vital forces insisting beyond emotion—that can serve to drive us toward movement, toward thought and extension'.[67] Or as anthropologist Kathleen Stewart explains, affect is about '… being sentient to the world (one is) in. It is a matter of literal contact, exposure to rhythms, interruptions, bodies, pacings and *relations of a territory*'.[68] The unfurling of affect in Gill's photographs is prompted by a simultaneous gathering of various constituent parts—bodies, time, architecture—and their ever-changing mutual relationships marked by 'rhythms', 'interruptions' and 'pacings'.

Yet these same photographs are equally disturbing. They are beautiful, curious, enigmatic and intimate even while they depict something which is putrefying, crumbling, degenerating and corrupted. The photographs raise the spectre of architecture caught up with, and sullied by, things and processes that do not belong. Anthropologist Mary Douglas argues that this condition worries us because they unravel something that we can still recognise; 'pulverizing, dissolving and rotting' eating away at an established 'clarity of the (architectural) scene' for which their 'out-of-place' presence ultimately impairs.[69] These photographs trouble because they show not unfathomable ruins but fragments that we can distinguish, even interiorise and inhabit. We might imagine the provenance of the displaced panes of glass in those half-built houses at the ransacked Port Dickson suburb; we might recall from our past experiences similar layouts of rooms and furniture as those found in the Gill family bungalow; we might even recognise the emptied-out school building or the mossy columns standing in an open field of tall grass. We can still make out the windowpanes, the shape of the room, the furniture and the columns. These have been inexorably altered by weather and time but mainly neglect hastened by human will. The power of these photographs comes from us wanting to look in, to get into those spaces and to fill in the blanks and gaps which they seem to hint at. This compulsion echoes Gill's description of her own fascination in wanting to look in:

> All I can do is go and look. All I can do is bring back those records of what I looked at. The political charge is almost a separate thing. … I'm not setting out to make an argument. I'm saying, 'Here they are, I went to look at them'.[70]

This same unsettling interiorisation of architectural decay is what Gill brings to her series of seventy-six unpublished photographs of the abandoned Station Hotel. The untitled and unpublished work systematically records the thirty-four hotel rooms looking from within and from without each interior space, its corridors, hallways and staircases. The photographs show these hotel rooms and their surrounding spaces in abject disrepair (Fig. 3.11). As sunlight rushes into each space, we see the whole architecture disassembling in time. The rooms are enveloped in silvery dust, gossamers spin in the air, flowery stains of damp pattern ceiling boards, animal droppings make tiny ink blots on walls, concrete floors breathe away from the carpets that once buried them, the rectangular shapes of absent aircon units mark walls, leftover objects like ceiling fans and spring-coiled mattresses once purveyors of comfort now threaten with collapse or disease.

In these photographs, we witness the immanent undoing of architecture: the building as object is returning into the building as flows; from immutable forms into fragile material entities; its previous 'categorical status' as the grand hotel already elapsed.[71] In the time-ravaged hotel rooms, almost all traces of architectural order have vanished. We can only just about make out the edges of each space. The vivid colours which illuminated and clarified the murals in the station, have a different effect in these rooms. The colours—hues of seagreen, turquoise, brown and mustard—are the result of an architecture dissimulated by fungi, mould, rot, termite and dust. Through doorways and openings, we catch glimpses of the celebrated murals (Fig. 3.12). Yet against these crumbling visual referents, the murals also appear to be biding their time.

Decay is a process where things tether at the edge of legibility.[72] In the photographs, the hotel rooms are barely legible for what they were. It is Gill's persistent framing of these spaces against the colourful murals which gives them a recognisable, if also static, reference point. In doing so, the artist is reiterating how architecture is habitually understood and traversed through its plan and section drawings, always as a sequential procession from a main circulation spine into its branching subsidiary spaces. Every photograph has the same frame, the same aspect ratio. The camera's immobile gaze obeys the limited posture which the urbanised body is expected to take. However, the frame does not discriminate what is caught within. Its objective is not unlike

3.11 Gill's unpublished photographs of the abandoned and forgotten Station Hotel capture the remaindered.

an unbiased orthographic projection. It is like Gill declares, only to prove 'Here they are'. Such disciplined optics are at odds with the decaying space that the frame scrutinises.

Indeed, geographer Tim Edensor argues that the ruin evokes a 'differently performing body' which must react to the ruin's sensual stimuli.[73] This sensuality ultimately challenges how the 'modern body' has been conditioned to 'comport itself in the city', to destabilise how it should 'apprehend and sense the city' beyond the normative criteria of what is 'efficient' or 'healthy'.[74] Furthermore, the narrative thread of the ruin is unlike that of the monument. This narrative is likely conjured in pieces, or as Edensor argues, the ruin:

> … foreground(s) the values of inarticulacy. The disparate fragments, juxtapositions, traces, involuntary memories, inferred meanings, uncanny impressions and peculiar atmospheres cannot be woven into an eloquent narrative. Rather like the nature of a ruin, the stories about it must similarly be constituted out of a jumble of disconnected things, occurrences and sensations. Ruins are disarticulated spaces and language can only capture their characteristics through halting speech. Bits of stories suggest themselves and trail away into silence. As an encapsulated narrative, the telling of the ruin's tale from beginning to end is impossible, for such a story must be open-ended.[75]

The paragraph above uncannily proposes how Gill's two pieces at the Tanjong Pagar Railway Station might be read as an architectural document, a kind of 'historical evidence' where the caveats for reading are made transparent from the start. Yet unlike a historical archive where the principles of 'historical truth' are assumed or produced through 'official and expert memories' or a 'subaltern account',[76] the guidebook and photographs question the fixed agencies of architectural history-making and telling. The reader is invited to look, and to sense that what they are looking at is contingent. In doing so, the reader is implicitly collaborating in a narrative of the architectural remainder that is neither imperious nor causal but fully conditional on the past being undecided. Gill's interventions acknowledge that in between fact and affect, architectural history is made up by 'suppositions, fantasies, desires and conjectures'.[77] The focus on decay is strategic. As spaces suspended in between rust and ruin, we cannot help but see the station and hotel as 'remainders and reminders'.[78] Their varying degrees of ruination lead 'into a labyrinth of ambivalent language—no longer, not yet, nevertheless, albeit—that plays tricks with causality'.[79] The stories these two works tell are neither linear nor seamless. Their histories are multiple and varied; incoherent and often difficult to describe. Their halted mode of telling strikingly mimic the affects of the Singaporean-Malaysian diplomatic stalemate, for which the station inevitably became collateral damage (Fig. 3.13).

3.12 From within each hotel room, the murals in the ticketing hall are visually referenced.

Arguing against the urge to flatten out historical narrative into a 'linear chronological progression', DeSilvey pioneered a mode of 'anticipatory history' which attempts to rewrite a history through the effects of transience and time.[80] Proposing a shift from critique to collaborative mode, DeSilvey asks, 'What kind of cultural work might be required to give time back to a timeless landscape, and to open up an appreciation of the past not as static and settled, but as open and active?'[81] This question is also what the guidebook and the hotel photographs prompt. What DeSilvey and Gill argue separately but unequivocally is that architecture (or space) is constituted through a complex constellation of forces. Yet the risk of any monument or heritage site is that these often discontinuous, contingent, contradictory and competing narratives are reduced to commodified foundations that will not account for what is transient or circumstantial.[82]

The station's official architectural narrative, which is already in-the-making as this chapter was written, will likely overemphasise its Art Deco and Neoclassical lineage, its colonial roots, heroic political past, and certainly timely rescue from its impending decay. Yet, the anticipatory nature of the architectural narrative for a monument like the Tanjong Pagar Railway Station suggests that a different representational strategy is needed to restore its histories. And if we wish for a preserved architecture which is open to life and change, ways of narrating, picturing, showing, documenting and writing about it must be more open-ended and speculative. Gill's *Guide to the Murals* and hotel photographs have reappropriated architecture's representational strategies—using the architectural guidebook and the purist emptiness of architectural photography. In each, the iconic strategies which also predetermine a narrative thread is challenged. Text, image, light and colour are deployed to bring out movement, change and transience. These artistic-architectural representations destabilise the monument's production, suggesting an endeavour multiply sited, jointly authored and still persisting.

Although conceived as two separate works, a dialogue between the guidebook and hotel photographic series is viscerally enacted across slippages between the static geometries and facts in the architectural text versus the transient forms, subjects, materialities and fables in the images. The artist herself is still reluctant to put the hotel series out there in the world.[83] Perhaps, this is Gill's gesture of personal closure for an expired space for which she has affection—keeping the photographs as a bottom-of-the-drawer piece and sustaining its lifespan as work-in-progress until she is ready to let it go. Her approach emphasises an embodied and occupied materiality about the building which can indirectly tell us 'stories about power, agency and history that we could never grasp from more direct forms of representation'.[84] It interrogates the systems of evidence-gathering, meaning-making and history-telling which must be transformed when the architecture's historical trajectory is under dispute; when there are obvious gaps and silences in the building's past. It draws the uncomfortable

3.13 The corridors of the Station Hotel encircle the ticketing hall below.

outlines of an affective architectural history by affiliating with decay, thus disrupting authoritative academic language.

One final note must be made about the centrality of the murals in Gill's two works. The artist refers specifically to the murals rather than the station in the title of her guidebook. The vividly coloured murals have outlasted the hotel, and they would have outlasted the station. Held aloft, like cathedral windows, they are the common referent for the guidebook and the hotel series. To get anywhere, it is to the murals that we must turn. Their prejudices of racial hierarchy and labour distribution in a bygone colonial age are now muted into luminescent saccharine colours, made palatable by nostalgia or just time passing. They are decorative, innocuous, photogenic and fetishised. However, like relics, they can be delusory and discomforting. In these works, the murals, and not the station, are perhaps really the lasting thing, 'the monument' in its most conventional definition. The guide talks about the murals, but in between the lines, and between text and image, Gill is letting slip that what is remaindered, residual, and leftover needs to be looked at, and looked into. •

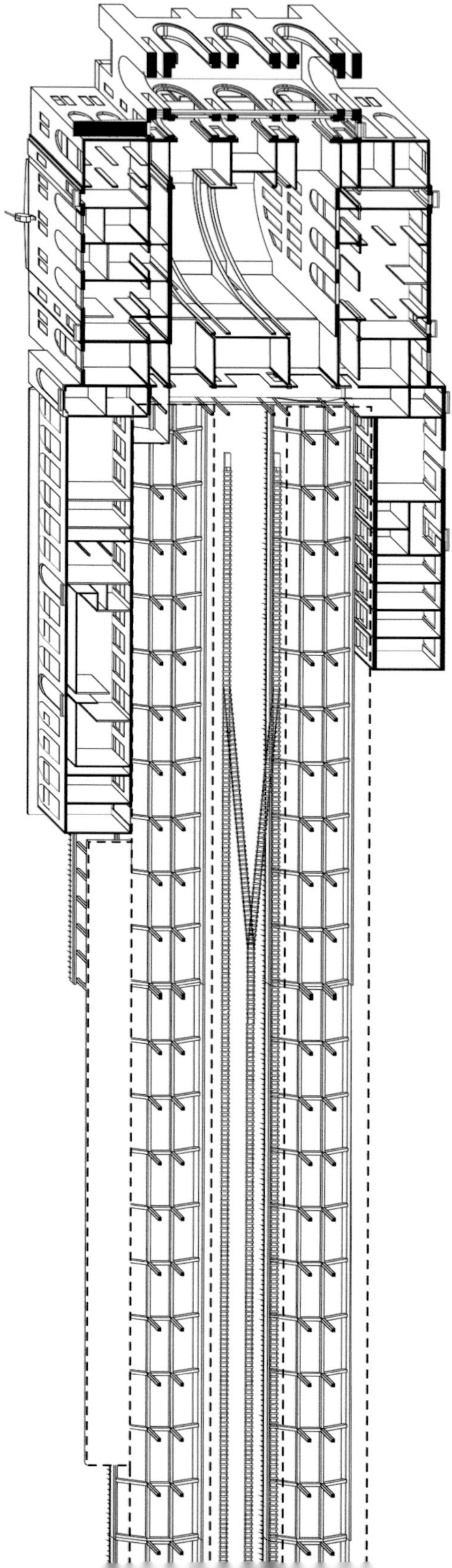

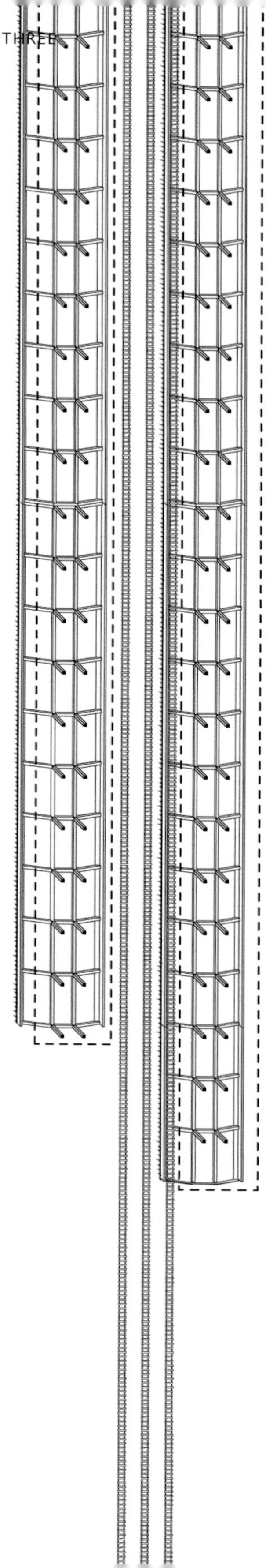

Part Two: Block

4 • Keeping Cats, Hoarding Things

Situations in the housing block

In January 2013, an elderly 'rag-and-bone' woman who collected discarded items for repurposing and resale was found sleeping rough on one of the staircases in Block 419, a public housing building located in the eastern Singaporean neighbourhood of Bedok North. More remarkably when questioned, Madam Ho who was in her 60s, revealed she owned a flat in the same block. She could no longer enter her unit because the items she had previously collected now obstructed access.[1] Madam Ho's neighbours had taken issue with her compulsive hoarding habits, especially when these items overtook the shared common corridor. The woman's 'home' was subsequently reduced to a shopping trolley filled with her belongings. She lived off meal vouchers and sought assistance from local community services.

In the recent decade, the number of hate crimes and accidents involving stray or street cats have risen in Singapore's public housing estates. Street cats have suffered strangulation, were thrown from height, had their limbs mutilated, their bodies slashed, and some were found caught in glue traps laid for rats at the void deck of the housing block.[2] In April 2022, 13 street cats were rescued from a smoke-filled flat when the Singapore Civil Defence Force were alerted to a fire in Bedok North.[3] Cat ownership is prohibited in public housing flats. There are no regulations which protect the animals even though cats commonly roam free in many housing estates.

How do these domestic, and perceivably, nuisance situations—the plight of the hoarder and the fate of street cats—implicate the architectural discourse of Singaporean public housing?

What if it fails?

The architectural critique of public housing either revolves around its socialist ideals, aimed at better living standards and new modes of living, or dwells on the compromises made to insurmountable social problems and urban congestion. Public housing strategy is largely hierarchical, sometimes utopian and often myopic. Although public housing could be radical, its failure becomes a byword for architectural hegemony. Examples of successes and failures include radical modernist architectural plans of the Weissenhof Siedlung (1927) in Stuttgart designed by Mies van der Rohe, Le Corbusier, Walter Gropius and Hans Scharoun amongst others; the Soviet experiment of low-cost, prefabricated industrialised housing (the so-called *Khrushchyovka*) beginning in the late 1950s under Nikita Khrushchev's dictate to stem out all architectural excess and the creation of hundreds of suburbs or *mikrorayony*;[4] and arguably, the infamous high-rise public housing project from the last century, architect Minoru Yamasaki's Pruitt-Igoe in St Louis (1954), which was originally developed for a race-segregated middle-class community but in under two decades, became an architectural shorthand for dysfunctional mass housing with its social problems, racial violence, crime and urban decay.

The early social housing experiments in Europe and the United States stood at the frontiers of architectural production but their sociocultural conditions were hostile to mass housing. The Soviet example is resonant with Singaporean public housing given that the body politic of the Russian state was ideologically embedded in its rapidly disseminated blocks. However, with the shift of the urban economy away from production, the Singaporean example (the Housing Development Board flats, or HDB) is differentiated from the Soviet *Khrushchyovka* because while the Soviet experiment stopped in 1971, Singaporean public housing continues to be built, efficaciously regulated and popularly subscribed. It is the housing choice of more than eighty percent of Singapore's resident population. Thus, the HDB is an exceptionally successful non-Western condition but more significantly, a study of it offers a ground-up perspective of an architectural typology that, in both Western and non-Western examples, has tended to be read through and expressed exclusively as normalised components of state ideologies and urban planning technologies.[5]

In her 1967 article in the *Chicago Tribune*, architectural critic Ada Louise Huxtable who visited Moscow to examine its public housing on the 50th anniversary of the Russian revolution described the technological advancements enabling the rapid roll out of standardised apartments to be advanced but lamented 'it looks promising, measured at one inch to 100 feet, but becomes inhuman when measured out in footsteps'.[6] As Huxtable observes, the optics of public housing is deeply entrenched at the urban scale while simultaneously alienating and subjugating knowledges and experiences on the ground. An inhuman public housing. What she, and this chapter advocate, is a recognition of public housing's spatial complexity forged through the habits and decisions of its dwellers, and particularly, an acknowledgement that something else transpires; something physical or psychical is produced when architecture is occupied.

Cultural theorist Lauren Berlant names that productive impulse of human inhabitation as adaptations, adjustments and improvisations to glitches.[7] These constitute coping mechanisms for us to refigure our relationship to a transitional situation. The situation, Berlant explains, is a genre that pulls us into something emerging, something that is likely already amidst us but still nebulous, 'materials in which worlds are made and not forced'.[8] Situations press themselves upon us, they draw us in; they are like a *punctum*. As such, situations are often disturbing circumstances in which we are mired but for which we have no control.[9] Led into these, we become 'interested in its potential changes to ordinariness. When a situation unfolds, people try to maintain themselves in it until they figure out how to adjust'.[10] Berlant argues that situations are modes of a historical present in that they embody a shift from institutionalised history's continuities to offering a preview into 'something incoherent or uncongealed in the ongoing activity of the social'.[11] This definition of 'situation' brings us back to the illegal cats at the HDB void deck and the rag-and-bone woman's things at the HDB corridor.

Despite Singapore's efficient law enforcement, these domestic disturbances persist, overlaying its public housing's securely managed ideologies with ground-up situations and messy outcomes of a lived domesticity. I recount these situations to think through the litany of 'worry, ambiguity and concern',[12] which habitually pervade the 'public spaces' of public housing. 'Public spaces' refers to the shared spaces in the housing block which do not have a specific domestic function yet are adjacent, or proximate, to its living quarters. I use the term 'public' intentionally to signal the tension between two entities—on the one hand, what constitutes the private household realm of the flat where one has autonomy to freely act, and on the other hand, what is *not* part of but closely located to this private domestic environment.

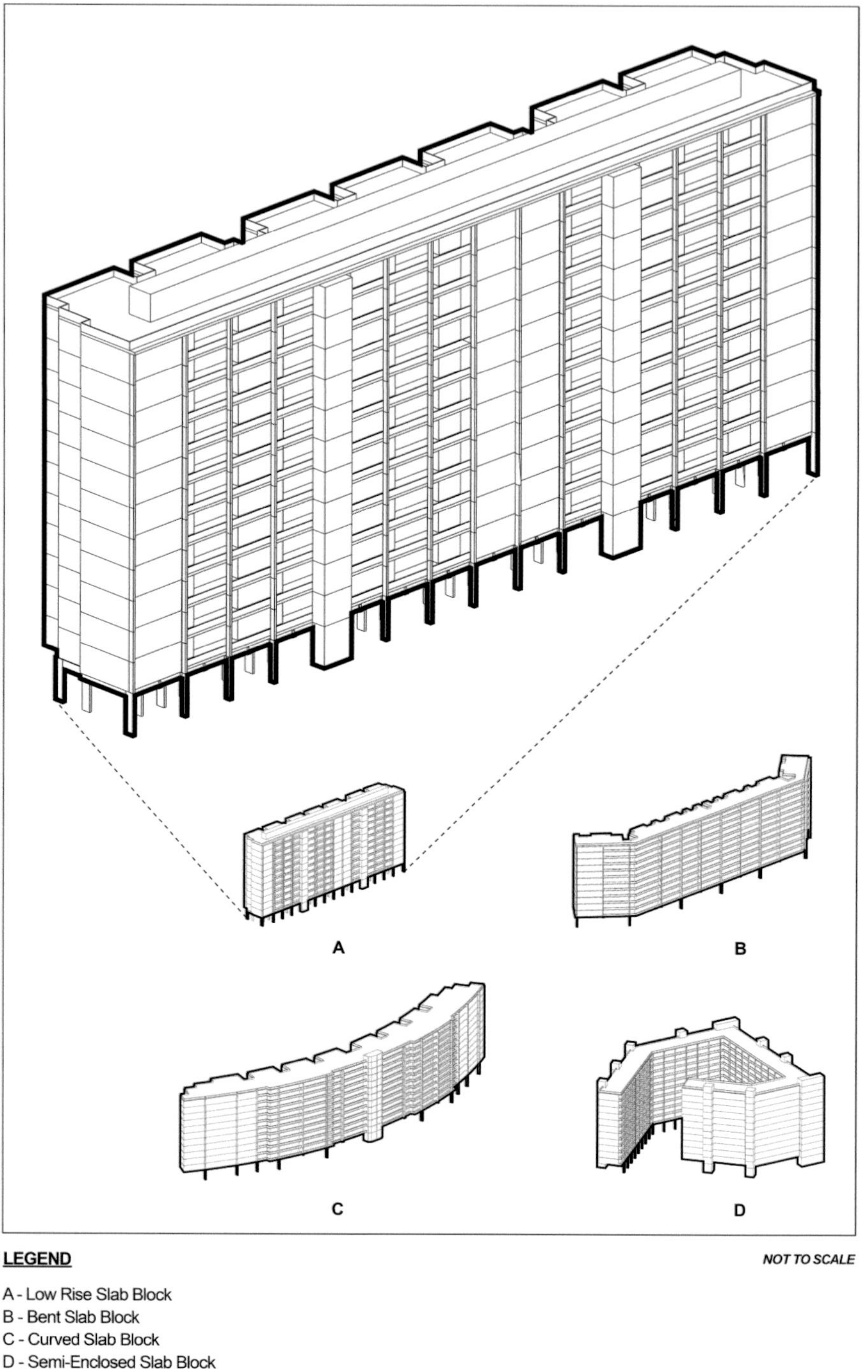

4.1 The HDB's distinctive (A) low-rise slab-block typology with void deck and common corridors, introduced in the early 1960s, with iterations of the typology conforming to site constraints: (B) bent slab block, (C) curved slab block, and (D) semi-enclosed slab block.

Two public spaces in public housing—the common corridor, a linear circulation space leading to the individual flats, and the void deck, an open public space at the ground level of the block—which constitute the primary elements of the early slab block designs, are of note. A first-generation iteration, the distinctive slab block typology was introduced in the early 1960s, popularised in the 1970s and last erected in 1985 (Fig. 4.1). It took on various shapes to conform to the urban exigencies of a range of housing estates. Its corridors have raised parapets which stretch over fifty metres in length and typically adjoin twelve to sixteen flats. The void deck is an open plan, porous ground level space with pilotis arranged in a grid configuration. This same column grid extends upwards into the interiors of the high-rise dwelling.

It is significant that the overall design of the void deck and the corridor remained relatively unchanged over a period of twenty years. Their purpose was to enact specific ideological intentions and solve practical issues. They were employed as part of a scalar solution to integrate nuclear families into larger communities in a bid to moderate the overwhelming scale of a new housing precinct. They were also important devices for healthy and hygienic living since both spaces provided through ventilation and generous open space in contrast to the densely inhabited single unit. With no specific function required by the dwelling units themselves of these spatial typologies, the corridor and void deck are, in effect, slack spaces with loosely defined purpose, accessible to residents and non-residents alike. If the block's public connective spaces were meant to forge community, can we accept that the corridor and the void deck's new definition of 'community' might challenge the normative model of belonging and sharing? Spatially, can there be more nuanced articulations of the intersections between private domesticity and public housing, particularly pertaining to the interstitial spaces of the corridor and the void deck?

Functionally ambiguous, the corridor and the void deck contrast the HDB's tightly regulated program of spaces and facilities. In its early years, the eponymous high-rise flats were constructed alongside new policy guidelines that set out to regularise and systematise nation-wide living standards.[13] The success of the HDB enterprise was premised on rigorous quantitative research focused on perfecting plot ratios, balancing densities and occupant distributions, maximising the returns of regional planning in terms of transport and infrastructural connections, and a constant fine-tuning of unit dimensions. Singapore's public housing's efficiency is in part a response to its first Prime Minister Lee Kuan Yew's bold social experiment to create a 'home-owning society'.[14] The brief to make public housing affordable for the masses and to build at speed without compromising quality, meant that the most effective construction methods and materials were procured at competitive cost, and finished flats were floated on an open market at controlled prices, coupled with heavy subsidies.

Aligned with this trajectory, findings from HDB-commissioned research and discourse produced within professional circles of architecture and urban planning focus on utility and efficacy. Topics discussed include urban planning at neighbourhood, precinct and regional levels for purposes of liveability, social cohesion and neighbourhood identity, unit plan modifications, façade and block designs, new standards of prefabrication and better systems of waste disposal, ventilation and vertical transportation, new configurations for new types of occupancies including single dwellers and the elder population. From the homogeneity of the extant discourse, there is a sense of public housing lacking a public.

In writing about the house as 'equipment' sustained by various institutional norms—'ownership, sexuality, kinship, family, lineage, technics, servitude, repression, civilisation, privacy, intimacy'—architectural theorist Georges Teyssot reminds us that a piece of equipment is first and foremost necessarily economic in nature: it produces a 'regime of normalisation' that 'establishes limits (of property for example) and exclusions (between the forbidden and the permitted)'.[15] Singaporean public housing architecture may be said to resemble such an instrument. It rehearses the pervasiveness of similar institutional limits and exclusions. It also exercises these norms by establishing what is 'proper', normalising what is specific or exceptional in its architectural design and intentions.

Given that a large cross section of the population dwell in these accommodations, is the strict application of a norm realistic? What happens when an architecture borne of quantitative calculus starts to show cracks in its social fabric? What happens when the block bears witness to exceptional situations which are neither representative of the norm nor its tolerated average? How do we evaluate the successes of these spaces if not by their efficacy? In summary, how do we deal with situations where the functionality of the block appears to *fail*?

Scholars have demonstrated that architecture is continually re-produced through occupancy.[16] The insertion of the occupant as co-producer, particularly in the Singaporean public housing context, is significant given public housing's stake in the national project. Adding a new dimension to critical architectural histories on Singaporean public housing,[17] this chapter emphasises the importance of occupancy, both human and nonhuman, to review the social and psychological impact of public housing architecture. It looks at the ground floor open void deck from the marginal perspective of its cats and examines the housing block's common corridors through obsessive hoarding. The production of architecture by occupation is organic, open-ended and experiential in ways that traditional architectural evidence can neither sufficiently record nor express. Anthropological, ethnographical and geographical studies of human-animal social/spatial relations as well as material cultural studies on the affective/infrastructural spaces of waste and trash enable an understanding of how occupancy transforms architecture. To begin with, why (again) are the two domestic situations—the plight of the hoarder and the fate of street cats—at odds with public housing architecture and its normalised discourse?

Drawing the normal in public housing

In 1960, the rise of public housing as a systematic political and social instrument in the newly independent nation state of Singapore combined pragmatic and ideological will. The initial impetus to solve a housing crisis in the late 1950s transformed into a convenient mechanism for citizen-making when in 1964, the newly launched homeownership scheme soon co-opted every willing citizen and homeowner into a binding and contractual relationship with the state. The former was transformed into stakeholders.[18] 'My primary preoccupation,' declared Prime Minister Lee Kuan Yew, 'was to give every citizen a stake in the country and its future. I wanted a home-owning society. I … was convinced that if every family owned its home, the country would be more stable …'[19] The state enabled such stakeholding[20] through generous housing grants, rebate schemes, family bonuses for generational proximity between households and heavily subsidised prices for new-built flats that appreciated quickly in value once the property entered the open market. On their part, the new homeowners could lay claim to the country's then fledging economy, and they too rapidly evolved into citizenry who understood the criticality of vigour and discipline in a thriving labour force.[21] HDB ownership facilitated 'the expansion of commitment to the prevalent social order by the development of personal stakes in its survival'.[22]

Thus, the home ownership ideology forged a more enduring if also complex relationship with the state wherein basic shelter was guaranteed only if proper occupancy requirements were met and maintained, including uninterrupted servicing of the monthly mortgage. However, home ownership involved more than the citizen's financial stability. Public housing was a tool to manage the population, and here, the national imperatives of birth control, family planning and racial quotas weighed heavily upon home ownership regulations.[23] In 1973 for instance, in a bid to engineer smaller families amongst the less highly educated population, one of the spouses who included a female work permit holder married to a Singaporean had to agree to 'sterilisation after birth of the second child or lose government housing subsidy and other concessions'.[24] Sociologist Chua Beng-Huat thus argues that public housing is framed within an 'explicit ideological language' intent on defining the 'proper' Singaporean citizen.[25]

Consequently, the achievement of Singaporean public housing is neither the eradication of squatter housing and disease-ridden environments nor the innovation of its architectural solutions. The achievement is how deeply public housing has infused itself into the national psyche. As of 2020, the HDB houses an estimated 80% of Singapore's resident population, many of whom are also owner-occupants.[26] With the vast majority subscribing to this housing programme, it is almost impossible to occupy an outside to this pervasive architectural milieu. Monopolising the Singaporean domestic environment and designed as a total environment covering a range of scales and spaces from the individual flat to the block, the larger neighbourhood and the greater precinct, this architecture ultimately shapes the resident's worldview. It is all encompassing.

In its capacity to physically and psychically hold a nation, the HDB flat is incomparable to any other architecture both in Singapore and beyond its shores. It is simultaneously the most common *and* the most aspirational of architectures. Unlike social housing elsewhere which tends to be stigmatised, the HDB block houses a national microcosm: 'financially-challenged families, solid blue-collar families, and also the upper middle class and young upwardly-mobile professionals'.[27] The citizen who qualifies for such housing is shaped by public campaigns and programmes targeted to alter their social behaviour and personal disposition with regard to productivity, public hygiene, neighbourliness, family planning, energy use, moral values and speaking Mandarin, amongst others. From 1968 to 1982, 66 campaigns were launched, and many of these found their way into HDB dwelling etiquette, conditions of grants, ownership policies and regulations, 'In the home, citizens are instructed on family size, the timing of household formation and childbirth, the language they ought to speak, are compulsorily required to save a fifth of their income, and since 1996, to financially support their parents in old age'.[28]

Yet ironically, in responding to the nation's utopian dream of housing the 'good life', the architectural motivation is premised on efficacy. HDB's founding chief architect-planner Liu Thai Ker describes a 'single-minded pragmatic approach',[29] which focused on 'improved sanitation, improved floor space, improved floor plan and improved amenities rather than on an ideal approach or an ideal building plan based on sociological studies'.[30] This purposefully neutralised and function-driven approach differentiated the architectural outcomes of the Singaporean model from the idealisms embedded in Westernised conceptions of modernist housing projects. Le Corbusier declared that modern architecture's efficiency stood for a particular ethics and morality, '*Standardization, mass production, efficiency*: three connected phenomena that rule contemporary activity pitilessly, that are neither cruel nor atrocious, but on the contrary lead to order, to perfection, to purity, to liberty'.[31] Thus, for modernism, the raised standards of living and their modes of efficiency were not ends in themselves. These were means to a better 'life', a quantum which Liu openly declared as too vague a goal for a performance-oriented organisation like the HDB.[32] Indeed, Liu's pragmatism to modernist architecture, with its focus on holding the family unit together and creating homes for a stable workforce, was not only productivist but culturally and theoretically differentiated as an anti-Western approach to modernism. Ultimately, Singapore public housing stands out, apart from its pervasiveness, in how committed it was to using pragmatism to strategically perform neutrality.

The primacy of the collective workforce with the family as an economically and biologically productive unit steered the agenda of Singaporean public housing. The HDB's enduring architectural legacy of enforced neutrality and muteness permeates its design and programmatic approach (Fig. 4.2). HDB flats are, as geographer Jane M. Jacobs observes, ironically distinguished by their perceived 'lack' of design and/or designer.[33] The designs emphasise functionality of floor plans and cost-efficiency guaranteed by repetitive use of elements and employing mass production techniques and construction. The perceived differentiation between housing blocks are enabled by residents voting the colour of their blocks, an attempt to curb expected dissatisfaction of HDB's self-appointed neutral brand. Ironically, compared to its ideological ambitions and complicated ownership regulations, the architecture of the housing block seemed to have evolved purely out of a mechanical and objective process, restricted primarily by unit-costs and standards.

Thus, while the nation-building policies defining public housing were arguably utopian, ideological and sometimes even radical—with strategies including universal stakeholding, ownership regulations and grants premised on marriage, restrictions on non-familial ownership configurations and regulated racial quotas—the architectural approach was conversely conservative. Functionality and efficiency were readily adopted as the architectural baseline. In Liu's early papers on public housing, these parameters formed the basis for a new architecture for the masses. Liu's arguments focused on how variations in flat and block designs had 'functional objectives' determined by and affecting land use and its optimisation, planning controls, development processes and policies, facilities planning and distribution, new town planning and the compelling need to standardise land use distribution, density and layout of the HDB estates across Singapore.[34] Such an instrumentalised architectural rhetoric and functional design approach have ultimately influenced a perception of public housing architecture as one driven and justified wholly by pragmatism.

Juxtaposed against the radical social and moral reforms articulated in Singaporean public housing policy, the architectures of block, precinct and neighbourhood are conversely grounded in the quantifiable and the measurable. As the most visible physical expression of national policy, the architecture of the

4.2 Singapore's public housing embodies HDB's architectural legacy of enforced neutrality.

public housing block is purposefully neutral. Through its neutralised rhetoric and seemingly banal aesthetics, the standardised architecture of public housing achieves an almost *a priori* ontological status—one that is indisputable because it is a mechanical and automatic response to function per se. Its reason for being was to enable positive living outcomes including better sanitation, hygiene, fresh air, bigger flats and better interior layouts. But perhaps, the pragmatic character of public housing architecture has a performative role. Its objectivity provides the backdrop for a series of contestable ideologies on nation, society and family. Under the shadow of the block's neutral aesthetics and pure functionality, these contested ideologies were made considerably muted, and in time, naturalised. If the idiosyncratic policies provoked debate, the automated gesture of the architecture achieved the exact opposite: it normalised.

Architectural theorist Catherine Ingraham situates the 'normal' as entangled with 'propriety' and implicit in the concept of 'property'.[35] Delineating the premises of modern space on 'properness, normativeness, cleanliness, comfort, the cogito',[36] Ingraham traces the word 'normal' first to its Latin roots 'norma' that 'etymologically (refers to) a T-square, normal is that which bends neither to the right nor the left',[37] and then, through historian of science Georges Canguilhem's definition, where the idea of the normal becomes much more ambiguous, since it has the joint status of imposed value and indisputable fact. 'Normal … is that which is met in the majority of cases of a *determined* kind, or that which constitutes either the average or standard of a *measurable* characteristic'.[38] The normal is thus not a self-evident category but an artificially regulated one. It is necessarily bolstered by measurable quantities which redress the biases and exclusions which normalisation automatically installs.

In the architecture of Singaporean public housing, the 'etymological cohorts'[39] of property, propriety and the proper are productively developed. Personal stakeholding in such property is based on the prerequisite that a proper family unit is already in place, defined as 'one man, one woman, marrying, having children'[40] and further, that this basic unit will necessarily exercise appropriate kinship relations between family, neighbour and citizenry, with the three categories being non-exclusive. These values are shored up by an architecture whose variations of flat size, interior layout and public spaces in the block, have over the years, shifted in tandem with national family planning policies, community-centric campaigns and nation-building enterprises. Ingraham proposes that architecture's propriety is not however, limited to 'typologically appropriate buildings' but that it fundamentally serves as a 'housing for the proper inside other disciplines and other discourses'.[41] Similarly, the objectified and neutralised characteristics of HDB architecture both accommodates and normalises a value system founded on the heteronormative and economically productive family unit. Further, if 'the architect is still understood to be a kind of physician for the (normative) body in space',[42] then HDB architecture inevitably reinforces this body politic through its design and spatial logic.

'HDB existence is characterised', attests Singaporean poet Alfian Sa'at, 'by aerial threats, where a mouth expels its contents, a hand dispatches its cargo, before withdrawing back into a planar anonymity (which one hides the culprit, one wonders, looking up), where a primary accomplice to misfortune is gravity'.[43] Sa'at's account delineates a deeply affective space populated by sputum, discarded sweet wrappers and dirt. It leads us directly to where these tensions brew: at the edges of two colliding territories, where the housing block's public space frays against the flat's private domestic space. Sa'at reveals that the riskiest spaces are those just around the corner from his flat—on the common staircase and at the edges of the void deck. At this public-private threshold, interdictions, breaches and situations spontaneously unfold as if to reinforce the tripartite relationship of property, propriety and the proper. Yet instead of fortifying the normative, domesticity's excesses thrive on impropriety, thwarting alignment with housing's proper intents. The two boundary conditions within a housing block—the void deck and the common corridor—offer a counternarrative of the housing block from within its public spaces and their situations.

...What such open space might do

The rationale for open space in Singaporean public housing was historically driven by health. Miasmic discourse was common in the British empire and produced various models of 'hygienic' architecture, including the discipline of tropical architecture.[44] The problem of disease was a problem of space. Congestion and overcrowding in the city centre were endemic. In 1947, up to 100,000 people lived in slums.[45] The British colonial government perceived these slums as an uncontrollable rot that ate at the city. Occupants lacked legal rights, access to infrastructure and escaped municipal surveillance. Consequently, the 1955 Masterplan[46] sought to map and eradicate such squalor. By its final year of operation in 1959, the colonial administration's housing body—the Singapore Improvement Trust (SIT)—only managed to build 23,000 homes.[47] In actuality, the rehousing of squatters required ten times as many. Further, the housing crisis was explicitly presented as a problem of space—the slums bore a miserly 0.35 hectares of open space per person in comparison to the UK standard of 4.5 hectares per person. Thus, the 1955 Masterplan emphasised the necessity and urgency of open space as a baseline for hygiene and health. At the same time, it described the significance of open space's 'pleasantness'—an excessive if also unquantifiable value—because it was perceived that open spaces could enhance communal relationships.

Open space was thus conceptualised in public housing for hygiene and health.[48] Its pragmatism was also extended to forging communal ties. The naming of the two spaces—the ground level open space called the 'void deck' and the multi-level open space called the 'access balcony'—offered little in their suggestion of communality and flexibility of use. This suggests that motivations to delineate the programmatic possibilities of such space were nascent and perhaps, ambiguous. Certainly, the question of control, use and maintenance of these open, loosely programmed spaces would have posed issues to the HDB although these anxieties were not documented.

The earliest void deck was reputedly constructed in 1963 at Block 26 Jalan Klinik, an estate located at the fringe of the city centre.[49] It was little publicised and appeared more like a residual space between facilities on the ground level including the stair core lobbies, a provision shop and spaces housing building services. There were no void decks in the first HDB township of Toa Payoh when that estate was established in 1964. This changed in the mid-1970s when almost every block was found with one.[50] Similarly, the HDB annual reports did not refer to this space at all until as late as 1977-78.[51] However in 1968, a decade prior to the popularisation of the void deck as a vernacularised open space in public housing, its spatial potentials were highlighted in the *Singapore Institute of Architects Journal* (SIAJ).[52]

In the journal, the first significant mention of the HDB 'void deck' as a purposefully designed open space was made in reference to the Park Road development, a pilot project of the state-initiated, inner city core Comprehensive Urban Renewal programme. The programme sought to intensify land use, modernise development and clear urban slums. The void deck described in this article refers to an open access roof-top space on the third level of a city podium block (Figs. 4.3a-b). Here, the void deck functioned to separate the commercial podium below from the residential flats above it. The void deck on Park Road 'act[ed] as a buffer zone to promote cross-ventilation and lighten the appearance of the building' and 'serve[d] as play and recreational areas for residents'.[53] It housed a crèche with landscape and pool for children's play, a kindergarten, a reading corner and recreational area for adult residents. According to Alan Choe, then HDB's chief architect, the city was an 'unhealthy environment' that bred 'delinquency, crime, prostitution, dope as well as neurotic diseases'.[54] Thus, at its earliest stages, the void deck was employed to separate its residential units from the supposedly corrupting influences of the urban milieu. It functioned in the same way as the European urban planning concept of the *cordon sanitaire*, or a zone of exclusion, in and through which the damaging aspects of city life could be mediated or filtered out. Choe's understanding of the void deck as an urban zone of exclusion is significant because the void deck is here conceptualised as untethered to the block, though necessarily, part of the city. Its scale, anonymity and ambiguous boundaries lend the void deck being perceived and used in ways which are consistent with an urban space but these ways become conversely contentious viewed in the context of a residential block.

NOT TO SCALE

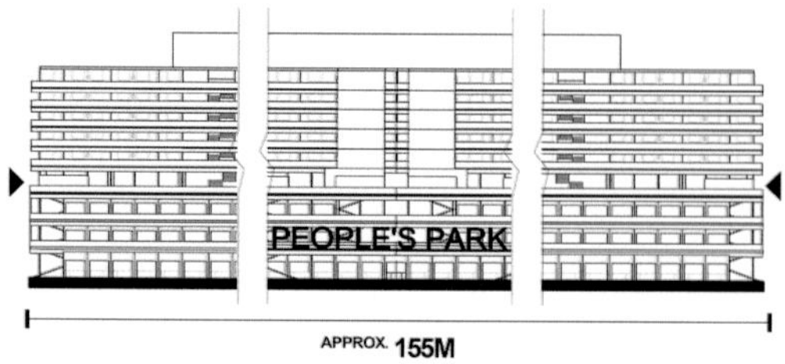

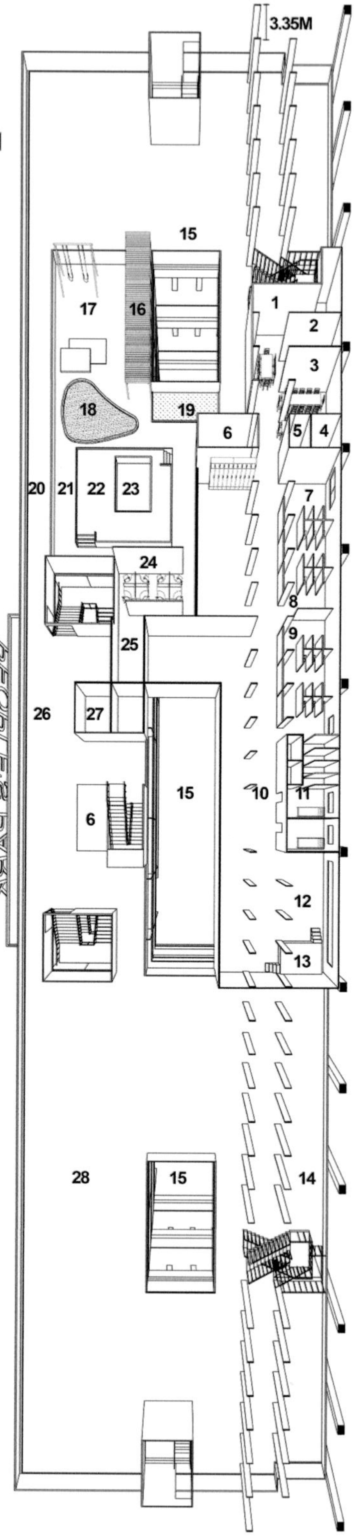

LEGEND

1 - Laundry
2 - Staff Dining
3 - Dining Room
4 - Changing room
5 - Shower
6 - Store
7 - Water Closet
8 - Boys' Toilet
9 - Girls' Toilet
10 - Staff Room
11 - Sick Room
12 - Classroom
13 - Stage
14 - Communal Area
15 - Air Well
16 - Timber Trellis Above
17 - Playground
18 - Sand Pit
19 - Turfing
20 - Timber Perimeter Wall
21 - Babies' Room
22 - Raised Resting Pavilion
23 - Wading Pool
24 - Office
25 - Entrance Hall
26 - Lift Well
27 - Motor Room
28 - Roof Play Area

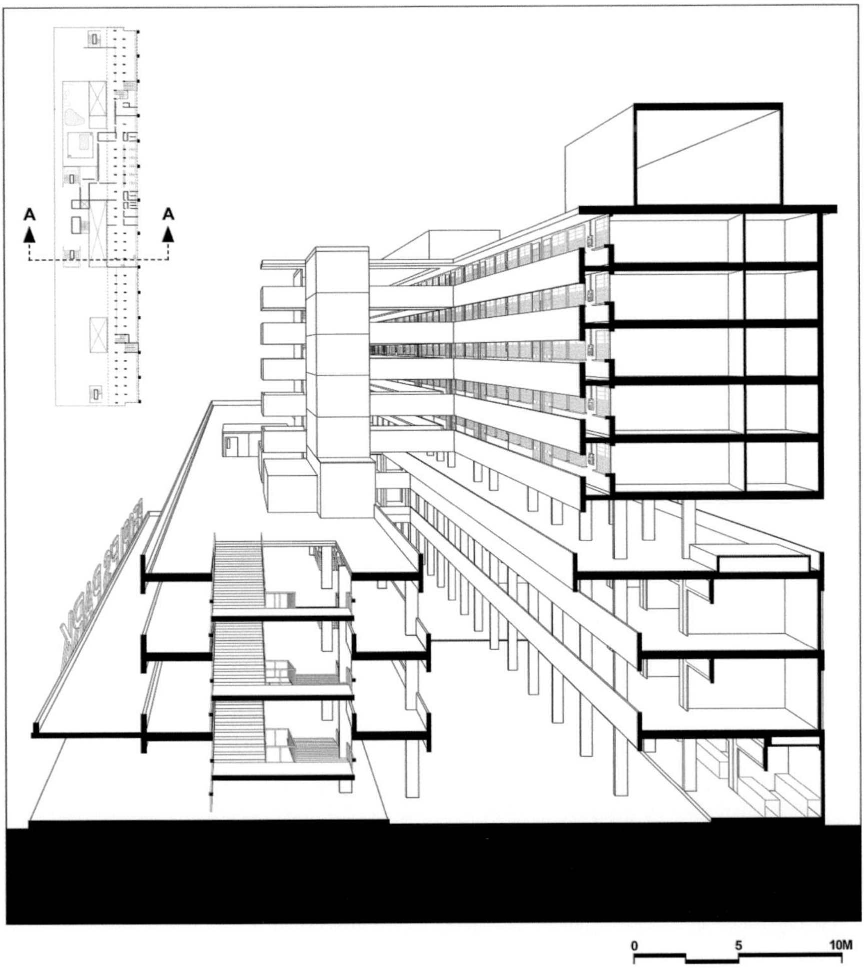

[Opposite] **4.3a** Void deck of the Park Road development, 1968.

[Above] **4.3b** Sectional perspective of the Park Road development, 1968.

4.4 Porous boundaries of the void deck.

However, when the HDB transferred this open space to the suburban housing estates in the 1970s, the same conditions of city-versus-flat no longer applied. The ambition shifted to understanding the void deck as part of a series of spatially defined and enclosed 'outdoor spaces'.[55] Yet, its morphology defied this simple re-categorisation. Defined by the extrusion of the structural grid of the domestic units above it, it repeated and reinforced the spatial rhythm of the high-rise

dwellings above. The void deck is porous to its surroundings but at the same time spatially independent of what is adjacent to it (Fig. 4.4). Surrounded by a raised apron that differentiates the block from the ground plane, the edges of the void deck are also marked by specific architectural points of exit for occupants and their waste—stair and lift cores open to one side of the block mirrored by refuse chutes and wastewater stacks located on the other end. The open space of the void deck is, as such, a space suspended between these circulatory modes of people and their discards. It has a vague sense of both belonging to, and not of, the block of flats. The void deck's 'public' character is ambiguously domestic and 'other' at the same time.

Consequently, its spatial and programmatic ambiguity came to be at odds with the rigidly circumscribed system of open space in public housing, which included playgrounds, town squares and hawker centres. Defined by clear physical boundaries and use patterns, these spaces reinforce the favoured orthodoxies of efficiency and pragmatism. While the urban planning mechanisms invested heavily on higher-order planning hierarchy in the town centre,[56] the proper usage of the void deck was subsequently left to interpretation and appropriation. Conspicuously populated during family gatherings including weddings and funerals, the void deck often emerges as a space rife for troublesome situations. These situations might even border on minor offences—'people sleeping overnight, parking cars, littering, gambling, smoking, drug taking and the presence—and occasional abuse—of cats'.[57] In this casual list, the void deck appears to accommodate both the excess relations of domestic space and an almost absurd catalogue of asocial behaviour, which is unsurprising considering the diversity of domestic life contained within the flats above it. Through its risqué associations and semi-regulated usage, the void deck shifted in its intended role from being the *cordon sanitaire* to becoming a salacious territory that needed to be kept in check, or avoided altogether.

A similar tension might be detected in the common corridor. Known in its early iteration as the 'access balcony', the common corridor runs alongside the front façade of the housing block. Tectonically, the corridor was established as an important architectural design element. In the early Singapore Improvement Trust[58] flats including the nine-storey blocks on Upper Pickering Street (1952-53) and the fourteen-storey Forfar House (1955), the access balcony or corridor acted as a distinct 'form-giver'[59] by providing the domestic high-rise its characteristically clean and striated modernist facades. Installed initially as a circulation passage, the access balcony was further developed to enhance social interaction between neighbours residing on the same floor. Adjustments to the design of the corridor dealt primarily with altering sightlines from the corridor into the private flat and experimenting with different dimensions to facilitate spontaneous socialisation. To address issues of privacy, the level of the corridor was lowered so it eventually sat half-a-metre below the flats that adjoined it. Through the years, there were several variations in the resizing of its width—ranging between 1.4 metres and 2.4 metres—such that the corridor became not just a site for moving to and from the flats but also a space to linger, encounter, engage and survey (Fig. 4.5).

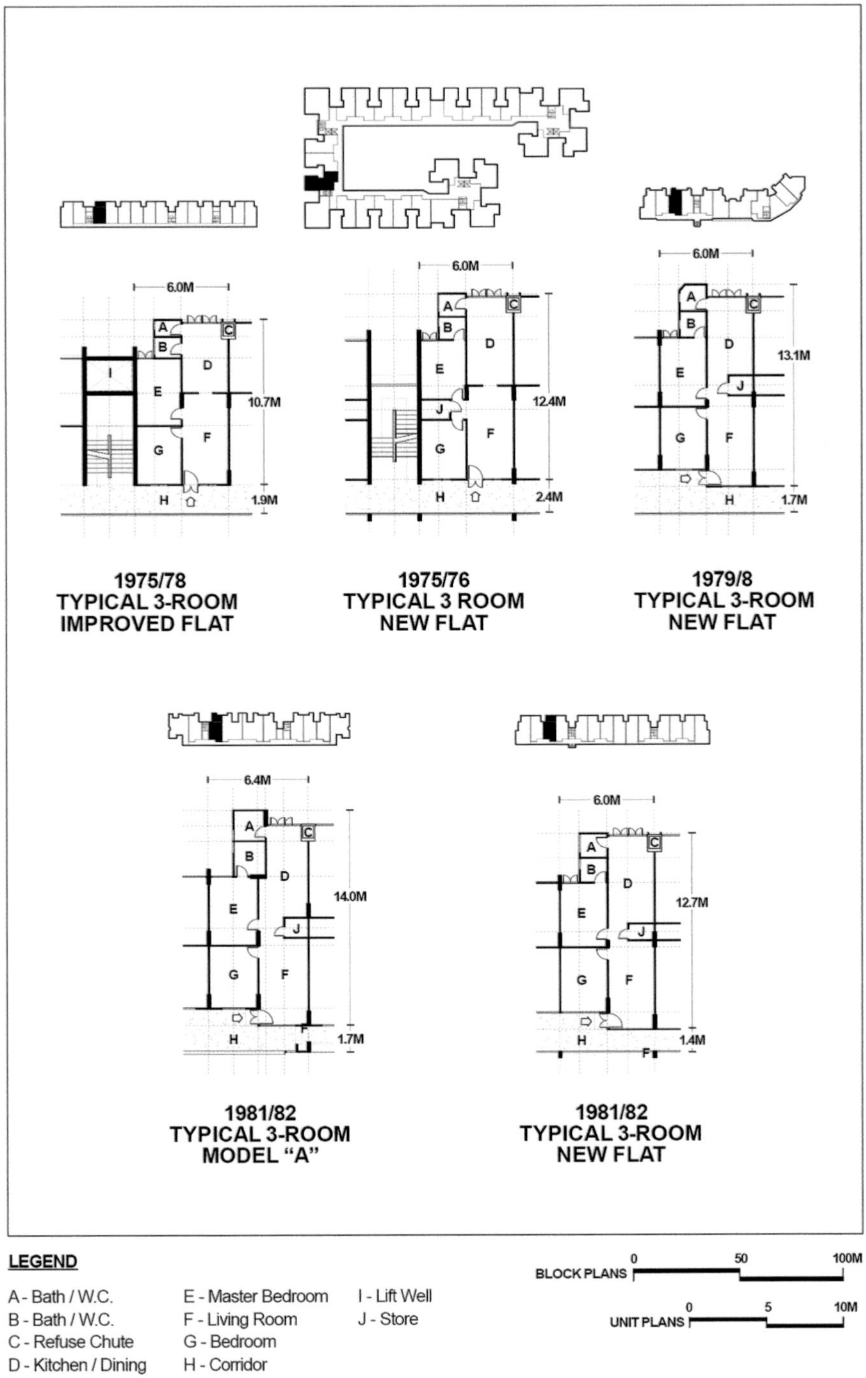

4.5 Resizing of the common corridor width in different housing layouts.

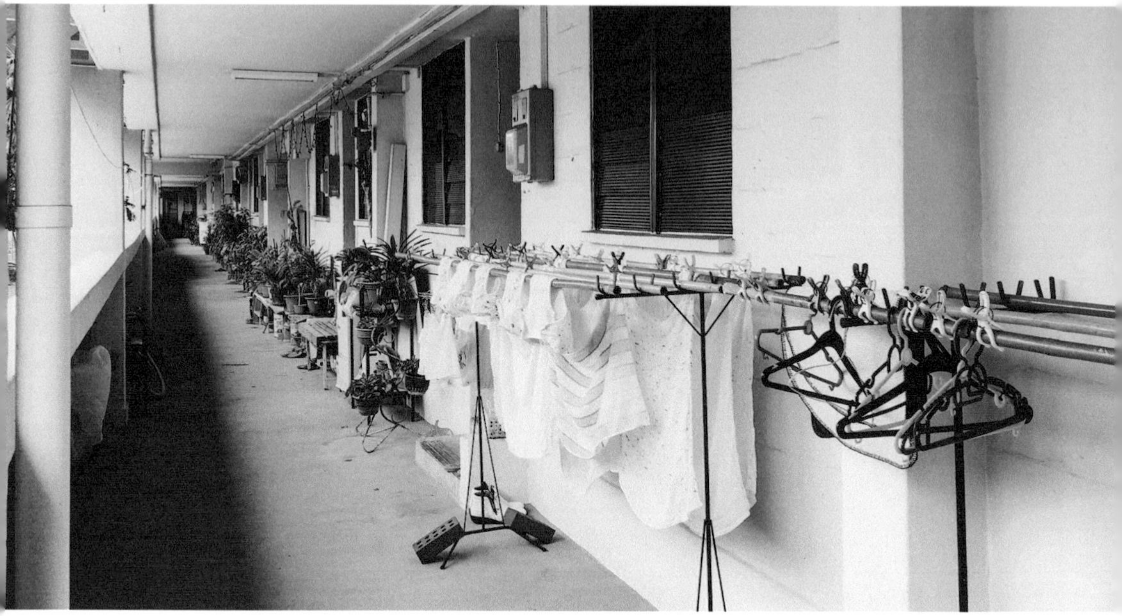

4.6 The common corridor is a panoptic and self-regulating space where irregularities become immediately visible.

Liu wrote about how the raised units overlooking the corridor would 'give the impression of houses along a street'.[60] Indeed the corridor was instrumentalised as a social condenser. It was constructed with an explicit aim of creating a community of neighbours who lived side-by-side but whose increasingly atomised existence rendered them strangers to each other. Yet, the provision of the corridor catered to more than space for the exchange of pleasantries. It had the double-edged function of interaction and surveillance—its visibility meant that comings and goings could be monitored by the 'houses along a street'. Whether or not such acts of surveillance happened, the corridor became a panoptic and self-regulating space where irregularities were quickly exposed: 'the extroverted corridor was a corroborator in the celebration of publicness and the suppression of privacy'.[61] As architectural historian Eunice Seng suggests, the corridor repositions publicness as demonstrative of that desired in the proper citizen-worker-homeowner-family member. The visibility of the corridor generates expectations of openness but at the same time reinforces specific standards undergirding the institutions of family and nation. At stake in this narrow space are complex issues of security, ownership and territory (Fig. 4.6).

Architecture can produce certain kinds of bodies just as specific bodies co-produce certain kinds of architecture. Philosopher Henri Lefebvre argues that spaces emanate particular 'energies' which are then played out through their users' relationship to those spaces, 'A body so conceived, as produced and as the production of a space, is immediately subject to the determinants of that space ... the spatial body's material character derives from space, from the

4.7 The loosely regulated HDB void decks operate as transitional public spaces.

energy that is deployed and put to use there'.[62] Lefebvre's provocation about space's energetic impulse describes the ambiguous conditions in the void deck and the corridor. The body, he tells us, reacts not merely to geometry and form but is galvanised by almost every quality coincident with that space including interactions, oppositions, centres and edges. In Lefebvre's statement, space is produced through and by these aspects, both what is quantifiable and what is not. Amidst the network of overdetermined public housing spaces with their preprogrammed national and familial expectations, the void deck and the corridor stand out in their spatial and programmatic ambiguity. They constitute public space in public housing but their proximate relationship through geometry, distance and overlapping boundaries with private flats give way to dissonance. Certainly, if the closely regulated dwelling and civic spaces of public housing were intent on producing the disciplined citizen-homeowner-worker, then the loosely regulated transition public spaces of the void deck and the corridor were appealing in their unintended liberalism (Fig. 4.7). It is significant that these two spaces were inadequately described, catalogued and understood through HDB's characteristic statistical enumeration. Through this efficacious mode, they appear feeble and excessive. Viewed through the categories of efficiency and use, they are perceived as defiant and bothersome.

Rather, these spaces seem pragmatic only when they were beset by some troubling situation. The corridor's parapet wall was raised to one metre when it was suggested that some had found it convenient to leap to their deaths from there.[63] The void deck was slowly filled up with more useful communal spaces like crèches, kindergartens, shared meeting rooms and elderly persons' corners to minimise unauthorised usage as harmless as football and washing one's motorbike, to more intolerable gambling and uncontrolled gathering of non-Singaporean residents.[64] Articulated in these ways, the void deck and corridor's effectiveness could not be gauged through functionality. Instead, these problematic situations highlight what is spatially different about these spaces, and that they are always encountered as contradictory: efficient and inefficient; clean and dirty; outside and inside. The situations also suggest that here, architecture as the efficient machine, an ethos upon which this public housing programme built itself, ultimately breaks down. It can no longer control the body, which 'is not only prone to but thrives on the tensions between propriety and impropriety'.[65] Considering the recent spatial pathologies exemplified by hoarding in the corridor and keeping cats in the void deck, how is this architecture—one built on a certainty of numbers, perfected through utilitarian calculus, conceived and constructed with the utmost respect for lineaments and intentions—unravelling?

Domesticating housing

In August 2012, neighbours in a Singaporean public housing block at Bedok Reservoir complained of a stench permeating their corridor.[66] The remains of a 76-year-old Singaporean man was subsequently found in a two-bedroom flat along the corridor. Rubbish collected over the past decade by the man piled floor to ceiling and obstructed the search for his body, which took at least an hour. In several cases, like Madam Ho's highlighted at the beginning of this chapter, the hoarder was driven out of their flat because it became inaccessible. Hazards posed to neighbours were just as disconcerting. In May 2015, workers from the Singapore National Environment Agency sprayed insecticide in a block at Eunos Crescent to destroy an active dengue cluster in the estate.[67] Subsequently, hundreds of cockroaches from a hoarder's two-bedroom flat invaded the corridor. A neighbour who posted a video of the cockroach-infested corridor on social media described how other flats along the same corridor were made to vicariously experience the perils of hoarding when swarms of insects and cockroaches turned up in their flats. Here, the corridor transformed into an undesirable conduit for infestation and disease. It becomes a threat to property, which together with propriety and the proper, constitute the architectural norm.[68]

Social scientist Neil Maycroft and architecture critic Sylvia Lavin[69] discuss hoarding as involving improper matter placed improperly in space. That is, the hoard becomes a situation of interest to architecture only when it fails to 'move along' to more 'marginal' areas of storage. It is the non-proper positioning of questionable things in non-marginal, visible domestic space that creates trouble. 'Sorting and classification have a spatial dimension: this goes here, that goes there. *Nontrash* belongs in the house; trash goes outside. Marginal categories get sorted in marginal places (attics, basements and outbuildings) eventually to be used, sold or given away ...'.[70] There is an alternate order to the hoard which has as its primary aim, continuous accumulation to fill up space: 'usually working from periphery to centre, sometimes carving out elaborate pathways or structures in order to enable functional day to day living, and often, more hoarding'.[71] Maycroft argues that hoarding breaks with storage conventions and challenges 'values' attributed not only to commodities but also towards space. In architectural terms, hoarding has a devastating impact on domestic space. It not only paralyses commonsensical spatial use but completely undermines its order and propriety. Hoarding violates the '"visible" space of the everyday' by desecrating the '"proper" ordering of objects in space'.[72] The hoard is precisely an architectural situation because it questions and breaks down accepted spatial boundaries and conventions; as Lavin puts it, hoarding is like a 'a critical resistance to the regimes of propriety enforced by architecture', immune to 'standard measures of use'.[73]

In the Singaporean case, the problem of hoarding is compounded by the close proximity between individual flats, and specifically worsened by the territorial ambiguity inherent in the long, linear and uninterrupted corridors abutting the front of these flats. Indeed, the spatial logic of hoarding thrives on ambiguity in organisation, with the hoarder actively transforming the constituency of space in their transformation of void (useable rooms) into solid forms (walls of trash). The fact that hoarding has become something of a pathological condition in public housing ironically rehearses the question of open space. Here, open space is no longer marked by its habitual use or by its efficacy. The open space of the corridor motivates hoarding; it encourages and is sympathetic to, a spillover from the flat into its outside (Fig. 4.8).

Nevertheless, the spillover corridor space is not just a lure for hoarders. It is habitually treated as a surplus space adjoining the private flat. At this private-public threshold, occupants store an assortment of objects including plants, footwear, religious paraphernalia and altars, bicycles, scooters and laundry. They use the corridor for temporary religious practices such as the Indian population's drawing of the *mandala* on the floor, or the setting up of small tables for offerings and prayers during key Chinese festivals. What is left in the corridor and the manner by which these objects are found veer ambiguously between being acceptable storage and unacceptable hoarding.

Unwanted objects are routinely placed in unprogrammed spaces which are also typically obscured from public view such as the attic, the storeroom, or the bottom of the garden. The loosely programmed common corridor is differentiated by

BAO XIANG
breeze
BAO XIANG
原生态
高标准

4.8 The porosity of the corridor is sympathetic to spillovers from a flat's interior.

its visibility. The corridor is prominently positioned at the front of the block. It is thus, a loose, self-regulating space. Its patent visibility presupposes a certain propriety, demanding an order that preserves this space as part of the block's front facade and entailing proper usage of such frontage. When such propriety is violated by the hoarder and hoarding, the corridor shifts from an efficient façade into an affective space, punctuated by an overtly corporeal existence that cannot be contained. Putrid smells, the organic growth of fungi and mould, infestations of cockroaches, insects and rats, peculiar deposits and stains and the slow degeneration of harmless consumables such as newspaper, books, clothes, rags, empty plastic containers and broken electrical goods into something more sinister and obscure, beckon at every turn. In the wrong place, what was once prosaic and domestic becomes abject, threatening and repulsive.

Similarly at the ground level void deck, the nightly prowl of illegally kept street or community cats confounds the differentiated boundaries between inside and outside, clean and dirty, private and public, legal and illicit. There are officially sixty thousand street cats in Singapore.[74] In 2013, the HDB received 260 pet-related complaints, of which 150 were cat-related. Since September 1978, HDB regulates what it calls 'HDB-approved' pets, which include smaller dog breeds, birds, hamsters and rabbits.[75] Cats have been excluded from the approved list because they are deemed ungovernable: 'It does not follow the grid system of the city. It does not walk on a leash and cannot be trained. It is unpredictable, nocturnal, transgressive'.[76] Pets are to be contained within a household, where they have a proper place. The dog, which is permitted in the HDB flat, is always accompanied and restrained when it moves outside the flat. Conversely, it is difficult to keep cats indoors. The cat disobeys any kind of spatial demarcation.[77] The HDB website states that 'when allowed to roam indiscriminately, they tend to shed fur and defecate or urinate in public areas, and also make caterwauling sounds, which can inconvenience your neighbours'.[78] As such, cats are still banned as pets in public housing.[79] Yet at any given block, illegally kept cats can be found resting around staircase landings, or strolling nonchalantly along the corridors and void decks (Fig. 4.9). Like the hoard, cats not only transgress spatial limits; they are regarded as a risk, a vector of contagion that may pollute the environment in their ability to carry in something improper into the flat:

> … 'putting the cat out at night' signals incomplete containment in the home, only a partial domestication, even though cats may be cosmetically modified to fit conceptions of the homely and domestic. Similarly, the cat-flap is a breach in the domestic boundary, and cats bringing mice or birds into the home may still be seen as polluters of domestic space. In this respect, pet cats are transgressive, breaking the boundary between nature and culture.[80]

4.9 An 'illegal' cat in a HDB void deck.

Considering that the void deck was historically installed to minimise contagion by the city's negative aspects, the perennial presence of cats in this space appears to be a mildly annoying breach. However, this annoyance reveals something particular about the spatiality of the void deck. The situation of cats loitering in the void deck—cats whose owners cannot be identified—exemplifies the ambiguous private-public boundaries that occupants regard and behave in the void deck. The void deck's completely porous and open edges are contradictorily matched by its exact mapping of the residential units' structural grid—one gesturing to unregulated access and the other a physical reminder of every unit's fiercely guarded privacy. It is a space that feels familiar and alienating at the same time. This spatial ambiguity gives anonymity to the stealth cat owner, assuaged that their cats are roaming not far from home.

At night, cats congregate in numbers at the edges of the void deck, mewing until their feeder arrives with their feed. On reaching maturity at five months, if not sterilised, a cat is liable to mate and produce offspring. But perhaps more alarmingly, the cat 'has sex very loudly and uses its sexuality "irresponsibly," reproducing out of control'.[81] Often, these comings-and-goings happen within direct view of, or at, the void deck. The street cat produces an affective space of smells and sounds, which like the animal, cannot be properly contained within or covered up particularly in the open configuration of the void deck (Fig. 4.10). As a space that is contiguous with the exterior environment surrounding the block, the void deck attracts because it is unrestricted yet provides shelter and also in the cats' case, food.

4.10 The street cat produces an affective space of smells and sounds.

Yet not surprisingly, cats appeal to those sympathetic to their 'streetwise and sensuous liberty'.[82] The free-ranging cat colony at the void deck is inevitably linked to the presence of a volunteer feeder who comes nightly to feed the cats. Drawing from conversations and her own relationship with the Singaporean feline network, artist, critic and street cat enthusiast Lucy Davis argues that there is an obvious gender-bias pervading volunteer cat-feeders. Many of these 'nocturnal cat-feeders' are women who are usually 'single, divorced, gay, childless, or with children moved away',[83] thus departing from the normative mother/wife profile indicated for public housing tenancy. These 'cat-women' are not necessarily residents of the same block. Some drive to the cat-feeding spot nightly with a bag of kibbles in their car boot. Others come from neighbouring blocks. There are also cats belonging to residents who similarly roam the corridors and void decks. Although there is a high degree of tolerance for street cats congregating at the edges of the housing block for their nightly feed, there are invariably complaints regarding threats to hygiene. There have also been several recurrent cases of cats thrown from heights and subsequently found dead at the fringes of the void deck, causing much distress.[84] These occurrences may be attributed to the ban on cats in flats and animal abusers taking advantage of this rule.

The cat (alive or dead) at the void deck is a situation that emotes sympathy, affection, love, disgust, loathing or anguish. These different affects correlate to equally different perceptions of the void deck as public, collective, private, or domestic space, as well as what should or should not happen there and who has rights to such an open and undefined space. The cat's affective and embodied environment overlays and disrupts the void deck's anonymity. The presence of cats makes residents notice, and for some, worry about the limits of their property. The cat's body, not unlike the other bodies prohibited in the void deck—gamblers, smokers and lovers—upsets the order and propriety of this open space. It provokes petty offences by unlikely groups of people—the nocturnal cat-women who provide the feed, the young family with a sterilised but unsanctioned pet cat, the eccentric woman who harbours stray cats in her flat. It sometimes also incites tragic and abusive acts of violence like the spate of cats found dead on the apron of the void deck, ostensibly thrown from the upper floor. Through its cat situation, the void deck might be regarded as a corrupting site, but at the same time, the void deck also operates like a public space in its fullest sense—exposed to risk, chance, accident and event. The cat situation returns something primal, affective and visceral to the open and spontaneously liberating space of the void deck.

Towards an architecture of situations

Reconsidering public housing through the evidence of two bothersome situations shifts the perception of its architecture away from the normalised monolith of housing towards affective occupant-based space and time. In due course, such affective discourse might influence the multiple roles of public spaces in public housing's ongoing renewal programmes. Indeed, public housing's public spaces can, despite building regulations, remain fluid and myriad in their spontaneous constituencies and constructions, such that the liveliness already present in these public spaces may proliferate.

A situation, Berlant reminds us, unfolds in an affective present. A situation requires intuitive tracking to follow where things might go since situations are not fixed; they are atmospheres precipitating, formations in the making. Mating cats, sleeping ones, amputated cats, cats bleeding to death at the corners of void decks, cat-women slipping into the night, open car boots with bags of cat kibbles, armies of cockroaches, rats, stolen supermarket trolleys, piles of newspapers, old furniture hanging around, laundry drying in the corridor. Through these shifting configurations, relations between things, people and the world are being created in 'social time and practice'.[85] This is worlding—the coalescing of subjects, objects, routines, spaces and times into a situation. Yet, when we encounter these situations in their animated state, the representational modes available to us, so that we can recount, and account for them, are incommensurate with what we might sense. Thus, situations require us to critique and invent how we tell and how we inhabit. Our language needs to articulate what we sense, 'a poetic of immanent world making'.[86]

Keeping cats and hoarding things, though bothersome, are two amongst many more situations which public housing architecture typically hosts. Through these situations, the interstices of public housing are continually reproduced via occupancy and use, with occupant-produced interstices unfolding in parallel with HDB's designs. Also, that cats and mild hoarding are still largely tolerated is a form of recognition that public housing occupants are active agents, not without a strong sense of stakeholdership, and duty of care. The discourse of situations undercuts public housing's long-term discourse of efficiency. It presents conceptual reorientations and new terms of criticism, which ultimately rewrites the historical present of public housing as a multiply-authored architecture fully alive to its occupants.

The public housing sensed through situations, and its 'poetic of immanent world making' segues into the subject of the next chapter. Discussing filmmaking as an architectural research methodology, the discussion moves into the HDB interior, using the creative practice of the essay film as a critical expression to depict how we dwell. Film is a mode of seeing, of sensing and in *03-FLATS*, of making sense of the worldings taking place in the domestic situations of three single women owner-occupants. In the following chapter, you the reader, may sense another kind of situation, an impasse, an anxiety of one's existence in a heteronomous culture, not unlike the disregard for certain things and practices delineated in this chapter, giving you pause to consider whether, or how, housing might accommodate difference.•

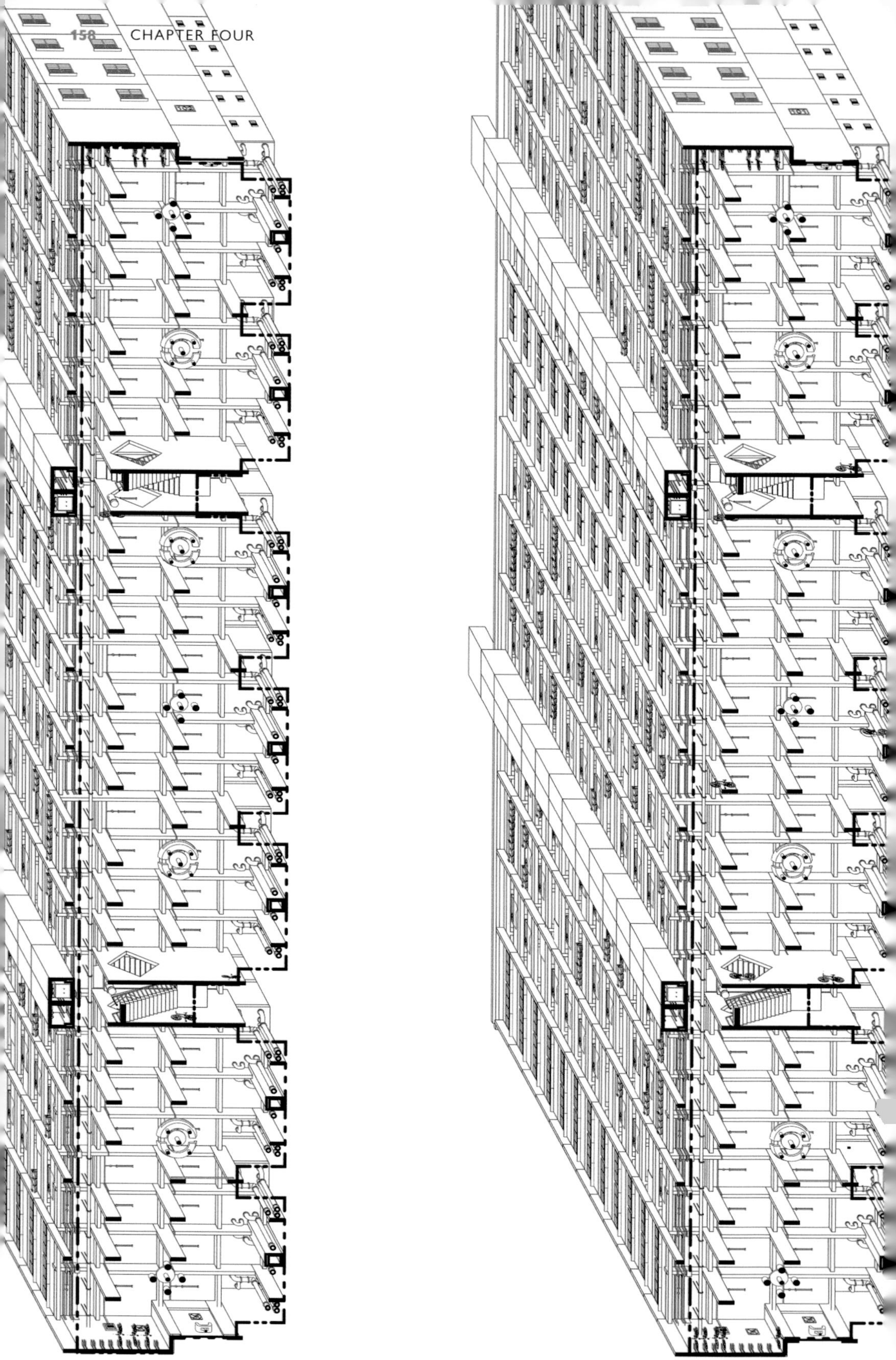

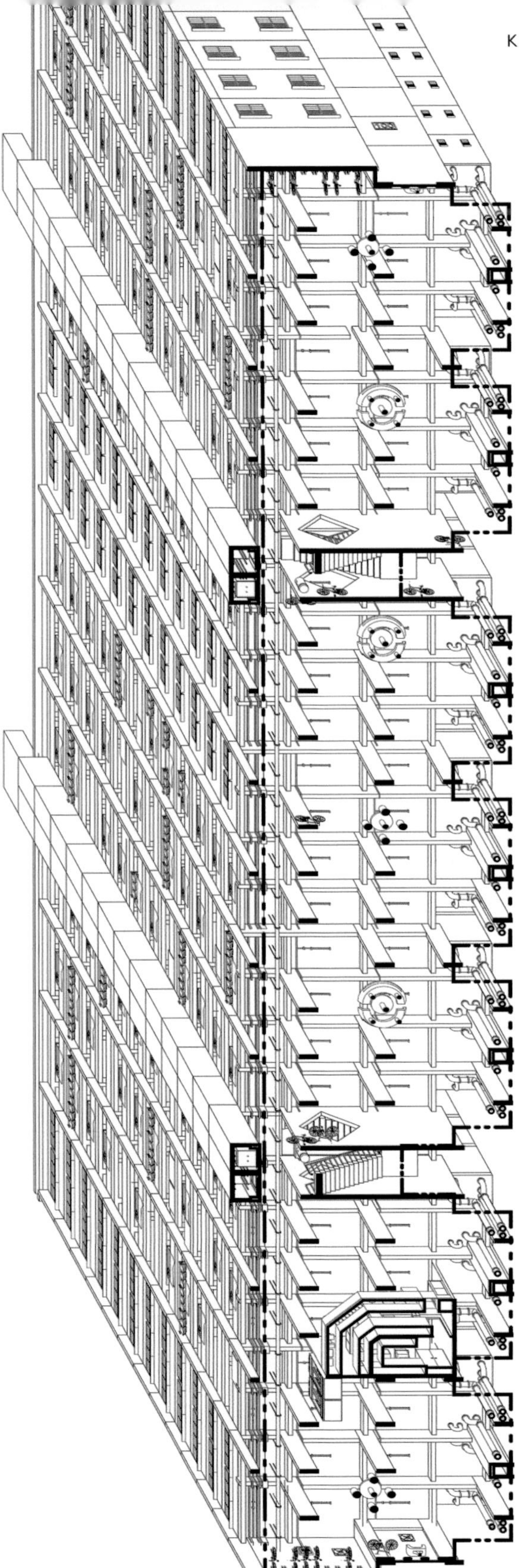

THE THIRD ARCHIVE

5 • Anarchiving Public Housing

Lilian Chee and Lin Derong, 2020.

Situated in three shared spaces of the Singaporean public housing block—the common corridor, the stairwell, and the ground level void deck—*Anarchiving Public Housing* reimagines this housing typology as a densely woven tapestry of occupants, their spaces and their domesticities. Extrapolated from events recorded in local news and popular fiction, the drawing makes visible individual desires and tactics, documented in both sanctioned and illicit domestic practices. Encountering architecture in-between actions of domesticity, the collage-axonometric shows the high-rise block distorted in scale to accommodate the fine-grained experiences of everyday life. Positioned in-between the genres of an infographic and a graphic novel, the drawing visualises the influence of sub-spaces and stories produced through spontaneous domestic action and exuberant imagination. Domesticity nests, appropriates, and/or alters the repetitive geometries of mass housing. Following Henri Lefebvre's schema of the social spatial production, the block corresponds to the ideological 'conceived spaces' thought up by architects and planners, while these private-public spaces in the block vary in their production of space, vacillating between the 'perceived spaces' of collective urban living, and the 'lived spaces' marked by utopian intentions, or disobedient manoeuvres.[1] Here, the architectural drawing, normally disciplined by the demands of the block (the instrument of 'conceived space'), brings together what visual representations of housing rarely accommodate. It speculates on the affective encounters experienced through these occurrences, picturing how they ultimately shape and/or unravel perceptions of housing's architecture.•

1. Henri Lefebvre, *The Production of Space*, trans. Donald Nicholson-Smith (Malden, MA: Blackwell, 1991), 1–67.

#01-01
#02-01
#01-03
#01-02
#03-01

#01-01 | GOLDEN STAIRCASE

In March 2017, eight steps of a bare concrete staircase on the 20th storey of a public housing block in the Balestier estate were covered in 400 sheets of 21-karat tissue-thin removable gold foil by a then third year fine arts student. The initially anonymous installation sparked controversy. A week later, she was asked to scrub it all off. She left behind eight gleaming concrete steps, and a 3x3 square inch piece of gold foil on the right hand corner of the lowest step as a memento.

Priyageetha Dia, *Golden Staircase*, 2017.

#01-02 | MURDER

'The body was found at about 5.30am, and by the time the police came there was something of a crowd around it… So I closed my eyes and thought about the man, the unfortunate man who stumbled upon our corridor, and got killed, seven stab wounds, no relatives found, dying words unrecorded, nothing to answer to but a ledge that blocked his view of dawn, and the fast-fading stars.'

Alfian Sa'at, 'Corridor,' in *Corridor: 12 Short Stories* (Singapore: Ethos Books, 2015), 51–64. Here: 54, 63.

#01-03 | SUICIDE

'It's difficult to explain what happens next. I jumped over the parapet wall, to my death, I think. The last thing I see is a pair of cats rutting; they are affirming life, the exact opposite of what I am doing.'

James Toh, *12 Storeys* (Singapore: National Library Board and National Book Development Council of Singapore, 2008), 19.

#02-01 | BEDROOM

She is a resident who sleeps long-term on the HDB staircase landing beside the flat she jointly owns with her daughter, as they share a difficult, and sometimes abusive, relationship. The makeshift bedroom includes a mattress, pillow and a plastic bag of personal belongings.

New Paper, 'This staircase is her bedroom,' September 13, 2010, https://www.asiaone.com/News/AsiaOne%2BNews/Singapore/Story/A1Story20100912-236782.html.

#02-02 | SMOKING

A smoker was fined S$200 for smoking at a HDB staircase landing. The National Environment Agency explains that 'common areas in residential buildings, such as common corridors, void decks and stairwells are off-limits to smokers.'

Belmont Lay, 'Someone caught smoking at 10.14pm at Toa Payoh HDB flat staircase landing fined S$200,' *mothership.sg*, October 21, 2018, https://www.mothership.sg/2018/10/fine-smoking-hdb-staircase/.

#01-04 | GARDENING

'The hating started one day when my grandchildren decided to play with water pipes. They had just come back from school and were waiting for me to open the door when the younger one decided to twist the tap that fed water into our neighbour's house… I opened my eyes and then turned around to find my neighbour looking at me from her window. I looked down at her still-surviving plant, the branches stripped bare because my grandchildren had torn off all the flowers one day.'

Sa'at, 'Corridor,' 56, 63.

#03-04
#03-02
#03-06
#02-03
#03-05
#03-03
#02-07

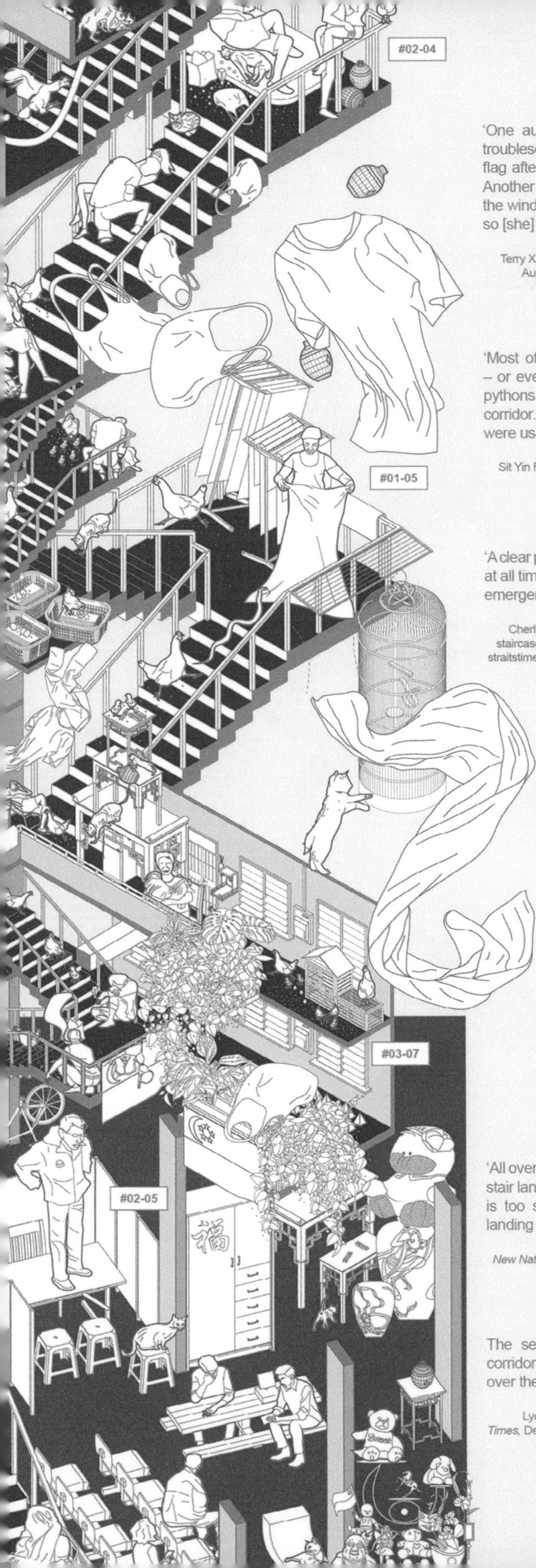

#03-01 | FLAG

'One auntie hanging her clothes out said, [it is] very troublesome to put the flag out and she does not have the flag after returning the flag back when it discoloured. … Another reason … was that the flag always fl[ys] up when the wind blows and [it is] pretty irritating to see it that way, so [she] decided not to hang up the flag.'

Terry Xu, 'The national flag, to hang or not to hang,' *The Online Citizen*, August 1, 2013, https://www.onlinecitizenasia.com/2013/08/01/the-national-flag-to-hang-or-not-to-hang/.

#03-02 | PYTHONS

'Most of the tenants on the floor refused to talk about – or even admit to knowing – that two metre-long baby pythons are lying in a makeshift cage at one end of the corridor. But a 70-year-old woman in the flat said they were used in "sacrificial offerings".'

Sit Yin Fong, 'Mum's the Word on these Pythons in HDB Passage-way,' *Singapore Monitor – 2nd Edition*, August 15, 1984, 2.

#01-05 | LAUNDRY

'A clear passageway of 1.2m clearance is to be maintained at all times, ensuring an unobstructed escape path during emergency situations.'

Cherlynn Ng, 'Resident advised to remove laundry at Anchorvale Link staircase by Town Council,' *Stomp*, January 8, 2018, https://www.stomp.straitstimes.com/singapore-seen/the-staircase-landing-is-not-your-private-area-even-if-you-have-loads-of-laundry.

#02-03 | CLUBBING

Some youths were seen clubbing in the HDB staircase during the recent pandemic circuit breaker, 'swaying to music, in front of a flickering light.'

Beatrice Del Rosario, '"Circuit Breaker Fever": Youths Go Clubbing in HDB Staircase,' *msn.com*, April 29, 2020, https://www.msn.com/en-sg/news/singapore/circuit-breaker-fever-youths-go-clubbing-in-hdb-staircase/ar-BB13lmlI?ocid=ob-fb-ensg-399.

#02-04 | LOVERS

'The naked young couple looked dead as they lay motionless on the HDB staircase landing. But they had, in fact, fallen asleep after having sex.'

Chan Seet Fun, 'The Naked… But not Quite the Dead,' *New Paper*, July 19, 1997, 12.

#03-03 | KITCHEN

'All over Singapore …, HDB kitchens are spilling over onto stair landings and common corridors. … when the kitchen is too small and the cooking paraphernalia fiddly, the landing or corridor outside becomes a handy extension.'

New Nation, 'When the Stairs Become a Kitchen,' February 10, 1980, 29.

#03-04 | SEVEN REFRIGERATORS

The seven refrigerators were lined up along a HDB corridor, causing one neighbour to get hot under the collar over the clutter.

Lydia Lam, 'Simei Corridor Fridges Removed after Warning,' *Straits Times*, December 21, 2017, https://www.straitstimes.com/singapore/simei-corridor-fridges-removed-after-warning.

#02-06 | HAMLET

The HDB void deck contained 'household staples such as rice cookers, refrigerators stocked with fresh groceries, a television, a sink and potted plants.'

Yeo Sam Jo, 'Yishun residents turn void deck into 'kumpung central',' *Straits Times*, March 24, 2016, https://www.straitstimes.com/singapore/housing/yishun-residents-turn-void-deck-into-kumpung-central.

#02-07 | WORKSHOP

He used his void deck as a workspace for four-and-a-half years. From a practical standpoint, the void deck's 'concrete table[s are] large space[s],... conducive to... leather crafting....' He would have registered the void deck as his work address if that was permitted.

Kenneth Chan (Director), *In Between: Into The Void* (Singapore: Viddsee Original Production, 2019), https://www.viddsee.com/video/in-between-into-the-void.

#02-08 | SOCCER

'With their slippers as goalposts, four boys play soccer with a plastic ball....Their field is a void deck, whose floor coats their soles with ovals of grey. When one of them scores a goal, an imaginary stadium roars euphorically.....'

Alfian Sa'at, 'Bukit Batok, 5 PM,' in *Malay Sketches* (Singapore: Ethos Books, 2012), 129.

#02-09 | GAMBLING DEN

From about 9pm to 3am every night, up to 20 people – usually senior citizens – gamble at the void deck's stone tables, playing different games, including blackjack, and placing bets up to hundreds of dollars.

Shaffiq Alkhatib, 'Chai Chee void deck turns into gambling den at night,' *New Paper*, March 7, 2016, https://www.tnp.sg/news/singapore/chai-c7hee-void-deck-turns-gambling-den-night.

#02-10 | SKATEBOARDS

The void deck contained '[s]kateboarding obstacles – consisting of a rail, ramp and some wooden boxes,' which have been 'glossed with layers of graffiti spray-painted over the years.' The police continually patrolled the area and took down the names of skateboarders. He responded by setting up rules to regulate how the space should be used by the skateboarders.

Reclaimland.sg, 'An agreeable anarchy,' accessed November 28, 2020, http://www.reclaimland.sg/rl/?p=677.

#02-11 | EXERCISE

Exercise equipment was set up on Astroturf, which was evenly spaced across the length of the void deck.

Stella Soon, 'Singapore's Most Spoil Market Void Decks That Prove They Don't Need Metal Railings,' *thesmartlocal.com*, March 29, 2017, https://www.thesmartlocal.com/read/unique-void-decks/.

#03-05 | JUNGLE

Almost 40 pots of tall leafy plants formed a jungle along the corridor outside a flat nicknamed the 'HDB Jungle House.' A neighbour living in the flat below claimed that he had to trim the plants that blocked his window, and that these attracted ants into his flat.

Salome Ong, 'Plant-loving Neighbour Causing Obstruction and Fire Hazard with "Jungle House," say HDB Residents,' *Straits Times*, May 27, 2019, https://www.straitstimes.com/singapore/welcome-to-the-jungle-hdb-residents-say-plant-loving-neighbour-is-beyond-be-leaf.singapore/circuit-breaker-fever-youths-go-clubbing-in-hdb-staircase/ar-BB13lmlI?ocid=ob-fb-ensg-399.

#03-06 | DECORATION

For the past few Chinese New Year festivities, she decorated the common lift lobby, stairwell and corridor with the most elaborate decorations.

Nyi Nyi Thet, Zhangxin Zheng, and Fasiha Nazren, 'S'pore Lady once again Decorates HDB Corridor, Creating Awesome CNY Atmosphere for Neighbours,' *mothership.sg*, February 5, 2019, https://www.mothership.sg/2019/02/cny-decorations-bedok-aunty-linda/.

#02-05 | CUBICLE

For 27 years, he conducted his Meet-The-People sessions at the void deck in a 'makeshift cubicle' made of light aluminium panels.

Icemoon, 'Chiam See Tong's Legendary Makeshift Cubicle at the Void Deck,' *Second Shot*, May 18, 2011, http://www.2ndshot.blogspot.com/2011/05/chiam-see-tongs-legendary-makeshift.html

#03-07 | FARM

A complaint was made to the authorities that 'chickens and ducks were reared in boxes under flower pots and posed a health hazard, in addition to the dirt and flies… Time and patience are required to educate these families to accustom themselves to living in flats.'

Straits Times, 'Chicken "Farms" at Kallang Flats: Board takes Action,' April 27, 1966, 8.

#03-08 | CATS

A parliamentary question was raised to the Ministry of National Development as to whether it was timely to remove the ban on keeping pet cats in HDB flats, and instead regulate cat pet-keeping to minimise neighbourly disagreements.

Ministry of National Development, 'Written Answer by Ministry of National Development on keeping pet cats in HDB flats,' January 6, 2020, https://www.mnd.gov.sg/newsroom/parliament-matters/q-as/view/written-answer-by-ministry-of-national-development-on-keeping-pet-cats-in-hdb-flats.

#03-09 | TOWER

He erected a 'pagoda-like tower of psychedelic toys, figurines and knick-knacks' in late 2016 at a HDB void deck. Its colours and familiar figures cheered up the occupants. Despite the tower's fans, the Town Council instructed that the structure be dismantled, citing fire-safety concerns.

Melody Zaccheus, "Artistic' tower of items at HDB void deck trashed,' *Straits Times*, March 2, 2017, https://www.straitstimes.com/singapore/artistic-tower-of-items-at-hdb-void-deck-trashed.

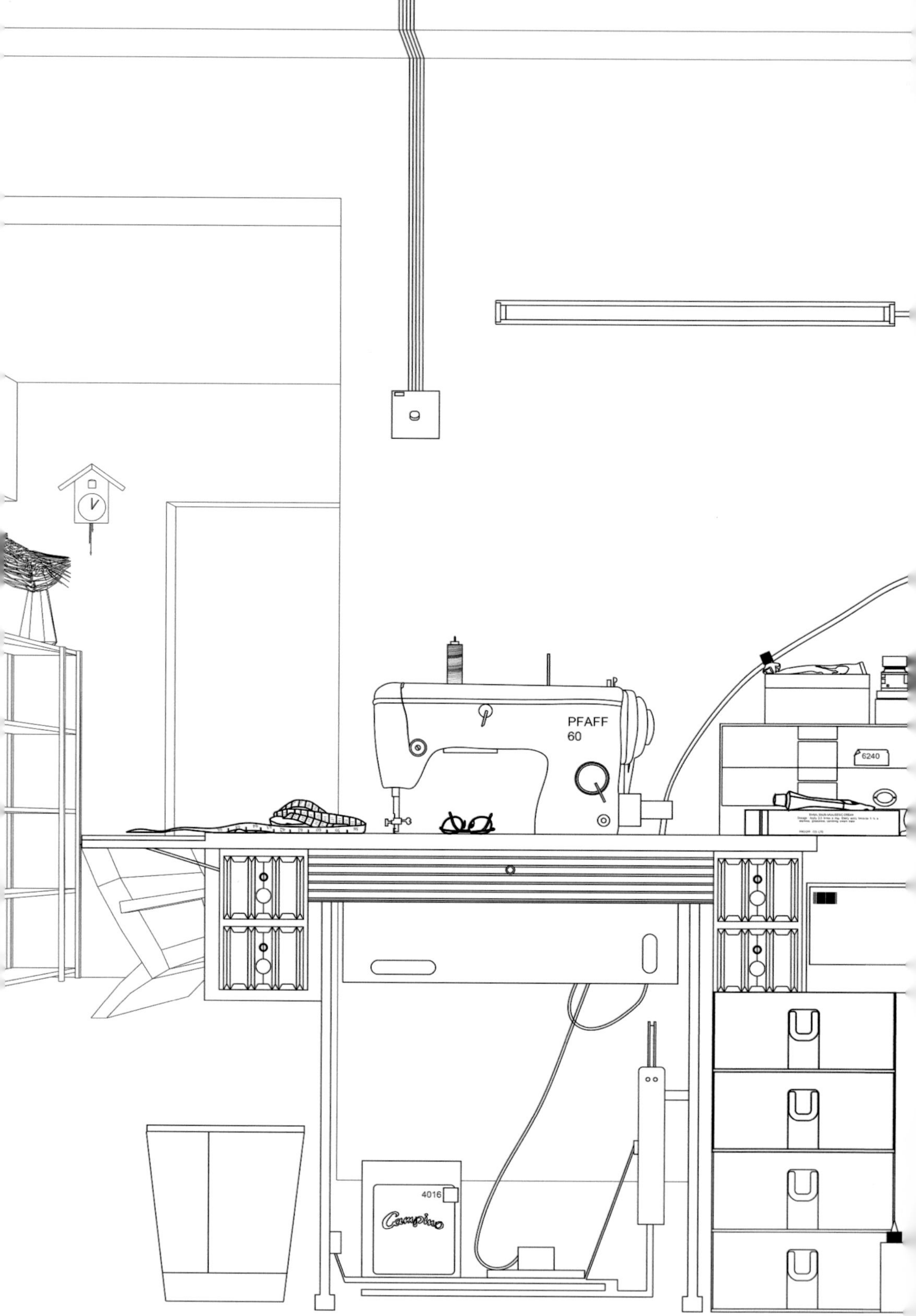
PFAFF
60
6240
4016
Campino

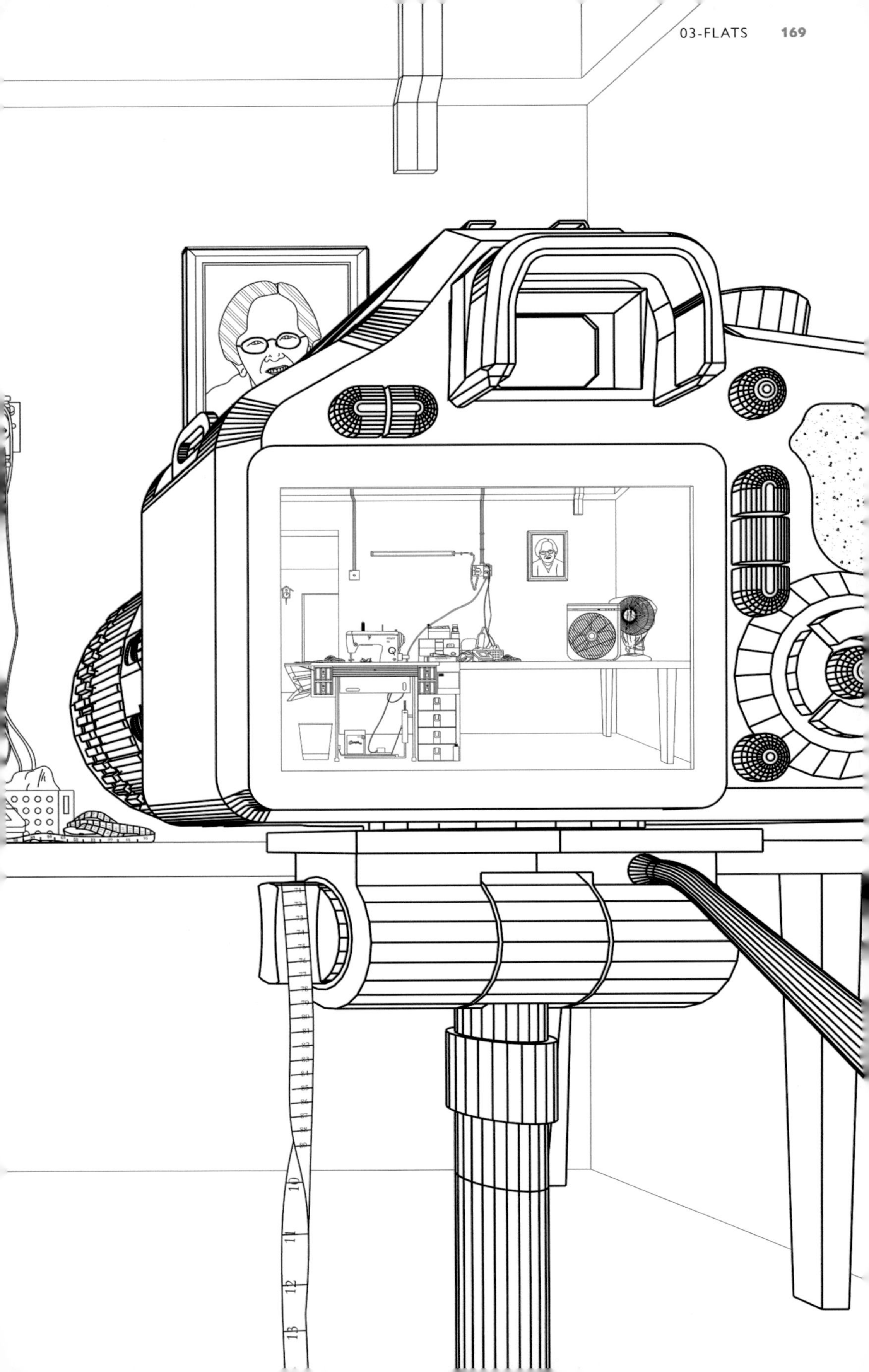

6 • *03-FLATS*

Architecture filmmaking, disciplinary questions

The long and narrow corridor overlooked by ten flats is mostly empty. Dust particles glitter in the sunlight; cobwebs encase plastic wastepipes. Carelessly placed potted plants, the odd plastic stool, a threadbare dining chair and an unpaired slipper left behind by someone. Occasionally, the silence is broken by the rustle of leaves. It is a warm afternoon. There are no children playing, no housewives chatting. At one end, the corridor bulges into a lift lobby. The lobby walls are covered with ceramic tiles, their cream and pink hues fading. A middle-aged Malay woman fusses over her cat. It lies belly-up on the cool concrete floor, taking comfort from the midday heat.

A portrait of a smiling elderly woman in a green collared shirt presides over a small bedroom. Placed against the same wall, is a black Singer sewing machine. A half-finished cotton shirt with small blue and purple flowers, hangs on the back of a plastic chair facing the machine. The adjacent wall is tidily lined with plastic drawers. A container of 1970s Lego blocks is kept above one of these. The corridor-facing room is dim even with its large window. A fluorescent light is switched on in the day. On the wall opposite the machine there is a framed jigsaw of a red tulip field. Two wooden elephants, a porcelain sheep, miniature Dutch clogs and two plastic goldfish sit incongruously on another column of plastic drawers. At the machine, another elderly woman is stitching a child's pyjamas.

In a much bigger room, a discarded signage saying '02-11 storey' sits atop a large whiteboard covered with photographs, newspaper cuttings, postcards, old envelopes, maps and handmade drawings. The cabinet under the whiteboard is stacked with boxes of charcoal sticks, masking tape, paintbrushes, chalks, pencils, pens, boxes of slides and a pile of folded newspapers. There is a large wooden table, with its top wrapped in clear plastic sheet. An open window faces the sea in the far distance. A storm is brewing. The papers on the whiteboard flutter in the wind. On the wall behind the table, there is a big rectangular frame marked out with masking tape. The edges of the tape are blackened. In this crowded room, a young woman stretches out a piece of canvas covering almost half her floor.

Blind spots and misalignments

The architectural essay film *03-FLATS*,[1] from which the scenes above are derived (Fig. 6.1), is a counter-representation of Singaporean public housing. The film highlights the interior spaces, household objects and detailed domestic routines taking place in flats belonging to three single women owner-occupants who live alone. It visually essays an intimate corollary to public housing's widespread reputation as an urban planning phenomenon,[2] for which explicit prioritisation is nuclear family living. *03-FLATS* materialises aspects of domestic life that are excluded from public housing's architectural narratives—the gendered occupancy of such spaces, the types of alternative labour which go into homemaking, and the nuanced ways in which the single women rearticulate their sense of self and space around an accommodation type designed for the nuclear family.

Stemming from an investigation of how domesticity might be visualised in architecture, the essay film is a research collaboration conceptualised by myself as principal investigator, in collaboration with the Singaporean filmmaker Lei Yuan Bin. It negotiates two disciplinary modes of visualising public housing architecture—from an architectural perspective of the flat shaped through everyday routine and use, and through a cinematic framing of the three women. The conceptualisation, making and reflective phases of the essay

[Opposite] **6.1** *03-FLATS* highlights how space is produced through the domestic practices of the three single women owner-occupants.

2

film advance the theory and practice of affective discourse; in this case, by making visible how the tenuous agencies and spontaneous production of space through domesticities intersect with the national agenda of public housing. As I have previously argued, 'domesticity' differs from 'housing' in scale and methodology.[3] Domesticities are concerned with processes, rituals and routines of homemaking. Domesticities, while mostly small-scale and interiorised, may also inflect wider national concerns of boundary-keeping and exclusion. As such, domesticity is a powerful analytical category, capable of crossing scales and agendas, including depicting what is affective to one's sense of agency.[4] In *03-FLATS*, domesticity becomes the primary angle around which the architecture of the flat, and the housing block, are observed and recorded.

Since its completion, *03-FLATS* has shown in major world cities, was adopted into university reading groups, and has won film accolades.[5] As such, the film has crossed geographic and disciplinary boundaries. It exists as an academic output but has also successfully reached out to a wider public.[6] This chapter examines the role of filmmaking in architectural research by traversing the difficulties and opportunities that unfold when one conducts research at the intersections of architecture and film's disciplinary boundaries. I conclude with thoughts about the struggles of creative research work, emphasising the importance of reflexive and affect-driven post-project writing as a companion to creative practice. Differentiated from the previous chapters, this piece of writing is an exegesis not only on the technical process of architectural filmmaking which implicates my personal role in conceptualizing the film, it also elucidates the fraught nature of using creative practice as a research tool, and the intellectual struggle of 'writing-up' an open-ended creative outcome.

To begin, some remarks are necessary to situate the continued political significance of the housing programme in Singapore, and why an architectural filmic study of three single women in their flats came to be. The contributions of the state-controlled Housing and Development Board (HDB) programme—which has produced and managed Singapore's public housing stock since 1965—to social progress, communal cohesion and congenial living are well-rehearsed.[7] The architecture of public housing does not merely fulfil the mundane day-to-day functions of domestic life (Fig. 6.2). For more than half a century now, it has served as an efficient and effective state instrument. As housing stock produced at subsidised prices primarily for middle-class to lower income nuclear families who are in turn obliged to draw a regular salary to service their mortgages, the HDB flat is implicitly embedded in a social, financial and political network. It reproduces a panoply of institutionalised functions including defining acceptable domestic standards, mandating regular paid employment, determining property rights according to eligibility by marital status/age/familial affiliations, and as such, also indirectly maintaining modes of surveillance and sexual propriety.[8] Additionally, the definitions of public/

private boundaries, social conduct and hygiene practices, preferred modes of socialisation with neighbours, home management and decoration, as well as blissful family life are unequivocally described, and often prescribed, through both official and unofficial channels including handbooks and newsletters circulated by the HDB.[9]

The heartland neighbourhoods[10] featured in the film—Eunos, Queenstown and Sembawang—have each been recounted as examples of successful public housing ecologies in terms of racial diversity (Eunos), pioneer housing and planning principles (Queenstown) and regeneration of an industrialised area (Sembawang). They are thus, to some extent, a representative cross-section of Singapore's evolving and aspirational public housing landscape. The selection of the three single women occupants is, however, non-representative of Singapore's typical housing demographic. As single owner-occupants, the three women depict a growing segment of the population who have made singlehood a way of life.[11] Nevertheless, with the objectives of 'housing the nation'[12] and thus, prioritising families, ownership restrictions of public housing for singles is a fiercely debated topic.[13] Singles may purchase resale housing units, or opt for more affordable but smaller built-to-order single bedroom flats upon reaching the milestone age of 35 years. Physiologically, this age group also marks an important threshold where female fertility is imminently in decline.

Yet the image, or imagination, of an unmarried woman (more so than a bachelor) in the public housing landscape is largely absent, perhaps, even still unthinkable.[14] The implicit relationship between biology and housing provision[15] has ultimately translated into how women are also gendered in the visual archives, particularly in the iconic black-and-white news footage from the *Berita Singapura* series.[16] Produced by the state-sponsored Singapore Media Corporation in collaboration with the HDB in the mid-1960s to the early 1970s, these short reports bore nondescript titles such as 'A New Look at Housing' and 'Celebration Part 2' (*Perayaan* 2). They emphasised instances of calibrated urban planning and its progressive outcomes: the strategic clearing of 'slums' for modern housing, the smooth resettling of nuclear families into their new residences, the enviable convenience of living close by to public amenities including shops, markets, schools and cinemas and the pleasure afforded to the harried housewife who could be efficient even with burdensome household chores. She is shown shopping, getting her hair done, cooking meals, tending to her children. In the archival footage, the figure of the woman in the flat is inevitably one of the mother or the housewife. There was no question that she could be anyone else, or that the flat she was pictured in might not accommodate anyone else but herself.

Thus, the single woman not only represents a different social demographic; it is for *03-FLATS*, a conceptual figure of difference, someone who produces

6.2 The architecture of public housing is institutionalised and embedded within social and fiscal networks.

an alternative optics of home, differentiated from the assumed normality of a nuclear family. Consequently, the selection of three single women aged between 40 years to 72 years at the time the film was made, is equally a conceptual decision to reconfigure the architectural representation of HDB flats. It is an attempt to study closely how a single woman may, or may not, rearticulate her own sense of space and self in family-oriented housing.

The chosen women are emotionally attached to their properties and had spontaneously developed their own modes of homemaking. Madam Sim lived in the flat since 1973 and inherited it from her mother when the latter passed on; Amy bought the flat to move into a neighbourhood where she felt secure living and aging amongst friends; Ling Nah purchased her flat from her landlord after many years of renting that same unit as her artist studio. Further, the women were also known to both myself and the filmmaker prior to the project. Our extant relationships resulted in the occupants' ease when filmed.

The architectural-cinematic document is 90-minutes, edited multiple times from almost 100-hours of observational footage, through a feedback loop between a filmmaker and an academic. Apart from the ambient sounds and occasional conversations taking place in the featured flats, *03-FLATS* has no dialogue or music. Interior scenes systematically detailing a similar domestic activity in the three homes—for instance, cooking, cleaning or doing laundry—are interjected at various points by blackout screens featuring policy texts about housing quotas for singles, the privileges of public housing for the citizen and responsibilities of homeowners as national stakeholders. The film vacillates between the three voices—semi-ethnographical material focusing on occupancy habits; architectural details and spatial patterns of use and the policies, politics and proprieties of public housing life.

Filming took place intermittently over 8 months. Using mostly wide-angle shots, long-takes and deep focus sequences with limited cuts, the filmmaking process negotiated intricacies of interior spaces, domestic objects and routines, contrasting these against the scale of the housing block. A resolve to be non-obstructive meant equipment used was minimal. Lei acted as the lone camera person, without additional lighting nor sound equipment. Borrowing from a semi-ethnomethodological approach, systematic close observation was adopted across the three households where the occupants' usage, modification and perception of interior spaces were conducted through cyclical domestic practices such as cleaning, cooking, doing laundry and watching television, as well as considering three occasions which include preparing a flat for a festival, sewing a child's garment and making a room-sized canvas drawing (Fig. 6.3). The research adopted a pluralist methodological approach, combining the

6.3 Systematic close observation was adopted across the three households.

analysis of architectural representations, ethnographical research on the women and their household practices and considering the material culture of household objects. In editing the footage, the longitudinal filmic records of domestic life were periodically examined and discussed across the disciplinary lenses of architecture and film.

Geographer Doreen Massey describes the affordance of space as a quality produced through multiple interrelations—social, cultural, material, historical, political, economic—happening at multiple scales, from the global to the intimate:

> [S]pace … is a product of relations-between, relations which are necessarily embedded material practices which have to be carried out, [space] is always in the process of being made … Perhaps we could imagine space as a simultaneity of stories-so-far.[17]

Massey's provocation of space as a product of affordance, with opportunities for shaping and changing, comes close to how *03-FLATS* was conceptually structured, and how the film sees the affordances of public housing. There are several overlapping threads in the film which reflect the sometimes parallel but other times interwoven concerns of visualising a single woman living in her flat. The film holds these threads together, but it also tries to separate them so that they can be read independently—the single women and their aspirations; the public housing ideology; the struggle, banality and rewards of everyday domesticity. The three women's flats are thus as much made of the 'relations-between' these material practices as they are physical accommodations.

Yet, the strong affiliation with the three women resulted in the film being misread as ethnographical in approach, and criticism that disciplinary protocols for ethnographic representation and research were inadequately addressed. Its mis-categorisation as an ethnographical film raises questions about disciplinary autonomy and disciplinary orthodoxy, as difficulties which develop when working through a hybrid creative practice research model. Indeed, the crossings between architecture, film and ethnography enrich the possible disciplinary convergences, and divergences when conducting research on a subject simultaneously technocratic (public housing) and emotive (a flat of one's own). In many ways, the theoretical aspects of visual ethnography have become critical in my struggle to theorise and write about the film's roles in architectural discourse. The making of *03-FLATS* ultimately located the research practice across material, discursive and affective registers.

Filmmaking as architectural research

Film is a research tool which allows social and cultural intimacy between the subject and the researcher. At the same time, there is the weight of ethical responsibility and self-reflexivity since content framing and editorial responsibility are also undertaken by the researcher/filmmaker.[18] The contribution of the moving image in ethnographic and anthropological research have been extensively covered particularly in the recent two decades.[19] This section briefly reviews how film is employed as a visualising tool for fieldwork research located particularly in the built environment discipline. It follows by comparing the use of film in ethnography to emerging discourse about film in architectural research. The discussion reappraises the methodology of *03-FLATS*, enabling critical comparisons to be made with visual ethnography, while arguing why and how this work differentiates and distinguishes itself as an architectural essay film.

Anthropologist Roxana Waterson writes that the study of vernacular architecture was not picked up by anthropology until the early twentieth century because architecture was seen to be located in the realm of material culture and did not readily provide the abstractions of mythology, social structures and kinship which interested anthropology.[20] Waterson's own anthropological research on Southeast Asian vernacular architecture locates these abstractions squarely within close examination of the indigenous built environment.[21] Describing how anthropologist AnneMarie Fiedermutz-Laun was able to let viewers visualise the intricate, dynamic, organic, gendered, communal and vernacular intelligence involved in the building of mud-houses in Burkina Faso, Waterson advocates film as a rich resource for understanding the built environment.[22] Studying the mud-houses through film, the viewer gets 'a vivid sense of physical spaces and textures as well as insights into social relationships, the organization of work, the rhythms of everyday life and the pressures of change'.[23] Waterson's vernacular-based research uses film to communicate a spatiotemporally dislocated culture and environment. Through the footage, non-verbal gestures, subtle patterns of use and the spatial interplay of otherwise invisible power relations might unfold for extended analysis and study.

Moving to more urbanised environments including private homes where fieldwork opportunities may be limited, anthropologist Sarah Pink's method of 'sensory ethnography' aims to draw out both explicit and tacit knowledges. Sensory ethnography attends to 'the experiential, individual, idiosyncratic and contextual nature of research participants' sensory practices' as well as accounting for how cultural conventions, moralities and knowledge play a part in shaping one's understanding of a subject.[24] Pink examines how sensory perceptions intersect with the researcher's cultural upbringing and knowledges, challenging the primacy of vision in knowledge production. Sensory ethnography demands a heightened reflexivity by the researcher, who must pay attention to how her position—determined by power and sociocultural relations—might influence, enable or limit specific ways of knowing:

> Doing sensory ethnography entails taking a series of conceptual and practical steps that allow the researcher to rethink both established and new participatory and collaborative ethnographic research techniques in terms of sensory perception, categories, meanings and values, ways of knowing and practices.[25]

Pink's research on the 'sensory home' attended to the practices of laundry and housework. In these, sound, vision, smell and touch were recorded to gain an understanding of how meanings of home security and comfort were constructed, given value, maintained and translated as part of one's identity.[26] In her recent sensory ethnographic project exploring energy and digital living in people's homes (Low Effort Energy Demand Reduction project, or LEEDR, 2010-14),[27] house tours, interviews and housework re-enactments were recorded using digital video at the participants' homes. Like *03-FLATS*, Pink's LEEDR research and fieldwork were conducted within defined, constrained and routinised domestic environments. It also involved participation and long-durations—the researcher would spend up to two days at each participant's home. Yet, for both Pink and Waterson, the video camera is largely a documenting tool. The films are fieldwork processes, and they function as visual notes which provide evidence for a textual analysis. In anthropological research where film is used as a means of recording fieldwork occurrences, the possibility of such a film behaving as an aestheticised object in its own right is likely supplementary to the objectives of the research.[28]

Yet, does the production of a research film that intentionally engages its aesthetic value invalidate its role as research method and instrument? Can the research film be critical of its subject matter and performative in its aesthetics? I ask these questions from a dual vantage point: from a design perspective, where the creative work produces its own argument within its material modality, and from an academic position where film is analysed as a kind of text which begs interpretation.[29] I argue that creative filmmaking not only surveys the field, it produces specific ways of seeing which embeds forms of knowing and knowledge. There is additionally the dilemma of how an academic producing architectural discourse might position herself in relation to her own creative project. There is no easy way around this predicament, with the exegesis of the work itself a discursive process worth deliberating.

In her illuminating essay about architecture's relationship to fieldwork, architect and academic Suzanne Ewing expounds that architecture works creatively 'in, from and through' the field, in that rather than recording data existing in the field, architecture *produces* something out of the field.[30] It creates 'connections and relations between field and work', that is to say, architecture sees the field as a material environment in which there is a material context to produce new work, reconfigured strategies and relational tactics, '[u]nlike other disciplines, field/work in architecture always contributes and connects, whether closely or more indirectly or collaboratively, to *a form of architectural production*, rather than remaining autonomous as field data'.[31]

The 'field' in the case of *03-FLATS* is simultaneously physical, psychical and projective. It constitutes an area both larger and more intimate than the vicinity of the three public housing apartments. These flats are located within an efficient system of public housing which holds architectural significance in terms of their prefabricated construction, waste, laundry and lift technologies.[32] They are also ideologically situated within a nexus of family-centred state policies. Although documenting spaces in-situ, *03-FLATS* is conceptualised to react to, as well as raise questions about, its 'field'. The film *produces* something out of the field—the image of the single woman in a flat of her own is projected as a statement about gendered single living, individual agency and creative appropriation. The film strives to be performative and generative of a new discourse that shifts from the habitual writings about public housing's relationship with urban technology and social equality to questions about how this architecture might develop with an awareness of gendered differences and embodied precarity.

Architect and academic Igea Troiani alongside Ewing argue for the use of visual research methods that function beyond documentation and interpretation such that their extended role is 'to critique and perform a practice-based visuality'.[33] Acknowledging that the 'politics of subject focus' will differ between humanities-based disciplines including architecture, ethnography, geography, phenomenological philosophy and material culture, they propose that the architectural researcher's 'use of tools or devices for drawing, painting, photographing or filming [can] seek out explicit and emergent architectural knowledge'. Visual methods, Troiani and Ewing emphasise, 'allow space, its occupation, the body, the temporal and their social inter-relations to be given architectural attention'.[34] Moreover, the increasing availability of affordable and higher quality software for photography, audio recording, filmmaking and editing create an environment ready for what Troiani and ethnographer Alison Kahn call an '"undisciplined" corporeal publication'[35]—where the research medium is critically mined to materially express the subject it records. The outcome is a reflexive piece of work, with a keen awareness of how people in the field—subjects and researchers alike—influence and shape processes and outcomes. The 'undisciplined' nature of the work is a resultant of various methods contingently coming together out of pragmatic and critical need, much in the way that Pink has described the modalities of sensory ethnography.

In this, architecture has some illustrious precedents. The architects Robert Venturi, Denise Scott Brown, Steven Izenour and Reyner Banham opportunistically employed the spatiotemporal nature of film to depict the city's restlessness, monotony, vastness or anonymity. In 1968, Venturi, Scott Brown, Izenour and their thirteen students from Yale spent ten days on the road in Las Vegas making a film to gain 'an understanding of the automobile-oriented city and to find an adequate image for it'.[36] The insights from the film, which 'treated architecture from the perspective of appearance and phenomenon',[37] subsequently led to the publication of their influential *Learning from Las Vegas.* To produce *Reyner Banham Loves Los Angeles* (1972), Banham learnt to drive so that he could experience Los Angeles from its ubiquitous freeways. The use of film in both cases frames architecture as a phenomenon. Film shifts perception and reception of architecture from something static to one predicated on circumstance and encounter.

Yet, even with a knowledge of these precedents, the making of *03-FLATS* persistently produced periods of disciplinary anxiety. Decisions about whether the subject of public housing architecture could surge to the fore or might only exist through the lens of the three occupants; whether an audience could relate to a film without dialogue or narrative; whether film is sufficiently critical or already compromised by *a priori* decisions—ensued in highly charged debates. The film footage demonstrated a lack of substantive architectural material, at least in the modes I previously recognised. The recording of everyday domestic existence drew blanks when read against the critical techniques I was familiar with. Subsequently, the misalignment between theory and everyday practice became a source of struggle and apprehension.

The situation recapitulated a question asked by architectural theorist Mark Linder: 'How, when, or where, does, did, or could architecture make its appearance *other than as architecture?*'[38] Linder argues that the discipline of architecture can be made stronger, not weaker, if it can be flexible in its identity when negotiating with, being exposed to, and having some of its discourse and practice reconfigured by other disciplines whilst at the same time clarifying what is distinctly at its core:

> Because it continues to use proper disciplinary techniques, concepts and vocabularies and, at the same time, is open to the alterations that emerge when they make undisciplined appearances or appear in altered forms in other disciplines, transdisciplinary architectural work, whether by architects or others, will both intensify and expand the discipline.[39]

Linder's argument takes architecture neither as object (buildings or spaces) nor representation (drawings, diagrams or models) but as a material discipline which is borne through discursive configurations: 'Disciplines are discursive: in architecture, as in any discipline, work and knowledge appears through, and is produced by, its specific discourses'.[40] The discursive practice of architecture through transdisciplinary work is also advocated by architecture-art historian and theorist Jane Rendell. Apart from one's material and intellectual investment in a collaborative relationship, Rendell brings up the often unremarked emotional aspect of transdisciplinary work which can result in 'feelings of anxiety and ambivalence'.[41] The collaborators (as we experienced first-hand) have to deal with 'invested positions situated at the heart of institutional power structures' and in exchanging 'what we know for what we do not know', one is ultimately giving up 'competence and specialism for the fear of inability and the associated dangers of failure'.[42] In making *03-FLATS*, decisions about content, subject framing and voice were often tensely deliberated. It was clear that interdisciplinarity did not mean the collapse of boundaries between the filmmaker's and my disciplinary training. Quite the opposite: areas of specialisation and disciplinary values were made even more distinct, and at times, even appeared oppositional.[43] The subjective nature of creative practice also meant that negotiations were potentially beleagured since it was not difficult to perceive one's position being compromised.

In this, the long-duration footage became an important tool for re-examining disciplinary biases. It allowed an interpretation of 'images between the images'[44]—the identification of other modes of dwelling which existed at a very different scale and form, a sub-architectural mode, one might say—from architecture's customary documentation and narrative of housing. In particular, there was a shift to reconsider how domestic labour, which academic research itself has largely denigrated as unimaginative, unintelligent and trivial, could hold creative spatial consequences.[45] At Ling Nah's flat for instance, the open plan design, often associated with an uncluttered modernist aesthetic, allows her to survey and manipulate an extensive collection of artistic fodder lodged on multiple vertical surfaces in the flat (Fig. 6.4). Her open plan is full of stuff. In the film, we see how a moveable shelf/partition of artist books opens into a wardrobe naturally lit by a window connecting the wardrobe to the exterior. Another moveable 'wall' of magazines and journals unveils a commodious toilet with a chair at its entrance. Unlike photography, film captures change, movement, and transience as Ling Nah herself reconfigures these walls/partitions/shelves. There is a bodily logic to things shifting around—the movement responds to domestic maintenance, to the artist's creative labour, to moments of activity and leisure. It is space produced through occupation, but it is not possible to contrast this production as radical to what architecture produces through design. The transformation of her flat is not antagonistic to what exists. It makes do, and often, makes something else out of what is there.

Making, questioning, seeing and thinking through film and architecture made the process, even in its conceptual stages, acutely material and discursive at the same time. Yet, the trade-off of transdisciplinary collaborations is the way

WHAT LIES BENEATH

in which the static lines of a discipline open up at a very personal level—I had to explain unequivocally what architecture stands for, what the discipline values, how it sees or does not see, and what relationships can be built across the diagonal path of disciplinary exchange. It forced a manifestation of Linder's provocation—can architecture be recognised other than as architecture?

Making an architectural essay film

In many ways, the collaboration between an architecture academic and a filmmaker threw 'awkward light on habitual practices'[46] and forced reflexivity not only between the researcher/filmmaker with their subject but also forced upon them disciplinary retrospection and negotiation. However, this hybridised practice can, as Troiani and Ewing argue, critically modify architecture's often internalised practices,[47] and compare the temporal relevance of discursive and material practices—the former focusing on the past, and the latter having a 'constructive, projective and speculative use'.[48] The use of film for architectural research holds both these practices in view. The film simultaneously becomes an archive, while also advancing a projective view of what could be.

Extending the discipline's customary output of drawings, photographs and text, film is deployed to explore how much can be deciphered in the confined and repetitive domestic context, wherein details of use and change are much more subtle and fine-grained. The use of film allows for what geographers Gillian Rose and Divya Tolia-Kelly term the 'co-constitution of visuality with materiality'.[49] Rose and Tolia-Kelly emphasise that research needs to develop beyond analysing texts at a distance to include observing and engaging with actual material practices. They also consider what it means for research to engage with new media used to visually record such practices and objects which are produced or used by the subjects:

> … this approach thus draws attention to the co-constitution of human subjectivities and the visual objects their practices create. This is somewhat different from enquiries based on looking, seeing, analysing and writing text; instead, it considers the (geo) politics of embodied, material encounter and engagement.[50]

To make a film, edit the footage and to write about the film after it is made involves different authorial positions. This complicated process reinvigorates a rethinking of how we look, what we see/do not see, the subjects/things/

[Opposite] **6.4** There is a bodily logic to things shifting around; it is space produced through occupation.

practices which matter/do not matter and how these become reconstituted in a renewed visuality, or what Rose and Tolia-Kelly describe as 'how things are made visible', 'which things are made visible' and, 'the politics of visible objects'.[51] These questions are key to *03-FLATS*. Apart from its architectural focus, perhaps the more engaging parallel stories are of evocative domestic objects and subjects—Madam Sim's sewing machine, Amy's cat Dizzy, and Ling Nah's densely packed walls of ideas and memories; the self-portrayal of each woman. These aspects could not be foreclosed by prior research framing. Thus, the adoption of the interstitially situated and loosely experimental essay film mode was an inevitable response to preserve such open-endedness.

The essay film expresses an argument, a speculative opinion or an untested proposition, for the purposes of public scrutiny and debate.[52] The power of the essay film, as film and art theorist Timothy Corrigan explains, lies in its dialogical private-public structure encompassing 'a personal point of view as a public experience' that subsequently promotes a 'process of thinking'.[53] The essay film crucially embeds the private self within a public experience, to yield 'different experiential encounters or experiential concepts linking subjectivity and a public domain'.[54] Nevertheless, the definition of the essay film is wide-ranging if also sometimes vague. In fact, film and media theorist Nora Alter suggests that the essay film is not a distinct genre but that it can be distinguished by its polemical and experimental character given that it is 'transgressive both structurally and conceptually' and consistently 'self-reflective and self-reflexive'.[55] Melding fact and fiction, combining art and documentary genres qualities, the essay film foregrounds the personal opinion in dialogical conversation with its publics.[56] The 'in between' stance of the essay film which is 'transgressive, digressive, playful, contradictory and political'[57] differentiates it from the more informative and often, prescriptive, documentary mode. Film theorist Laura Rascaroli iterates that the essay film is first and foremost a form of cinema that thinks,[58] or in other words, it preserves the thinking process within its structure:

> The essay film constructs [a] spectatorial position by adopting a certain rhetorical structure: rather than answering all the questions that it raises, and delivering a complete 'closed' argument, the essay's rhetoric is such that it opens up problems, and interrogates the spectator; instead of guiding her through emotional and intellectual response, the essay urges her to engage individually with the film, and reflect on the same subject matter the author is musing about. This structure accounts for the 'openness' of the essay film.[59]

Corrigan describes how the essay film format is best articulated through an argument made in Theodore Adorno's paper *The Essay as Form* (1958). In this paper, Adorno argues that 'the essay is concerned with what is blind in its objects' and that 'it wants to polarize the opaque element and release the latent forces in it'.[60] Adorno further sketches an image for the essay form saying that in it 'concepts do not build a continuum of operation, thought does not advance in a single direction, rather the aspects of the argument interweave as in a carpet'.[61] It is in this carpet of interwoven arguments, voices and questions that the essay and its filmic counterpart preserve the fluid and contested nature of thinking. In the architecture studio, the essay film has also been used as a propositional technique for design exploration and making. Advocating it as an instrument for capturing thought in action, designer and academic Penelope Haralambidou suggests that the essay film provides us with a hidden perspective, showing us how we think: 'As if the camera has turned around, and is now recording the residue that film leaves in the mind, rather than the image of the world outside, essayist films often become elegies of the nature and structure of memory'.[62]

Significantly, several notable films have been structured around architecture/space. These include: *Empire* (1964) by Andy Warhol; *Jeanne Dielman* by Chantal Akerman (1975); *Vive l'Amour* (1994) and *Hole* (1998) by Tsai Ming-liang; *London* (1994) and *Robinson in Space* (1997) by Patrick Keiller; *Koolhaas Houselife* (2013) and *Barbicania* (2014) by Ila Bêka and Louise Lemoine; *If Buildings Could Talk* (2010) and *Cathedrals of Culture* (2014) by Wim Wenders. Visual studies theorist Giuliana Bruno observes how film can also materialise time through architecture and space—time is sometimes ephemeral, or ponderous, or banal. She cites Warhol who describes his film *Empire* (1964), an 8-hour, silent, black-and-white film on the Empire State Building in New York, as 'just a way of taking up time'.[63] On the other hand, there is time unfolding in cycles through slow bodily labour in Chantal Akerman's domestic interior of her protagonist Jeanne Dielman:

> Repetition, private time, the unfolding of ordinary temporalities, the rhythm of the everyday, the time when seemingly nothing goes on—all this has been radically called into question as a corporeal texture of woman's time in Akerman's work.[64]

Akerman's film *Jeanne Dielman, 23 Quai du Commerce, 1080 Bruxelles*, revolves around 3 days in the life of a widowed mother/housewife and part-time prostitute. The site of the film is in the flat that Jeanne shares with her teenage son. A doting mother, she maintains a meticulous domestic routine. Akerman follows Jeanne's domestic actions closely—preparing a veal cutlet, washing her sink, making coffee. Film theorist Ivone Margulies describes how Akerman's portrayal of Jeanne's household chores are given an equivalence to more dramatic events—she stabs a client in her bed but this does not affect Jeanne as much as when she discovers the coffee does not taste normal despite being remade a second time.[65] A critique of 'woman's work', 'Akerman converts the story's feminist psychology into choreographic spectacle, depicting housework, sex and family life with a gestural and directorial precision that renders them monumental'.[66]

Margulies argues that Akerman's *Jeanne Dielman* refuses abstraction or abbreviation. It negates the tendency to summarise Jeanne's domestic actions as 'housekeeping'. Instead, Akerman focuses on an excessive duration and iteration of a familiar domestic scene, creating a 'concrete description' of this scene in order to materialise on screen what is experienced bodily. Concrete presentation subsequently leads to defamiliarisation and questioning of the familiar by the viewer.[67] The film does not dismiss repetition as ending up in the same. It stays with the subject and has '... the loquaciousness of the speech of children or of the very old, in that its refusal to summarise, to use the single example that would imply the whole, is like those feverish accounts of events composed of a string of almost identical details connected by "and"'.[68] For this, Akerman enacts a 'literal approach'—the camera follows the preparation of a bowl of potatoes in *Jeanne Dielman* in real, literal time. On film, this feels longer than it really is since we are used to seeing a recurrent series, particularly household labour, depicted through a single event as a shorthand for daily maintenance.

Anthropologist Margaret Mead who argued for the excessive depth and detail enabled by visual source against traditional reliance on field note-taking, claims that footage taken by a camera 'that stays in one spot, that is not tuned, wound, refocused, or visibly loaded, does become part of the background scene, and what it records did happen'.[69] The long fixed take which Mead advocates here, and which Lei used for many scenes in *03-FLATS*, is one that Mead accords a particular sense of objectivity. Yet it is interesting how the long fixed take has also been used by Akerman to do the opposite, that is, to 'create the rigid, distanced, albeit subjective perspective'.[70] Her approach favours a keen sense of materiality, particularly details catching on to the 'literal flow of time' as the camera stays with one event in a room.[71] Its framing draws our focus to filmic texture and structure of the objects in Jeanne's flat, 'the dishcloth, her apron, the weight of plates, the soup tureen on the dining table—all become heightened synesthetic foci'.[72]

The flat is not simply identified by how the space looks because, as Massey writes, 'spatialities are constructed as well by sound, touch and smell—by senses other than vision alone'.[73] In an otherwise generic architecture, affective identification is facilitated by the portrayal of a 'thick space', mired in actual embodied encounters and concrete material definition of atmosphere, potentialities and relationships. Film's propensity to capture embodied presence also means that it is more attuned to 'smaller and more relational tactile experiences' over larger architectural objects and spaces.[74] In *03-FLATS*, such relational, tactile and smaller-scale emphases involve the multiple spatialities in a housing block.

These spatialities include the state's pro-family housing policies; the architect/urban planner's egalitarian concept expressed in repetitive blocks; the occupants' usage patterns and self-appropriation of the flats, their experiences, memories, social and cultural relationships enabled through and within these flats, and the nonhuman effects of building and domestic technologies, weathering elements and pets (legal and illegal) within and beyond the compound of the housing estate. The eventual filmic representation is not simply focused on either the design or the actions of the homeowner/occupant. Instead, it is an affective space of multiple practices (cooking dinner or cooking for a festive occasion, making a dress, making a drawing, fixing a television, installing a new cupboard), relationships (with pets, friends, families, neighbours) and domestic artefacts (the sewing machine, the souvenirs, the sofa, the framed jigsaw). The flat is seen as a frame for the ebb and flow of these components, and their affective iterations of home and belonging. The flat is temporally bound to cycles of use, whether repeated (cleaning the house, cooking, watching television) or exceptional (sewing a garment for a child, starting a new drawing, preparing food for festivities).

In a scene from Madam Sim's kitchen, she is making lunch, stirring the contents of a wok over the stove. The cooking routine is set in a backdrop of overlapping temporalities and spaces—from the mundane everyday (the wind coming through the open window, the flickering of the stove's flames, the steam rising out of the wok, the makeshift cardboard used to protect the flames from the wind) to the larger scaffold of the city beyond it (the hum of a passing train in the distance, and row upon row of gravestones in the soon-to-be extinct private cemetery onto which her kitchen faces) (Fig. 6.5). The scene's materiality summons 'the senses, memory, body and history' as 'part of the analytical process'[75] by inserting into a simply furnished kitchen, the frailty of the elderly woman's body, the air of inherited frugality in her makeshift but pragmatic domestic solutions, the neighbourhood's past and future (the cemetery facing impending exhumation) and its ambitions (the expansion of the mass rapid transit line). It allows the viewer to inhabit not just a physical space but to glimpse into the emotional and social life of an elderly, working class, single woman. In another scene, Amy's 65 square metre flat is temporarily transformed to accommodate eighty friends for post-Ramadhan festivities. Through acts of friendship, labour, care and affiliation, Amy changes an often neglected and sometimes hostile corridor into a celebratory space. She makes it habitable by removing the litter left in the lift lobby beside her flat, scrubbing the floors, putting up streamers on the lobby's ceiling and setting up tables covered with starched linen (Fig. 6.6). The transformed corridor embodies Amy's care.

At the same time, filming was structured around the women's acquiescence of when and what could be publicly recorded in their households. While filming

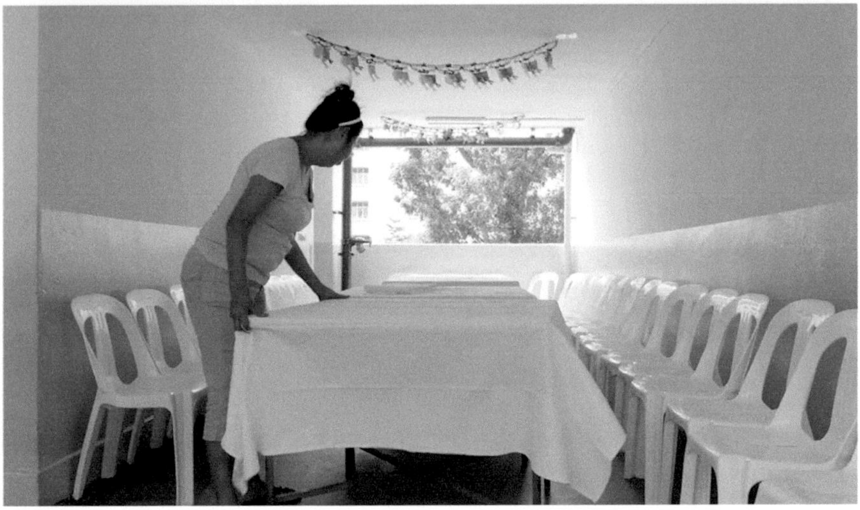

opportunities were initiated in relation to routine domestic work, equally the women would suggest filming particular events they felt unique to their household or defined them as the owner-occupant. These enactments included Madam Sim starting and completing a sewing project for her three-year-old grandniece on a sewing machine she has owned since 1973, an instrument enabling many of her personal accomplishments; Amy observing an installation of a cabinet from Ikea and later, shifting it around on her own with visible difficulty until it was satisfactorily repositioned; and Ling Nah fixing up a handed-down analogue television and watching a TV programme for the first time in her own flat (Fig. 6.7). These represent small, though not minor, acts of resistance as the women's relationships with these domestic artefacts were consciously constructed through ethics and aesthetics that challenged the status quo—maintaining a vintage sewing machine, accommodating an oversized cabinet in a tiny flat and choosing to reuse a second-hand television. These artefacts created relational spaces and opportunities of self-expression which were particular and available to the single woman occupant.

Yet the women's self-initiated enactment of their household chores for public viewing[76] are not inconsequential to research, nor do these compromise its integrity. Pink and ethnographer Kerstin Leder Mackley argue that enactments compel the occupants to recount, imagine and react to their familiar environments in reflexive and embodied ways. These enactments also give researchers and occupants insights into 'how the everyday is a site for both personal and individual transformation ... and of/for everyday improvisation'.[77] In her research on the making and perception of art practices, art historian Barbara M. Stafford argues that art practice is capable of expressing the 'visual, kinaesthetic, haptic, acoustic, olfactory, and visceral sensations'[78] lodged in the maker's body. Extending Stafford's argument to household enactments, Pink and Mackley advocate the self-reflexivity of these enactments since they require the occupant to represent who they are, their experiences, what they value or are good at, by simultaneously doing, remembering and imagining, in situ, what the activity really means to them:

[Previous Spread - Left] **6.5** A simply furnished kitchen and the frailty of the elderly woman's body, juxtaposed against the neighbourhood's past and its aspirations.

[Previous Spread - Right] **6.6** A hostile corridor is transformed into a celebratory space for Ramadan festivities.

[Opposite] **6.7** The women suggest filming particular events which define them as owner-occupants.

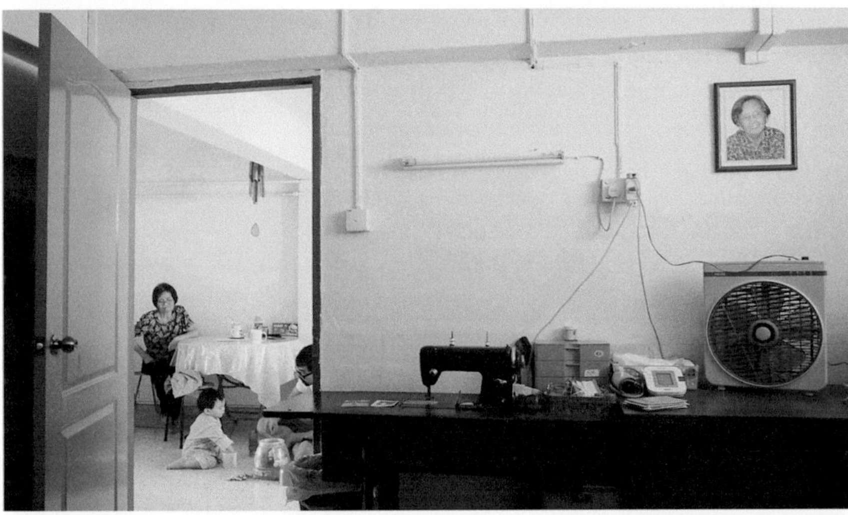

> [The] re-enactments [are] not about simply remembering something that one does or imagining what normal everyday life is like but as expressive ways of drawing on embodied resources and ways of knowing that are precisely part of who participants are, psychologically and physiologically.[79]

In this schema, I consider the three women my research collaborators (Figs. 6.8-6.10). Since the essay film gives each woman a space to produce her domestic story around independent living and selfhood, the collaborative space also shifts the autonomy of academic criticism. In their shadow, arguments about agency, resistance and disobedient domesticity required reframing and rethinking.[80] There was/still exists a struggle to 'write up' the research, since the argument and trajectory were not only nonlinear but sometimes also contradictory, as creative practice frequently is, but as academic research is often not seen to be. In the end, the writing-up process became the most trying phase of the project. It tested ways of advancing an academic argument without obliterating the project's intersections, complexities, productive detours and serendipitous alignments. The text following the film is an exercise in producing affective discourse.

Writing back…

'Writing up' the essay film is not uncomplicated. The film is located at the nexus of empirical research and creative practice. The interpretive text or exegesis of the creative work must do the work of both describing the relationality of the empirical and creative processes, and distance itself from the work, in order to articulate specific contribution to gaps in knowledge. Theorising creative practice as simultaneously engaging the production *and* consumption of knowledge, artist and academic Barbara Bolt argues that creative practice produces knowledge by implicitly requiring handling of 'materials, method, tools, and ideas of practice', and it is through such performative handling that we 'come to know the world theoretically'.[81] Bolt's position coincides with Paul Carter's argument of 'material thinking' which places the origin of invention within a social contract to repair, renew or renegotiate relations with the other.[82] Understanding *both* making and writing as material practices allows a dialogical relationship to exist between *03-FLATS* and its written expositions. In this sense, the exegesis does not annotate the creative work. Pairing with the film, this text is one point in a circuit relaying observations and ideas. The exegesis of the creative work does not reduce or paper over the complexities of the creative work. It enlarges its reach and relevance so that it becomes of interest to others.

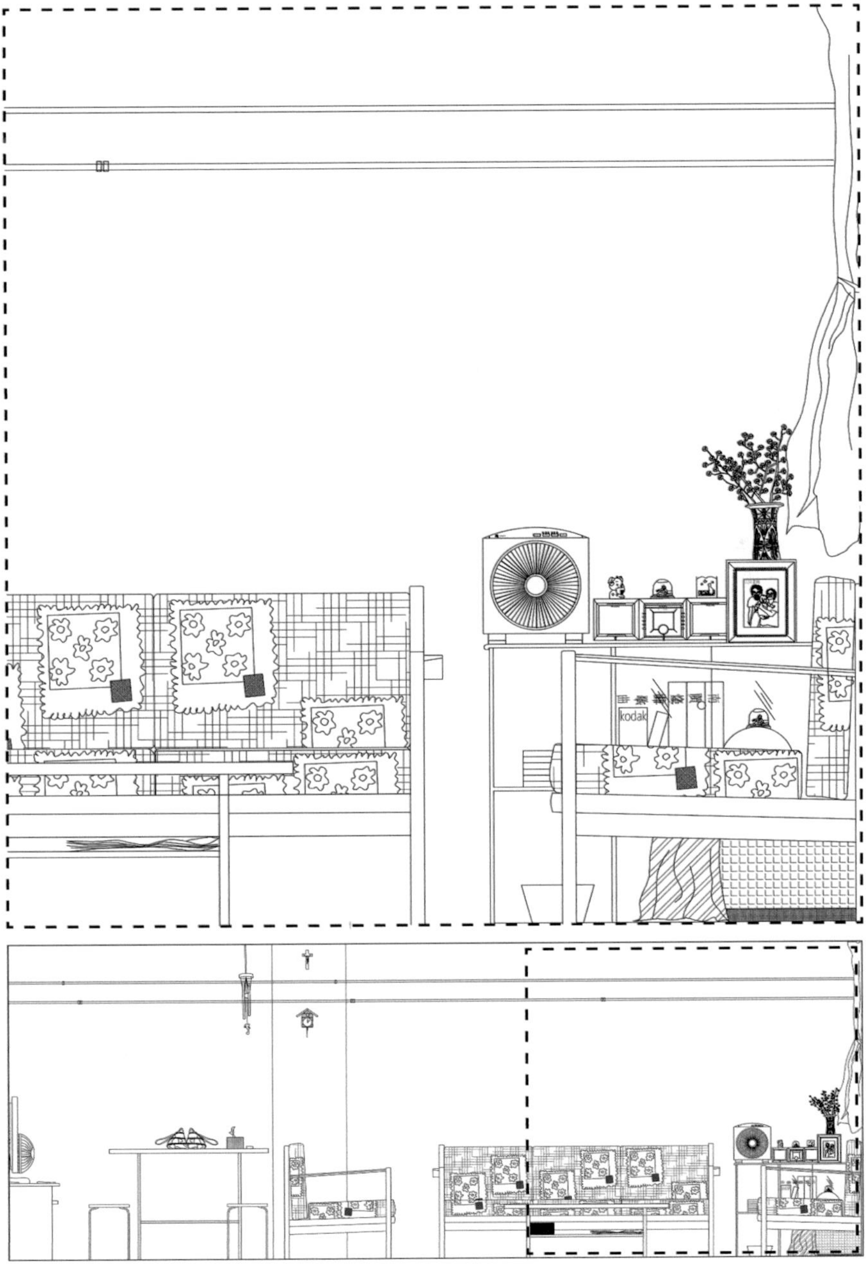

6.8 Madam Sim's living room, detail.

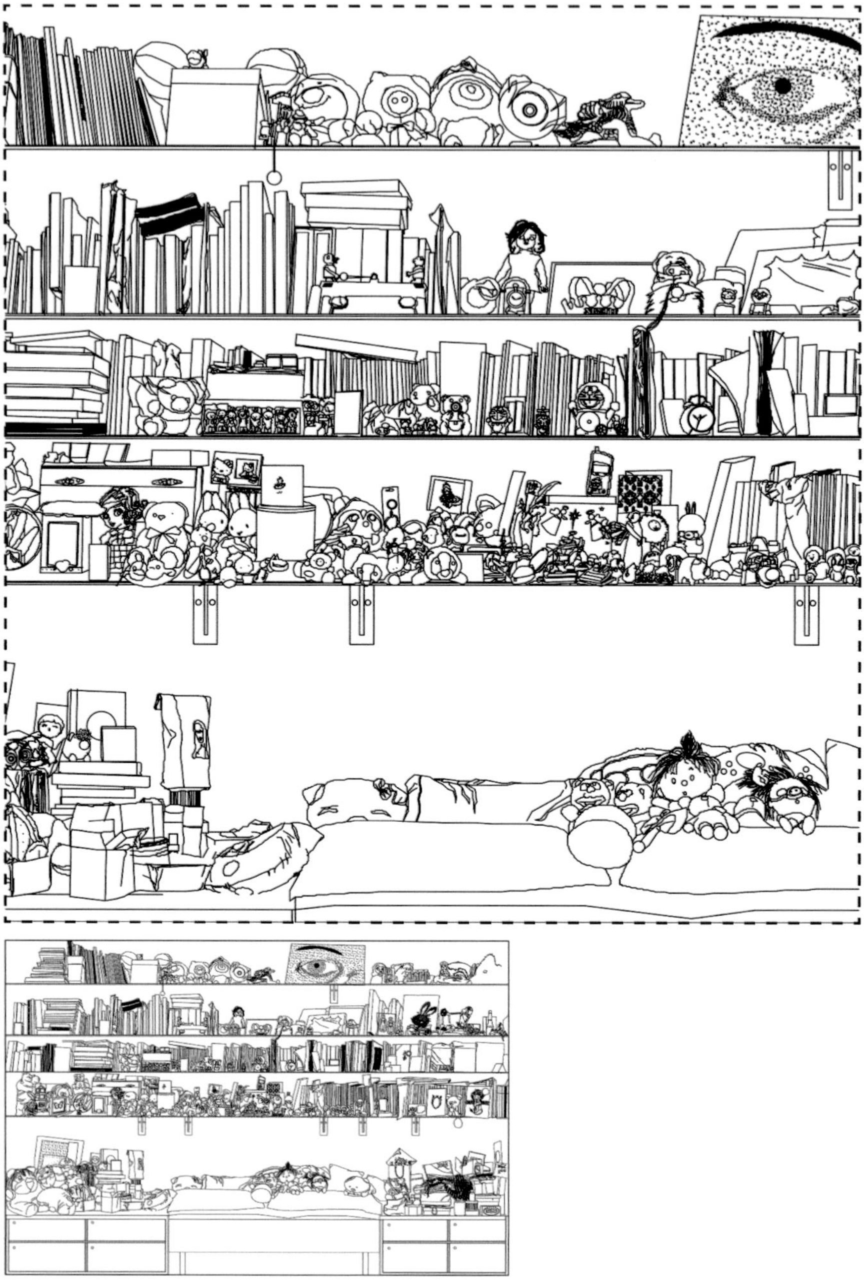

6.9 Ling Nah's bedroom, detail.

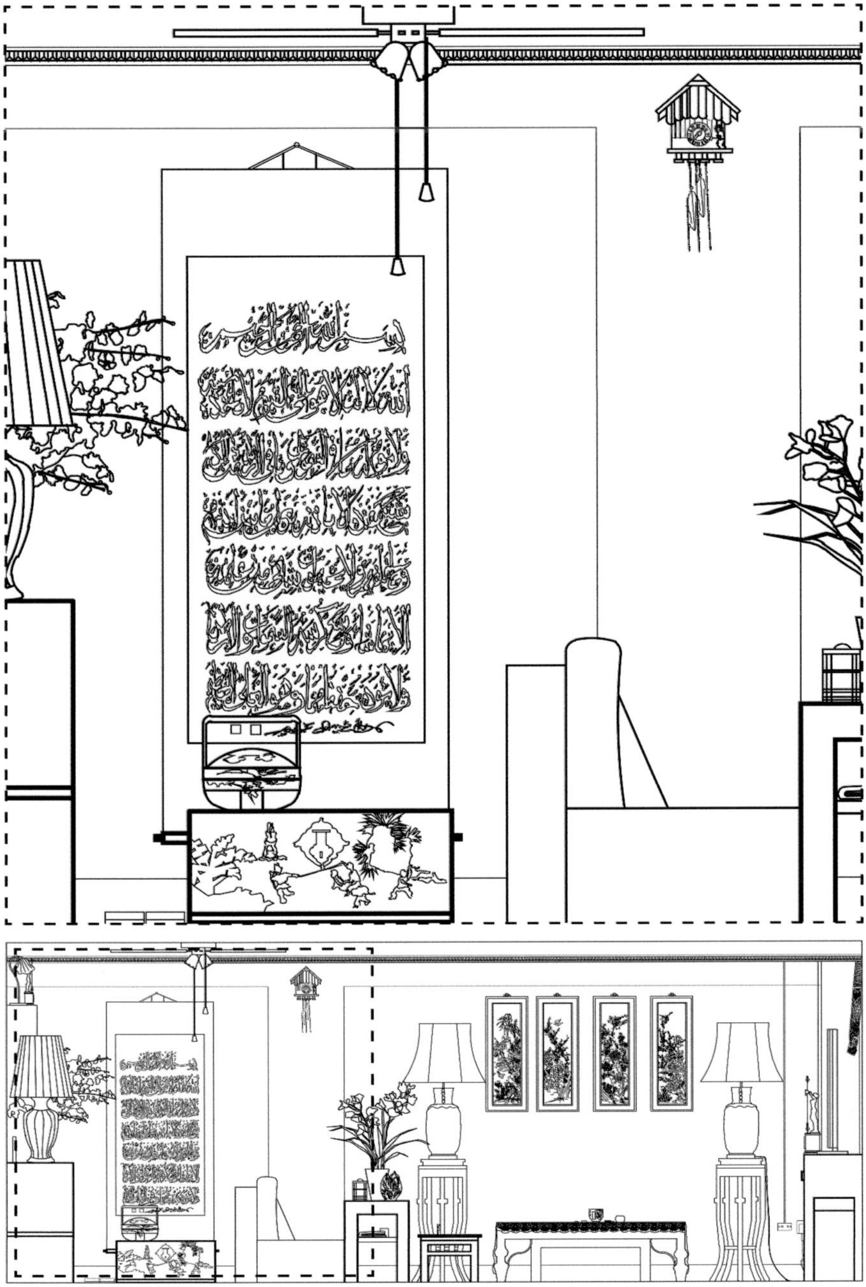

6.10 Amy's living room, detail.

Rendell, who has moved between the disciplines of architecture and art and has recently also been involved in public advocacy, articulates this conceptual traffic: 'I have been wondering how it is possible to be in two places at once, to hold alternative possibilities together, specifically creative and critical modes of writing, combining the analytic with the associative, intellectual inquiry with storytelling, remembering with imagining'.[83] It is a question not just of expanding the limits of architectural representation (in Rendell's, and my own case) but also of reworking such representational politics to confront the chasms between architectural writing and the visual, practice and theory, embodied evidence and researcher distance, to name a few.

Where the exegesis should be positioned—within a creative impulse, within a methodological process, or within the fraught in-between—is a necessary question that extends the depth of the research inquiry as it moves between creative and interpretive realms. While *03-FLATS* has shifted the architectural encounter with public housing beyond a 'research-based enquiry to become fundamentally a transformative "doing"',[84] its use of film as a research tool has been unhelpfully conflated with ethnographical fieldwork. Consequently, rather than using the film to theorise public housing, ethnographical theories were imposed upon the film with resultant critique focused on the impact of ethnographical fieldwork. The refusal to recognise film as the outcome of research itself rather than data collected has epistemological consequences. The understanding of film as data prioritises the written exposition as having to make sense of the data/film, while the inverse of film as research would demand that its exegesis engage and respond to the research. As Bolt argues, the written exposition for a creative piece of work is not just to describe, explain or contextualise the practice but 'to produce movement in thought itself ... to create "shocks to thought"'.[85] In other words, the exegesis continues the inventiveness of the creative work in a discursive form; it theorises out of practice.[86]

In writing about her experimental ethnographic films which have received diverse feedback from audiences both academic and non-academic, filmmaker and theorist Trinh T. Minh-ha reminds us that there is a fundamental difference between the film and 'the discussion around it or with the theoretical work that comes with it'.[87] There is a tendency in writing about a creative work, that one is pushed to explain the work, in other words, to lucidly say: 'This is what it is all about'. Trinh emphasises that it is virtually impossible to speak definitively about her films. Films have affective qualities that should be experienced rather than described in second- or third-hand. Instead, her position is to articulate 'nearby' or 'close' to these films, which is to say they are useful to inquire further about the subject, displace preconceptions and theorise new directions, rather than to simply illustrate or explicate what is already known.

'My intentions and motivations', Trinh discloses, 'are only part of what goes into the readings and theories advanced, which are also built in the unpredicted interactions among elements in the work and are often densely populated by other people's feedback'.[88] In saying this, she positions film and writing as being in a dynamic relay—moving fluidly between creator and audiences, and shuttling between vision, perception, feedback, description and critique. Similarly, the intentions and motivations for *03-FLATS* are never identical to how I can speak of the film today. Each category has evolved because the boundaries for perceiving and understanding the work also shifts depending on the audience's contexts, and my own reading which also continues to change with time. It is accurate to say that the film displaces the audience/researcher as much as it locates them conceptually and contextually. Sited at the edge of the architectural discipline, its function is not just to question the subject it portrays (single women in public housing) but also to probe internally into the framework of its investigation (filmmaking as research in architectural discourse and in filmic practice).

Troiani and architectural film scholar Hugh Campbell argue that the 'meaningful integration of film' in architectural research and pedagogy not only allows us to depict existing worlds or imagine new ones into being but is a way of 'expanding and exploring the languages and structures of the discipline itself and its interdisciplinary spatial and temporal conditions'.[89] The 'utility' of *03-FLATS* is not in making conclusive or generalised claims either about HDB living or about single women living in public housing. It has broadened, or disrupted, the *a priori* image, visibilities and architectural discourse, of Singapore's public housing. Through several competing and overlapping dimensions—bodily, emotively, materially, ideologically, socially, politically—the film presents this architecture as tentative, fragile and embodied. It elicits emotions and feelings relational to the space and to the inhabitants in those spaces, and in doing so, exceeds the objectified and static capacities of traditional architectural representation, whether visual or textual.

The making of an architectural essay film that sits at the edges of architectural discourse, filmmaking and ethnographic study pokes at disciplinary limits and their affordances. A piece of writing such as this one cannot replace the film nor fully illustrate its purpose and conclusions. Instead, I also write 'nearby' *03-FLATS* to draw out its difficulties, encounters, opportunities, behind-the-scenes collaborations and failings. This exegesis theorises from its own tentative position. It works through disciplinary affordances to make visible the gendered body, to thicken their descriptions and their situatedness in ways that can no longer be disengaged from the politics of space in which these bodies are located. •

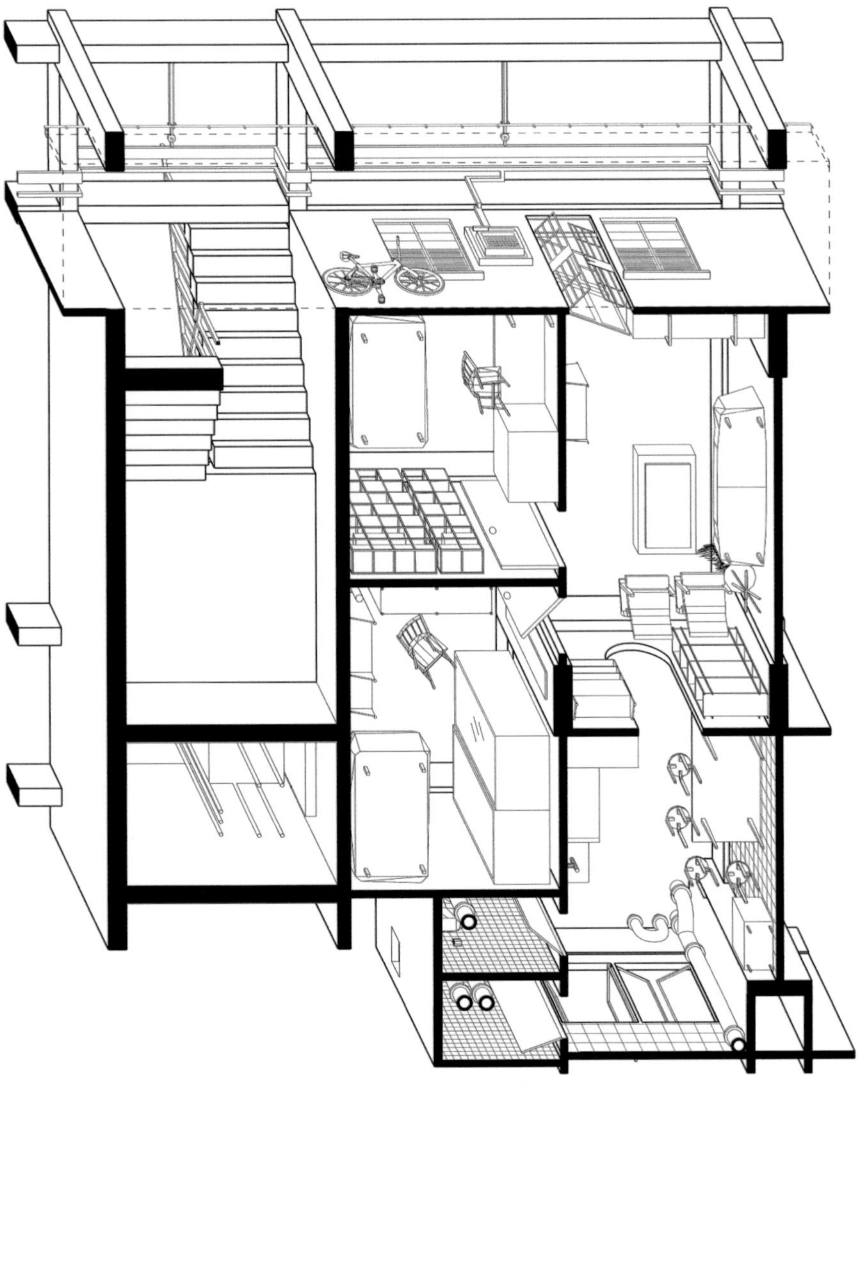

Part Three: Landscape

7 • In the Midst Of

Field notes at a cemetery

LIKE MANY OTHERS, I ARRIVED BELATEDLY AT BUKIT BROWN—there was already widespread concern for the 140-year old cemetery's slow but steady extinction. It had lain dormant for almost 40 years (Fig. 7.1). The cemetery had catapulted from obscurity into the citizenry's consciousness in September 2011 when plans for an eight-lane highway were first revealed. It rapidly became a popular subject of research by both specialists and amateurs.[1] In October 2013, the World Monuments Fund included Bukit Brown as one of 67 sites in the 2014 World Monuments Watch List.[2] The site featured prominently in documentaries, blogs, public debates, exhibitions and numerous publications.[3]

Located at the fringe of a suburb, Bukit Brown's insularity evolved from being untouched. Upon closure in 1973, a secondary forest took hold. 90 bird species, 13 of which were critically endangered or vulnerable, sought refuge here.[4] Undisturbed each year, the trees grew taller; creepers buried paths and tombs. Its presence waxed and waned with ancestral and religious rituals. Families came once a year for the *Qingming* (tomb sweeping) festival to pay respect to their deceased ancestors. The cemetery was also the site of ad hoc urban

7.1 4,000 graves in the 140-year-old Bukit Brown Cemetery were exhumed to make way for an eight-lane highway.

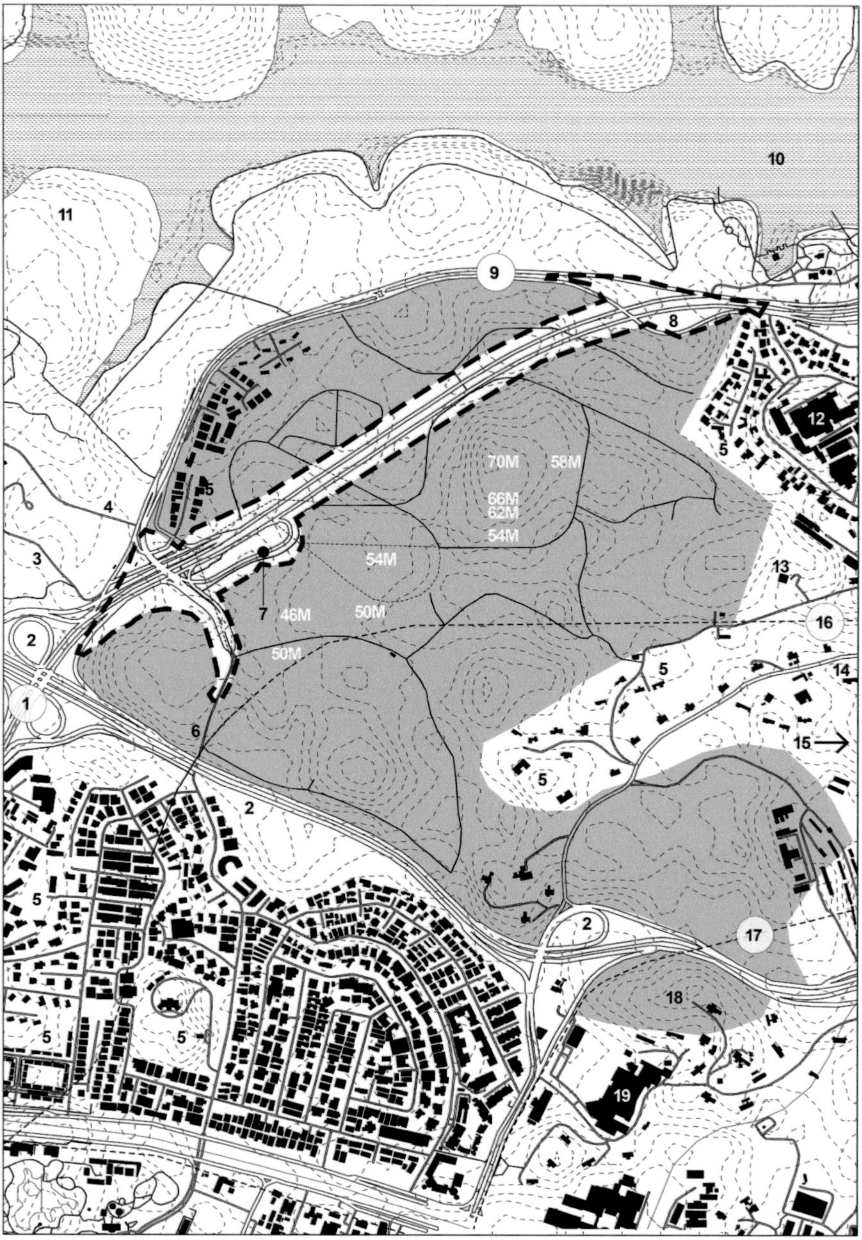

LEGEND

0 200M N

Graves Affected by Lornie Highway

Bukit Brown Cemetery Area Today

1 - Adam Road & Flyover
2 - Pan-Island Expressway (PIE)
3 - Adam Drive
4 - Sime Road
5 - Landed Private Housing Estate
6 - Kheam Hock Road
7 - Lorong Halwa (Old Access Road)
8 - Lornie Highway (2011 Announced)
9 - Lornie Road
10 - Macritchie Reservoir and Park
11 - Sime Golf Course
12 - Caldecott Broadcast Centre
13 - Omar Salmah Mosque
14 - Equestrian Federation of Singapore
15 - Singapore Polo Club
16 - Circle Line (Underground Train)
17 - North-East Line (Underground Train)
18 - *Kopi Sua* (Coffee Hill) Cemetery
19 - St. Joseph's Institution (School)

adventures organised by epiphenomenal groups such as the Asia Paranormal Investigators (API).[5] Allusions to hauntings and ghosts piled up as thick as the encroaching jungle.

Geographically, Bukit Brown (*Kopi Sua* or Coffee Hill)—named after the British landowner George Henry Brown who first cultivated the land as a coffee plantation in the 1840s—consists of five hillocks[6] forming a gentle, undulating topography. It was considered auspicious in Chinese burial geomancy because of its propitious aspect and elevated terrain (Fig. 7.2). Linked to several smaller burial grounds in the immediate vicinity, Bukit Brown is the largest Chinese cemetery in Singapore and outside of China.[7] The last recorded survey in 1922 put the cemetery at 213 acres, or the equivalent of 142 football pitches.[8] However, because of its hilly terrain and thick overgrowth, it is perceptually larger. Buried here are pioneer Southern Chinese immigrants of the mid-nineteenth century whose tomb designs manifest the material linkages between Southeast Asian Chinese, local Straits Chinese and European cultures, as well as prominent Southeast Asian supporters of China's 1911 Republican Revolution. Between 1942 and 1945, the cemetery was a World War II battleground. It served as a burial site for casualties and victims of the Japanese occupation. Prior to recent exhumation works, the cemetery reputedly contained more than 100,000 tombs, many of which were completely embowered by thick vegetation (Fig. 7.3).[9]

This cemetery has been recorded primarily through a history of land transactions.[10] At the turn of the twentieth century, migrant labour from Southern China flooded colonial Singapore's job market. As Singapore's population grew, unforeseen problems of providing burial sites for those not affiliated to existing Chinese clan associations intensified. Typically, the dead were buried in cemetery plots which belonged to groups sharing the same family surname as the deceased, or to members of the same dialect group and district in China.[11] Consequently, migrant workers who settled in Singapore without the requisite clan or dialect affiliations found themselves excluded from this privatised scheme. Unclaimed bodies were left in public areas such as the backlanes and five-foot-ways of buildings. Many were illegally buried. The Burials Bill, proposed in 1887, sought to control the widespread appropriation of land for non-developmental use; it was specifically drafted to license and regulate the use of burial land. This proposed legislation—not implemented until 1896 after several concessions were made to appease the Chinese residents—sparked concern amongst the wealthier Chinese who felt that their native land rights were repressed. At the same time, to cope with growing development, cemeteries were increasingly emptied out for other uses.[12]

[Opposite] **7.2** Topography and vicinity of the Bukit Brown Cemetery.

[Overleaf] **7.3** Timeline of exhumation and development of cemeteries in Singapore as of 2019.

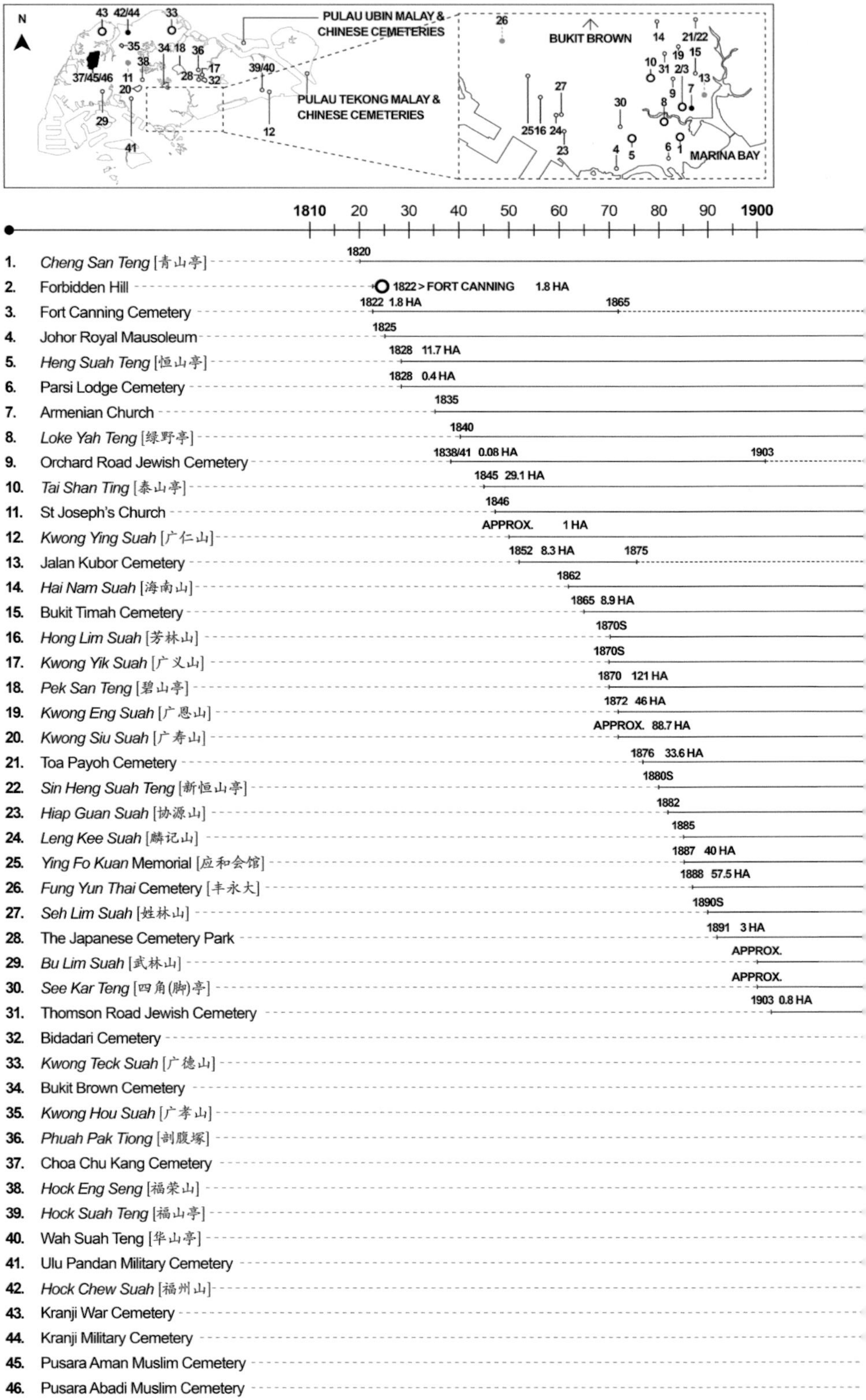
N
43 42/44 33
35 34 18 36
38 17
37/45/46 11 28 32
20 39/40
29
41 12
PULAU UBIN MALAY & CHINESE CEMETERIES
PULAU TEKONG MALAY & CHINESE CEMETERIES
26
BUKIT BROWN
14 21/22
10 19 15
31 2/3 13
9 7
27 8
30
25 16 24
23
5 6 1
4
MARINA BAY
1810 20 30 40 50 60 70 80 90 1900
1. Cheng San Teng [青山亭] 1820
2. Forbidden Hill 1822 > FORT CANNING 1.8 HA
3. Fort Canning Cemetery 1822 1.8 HA 1865
4. Johor Royal Mausoleum 1825
5. Heng Suah Teng [恒山亭] 1828 11.7 HA
6. Parsi Lodge Cemetery 1828 0.4 HA
7. Armenian Church 1835
8. Loke Yah Teng [绿野亭] 1840
9. Orchard Road Jewish Cemetery 1838/41 0.08 HA 1903
10. Tai Shan Ting [泰山亭] 1845 29.1 HA
11. St Joseph's Church 1846
12. Kwong Ying Suah [广仁山] APPROX. 1 HA
13. Jalan Kubor Cemetery 1852 8.3 HA 1875
14. Hai Nam Suah [海南山] 1862
15. Bukit Timah Cemetery 1865 8.9 HA
16. Hong Lim Suah [芳林山] 1870S
17. Kwong Yik Suah [广义山] 1870S
18. Pek San Teng [碧山亭] 1870 121 HA
19. Kwong Eng Suah [广恩山] 1872 46 HA
20. Kwong Siu Suah [广寿山] APPROX. 88.7 HA
21. Toa Payoh Cemetery 1876 33.6 HA
22. Sin Heng Suah Teng [新恒山亭] 1880S
23. Hiap Guan Suah [协源山] 1882
24. Leng Kee Suah [麟记山] 1885
25. Ying Fo Kuan Memorial [应和会馆] 1887 40 HA
26. Fung Yun Thai Cemetery [丰永大] 1888 57.5 HA
27. Seh Lim Suah [姓林山] 1890S
28. The Japanese Cemetery Park 1891 3 HA
29. Bu Lim Suah [武林山] APPROX.
30. See Kar Teng [四角(脚)亭] APPROX.
31. Thomson Road Jewish Cemetery 1903 0.8 HA
32. Bidadari Cemetery
33. Kwong Teck Suah [广德山]
34. Bukit Brown Cemetery
35. Kwong Hou Suah [广孝山]
36. Phuah Pak Tiong [剖腹塚]
37. Choa Chu Kang Cemetery
38. Hock Eng Seng [福荣山]
39. Hock Suah Teng [福山亭]
40. Wah Suah Teng [华山亭]
41. Ulu Pandan Military Cemetery
42. Hock Chew Suah [福州山]
43. Kranji War Cemetery
44. Kranji Military Cemetery
45. Pusara Aman Muslim Cemetery
46. Pusara Abadi Muslim Cemetery

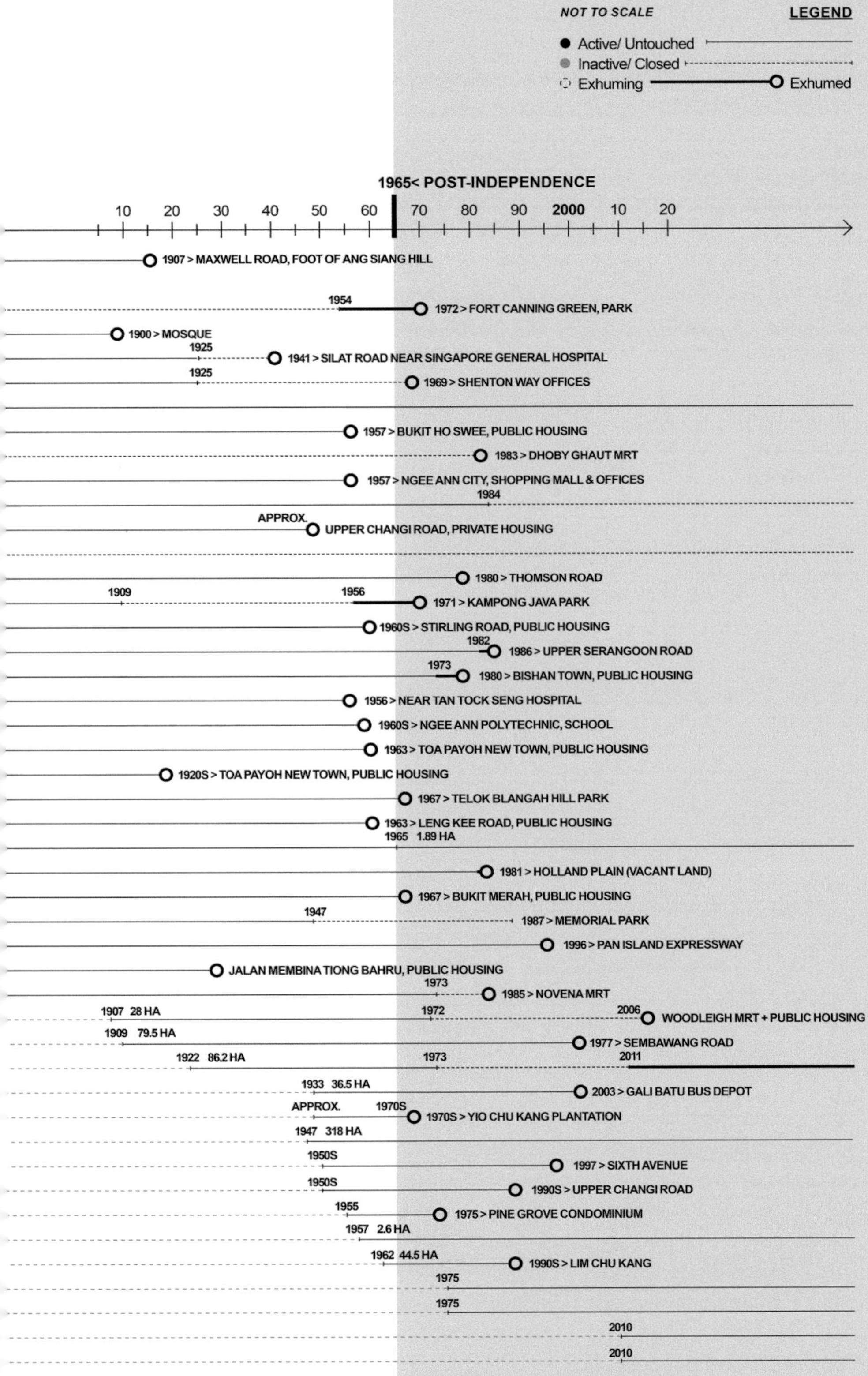
NOT TO SCALE
LEGEND
Active/ Untouched
Inactive/ Closed
Exhuming
Exhumed
1965< POST-INDEPENDENCE
10 20 30 40 50 60 70 80 90 2000 10 20
1907 > MAXWELL ROAD, FOOT OF ANG SIANG HILL
1954
1972 > FORT CANNING GREEN, PARK
1900 > MOSQUE
1925
1941 > SILAT ROAD NEAR SINGAPORE GENERAL HOSPITAL
1925
1969 > SHENTON WAY OFFICES
1957 > BUKIT HO SWEE, PUBLIC HOUSING
1983 > DHOBY GHAUT MRT
1957 > NGEE ANN CITY, SHOPPING MALL & OFFICES
1984
APPROX.
UPPER CHANGI ROAD, PRIVATE HOUSING
1980 > THOMSON ROAD
1909
1956
1971 > KAMPONG JAVA PARK
1960S > STIRLING ROAD, PUBLIC HOUSING
1982
1986 > UPPER SERANGOON ROAD
1973
1980 > BISHAN TOWN, PUBLIC HOUSING
1956 > NEAR TAN TOCK SENG HOSPITAL
1960S > NGEE ANN POLYTECHNIC, SCHOOL
1963 > TOA PAYOH NEW TOWN, PUBLIC HOUSING
1920S > TOA PAYOH NEW TOWN, PUBLIC HOUSING
1967 > TELOK BLANGAH HILL PARK
1963 > LENG KEE ROAD, PUBLIC HOUSING
1965 1.89 HA
1981 > HOLLAND PLAIN (VACANT LAND)
1967 > BUKIT MERAH, PUBLIC HOUSING
1947
1987 > MEMORIAL PARK
1996 > PAN ISLAND EXPRESSWAY
JALAN MEMBINA TIONG BAHRU, PUBLIC HOUSING
1973
1985 > NOVENA MRT
1907 28 HA
1972
2006
WOODLEIGH MRT + PUBLIC HOUSING
1909 79.5 HA
1977 > SEMBAWANG ROAD
1922 86.2 HA
1973
2011
1933 36.5 HA
2003 > GALI BATU BUS DEPOT
APPROX.
1970S
1970S > YIO CHU KANG PLANTATION
1947 318 HA
1950S
1997 > SIXTH AVENUE
1950S
1990S > UPPER CHANGI ROAD
1955
1975 > PINE GROVE CONDOMINIUM
1957 2.6 HA
1962 44.5 HA
1990S > LIM CHU KANG
1975
1975
2010
2010

The acquisition of the Bukit Brown Municipal Cemetery coincided with the colonial government's plans to reclaim an existing burial ground at the centrally located district of Tiong Bahru for housing in the mid-1920s.[13] Apart from space constraints, the Municipal Commissioners were dealing with what geographer Brenda Yeoh describes as a conflict between a capitalist colonial order over the value of land for development and a more esoteric native conception of the natural landscape as propitious burial sites for the deceased. The Chinese privileged 'round knolls and hillsides'.[14] Burial grounds multiplied so quickly that it began to appear that an entire 'Singapore seem[ed] likely to become a vast Chinese cemetery'.[15] In a 1917 Commissioners' meeting, Bukit Brown was publicly proposed as a viable option because of its competitive pricing and generous land size.[16] Two years later in 1919, and 13 years after the proposal was mooted for a public cemetery to deal with problems of overcrowding and illegal burials, the Municipal Commissioners introduced the Lands Acquisition Ordinance to compulsorily purchase a 213-acre site from the Seh Ong clan who sought to reserve the plot for their clan members. The new Bukit Brown cemetery with its small grave-plots and tight geometrical layout was initially unpopular. Modifications were made in August 1922 to account for preferences of larger plots sizes (20x10 feet for the general graves and 10x5 feet for the paupers' graves); favoured plot orientations facing East and South; and a path for circulation between every six rows of graves.[17]

Nevertheless, the cemetery's economic value was overshadowed by its atmosphere. Bukit Brown today remains notoriously inaccessible. Visitors have to trek through thick undergrowth, risking bites from mosquitoes, ants, leeches and snakes. Ephemeral rituals of animistic nature or those linked to temple-sponsored practices also found their way here.[18] Its obscurity furnished prospects for faction fights, robberies, murders, black marketeering and illegal swopping of burial plots.[19] The highway announcement in 2011 resulted in a flurry of organised activity antithetical to the wild character of the burial site. It was mapped, dug up, researched, photographed, narrated and written about at a furious rate. Lost graves and familial relationships were urgently traced and recorded. Bukit Brown's impending destruction haunted social conscience. Evanescent routes made during once-a-year visits were rapidly formalised as public interest grew.

Heritage groups made a case against development, emphasising the cemetery's cultural value. Historians argued it possessed rare material culture, evidenced in the lavish architecture of old tombs (Fig. 7.4). Similarly, natural history groups argued it was a sanctuary for endangered bird species. From the improbable coalition of birdwatchers, paranormal investigators, conservationists and heritage-conscious citizens, a groundswell of public support began to test the state's responsiveness towards the decision to obliterate 200 years of history with

[Opposite] **7.4** Detail of a tomb.

a highway. The intellectual construction of Bukit Brown took on the tenor of an Enlightenment project with enumerative knowledge—things were counted, catalogued, mapped, given reason and accorded proper provenance. The cemetery's otherworldly and uncanny atmosphere was notably absent from academic research where arguments were recast in terms of development versus conservation.

This chapter advances a different approach. Tracking the various activities, agents and objects central to the narrative of Bukit Brown at a time when the cemetery is legitimised as a historical-ecological subject, it focuses on the cemetery's unsaid, 'irrational' and mythic-historical attributes and narratives. The purpose is to return to the space of the cemetery itself, to reappraise how its physical and imagined features affect those who encounter it, as well as the perceptions, superstitions and stories which derive from, or embellish, its terrain. Geographer Mitch Rose argues that the work of culture is often impaired by relying only on what is known: '[Culture] understands the rituals performed, artefacts made, and stories narrated purely in terms of the forms of presence they are thought to be indicative of rather than the forms of presence they are endeavouring to affect'.[20] In other words, the legitimised historical narratives and identities can overwhelm or obscure other equally valuable accounts. Conversely, the open-ended character of Bukit Brown presents an opportunity to exercise a cultural engagement that makes its past relevant and proximate to the contemporary.

Field and spectre

To begin with, Bukit Brown must be unmoored from the polemics of development versus heritage. It requires a spatial perspective that views territory as simultaneously abstract, lived and imagined. One possibility is to engage the cemetery as landscape. Art historian W.J.T. Mitchell argues that 'landscape' invites us to 'engage in a kind of conscious apperception of space as it unfolds itself in a particular place', where we are compelled to 'look at looking itself'.[21] Overlapping with Michel de Certeau's conception of 'space' being more dynamic than 'place', and referencing Henri Lefebvre's triad of 'perceived space' (the 'space' of everyday routine and rituals), 'conceived space' (a 'place' planned and administered by engineers, city planners and architects) and 'representational space' (the lived space of inhabitants, which 'the imagination seeks to change and appropriate'[22]), Mitchell suggests that 'landscape' be situated within a triadic composition, analogous to 'representational space', thus becoming the third term—'space-place-landscape'.[23]

If we follow Mitchell's dialectical triad, Bukit Brown could be reframed through three competing spatial ontologies: 'land' (place), 'heritage' (space) and 'burial ground' (landscape). The first—land—is a territorialised place for productive infrastructure (highway, housing) that comes into being through planning and measurement (plot ratios, planning guidelines, land use policies, legal boundaries, soil types, monetised values). The second—an ecocultural heritage space—is a symbolic realm of meanings whose construction is cultural (historical pasts, cultural practices, architectural lineages, natural histories). The third—the burial ground—is a landscape where legible and visible frames of knowledge are challenged. A landscape conveys both an image and an ideology; beyond a specific appearance, it is ideological because it embodies particular ways of seeing, thinking and acting.[24]

When the cemetery was put at risk, an online archive of photographs spontaneously germinated (Fig. 7.5). Many of the photographs bear witness to the photographers' informal encounters with the cemetery. There are riotous mounds covered in vegetation and tangled greenery, aerial roots sprouting from gnarled ancient trees, and the barely-there outline of ad hoc paths covered with weeds and vines. The photographs show the cemetery in fragments—part vegetal, part topography, part architectural, part archaeological, part landscape. The separation of nature and culture, seen in the institutionalised knowledge of Bukit Brown, is absent. Landscape offers clarity through systematic hierarchy of fore-, middle- and background, with a subject positioned within these three distinct layers. At Bukit Brown, the same sense of depth and hierarchy are confounded by the equivalence of subjects which surge to the foreground or recede into the background without clear distinction: vegetation, tombs, evanescent paths, the remains of prayer items, receding hillocks, the low mist, the exhumed graves, the columns of the highway. These fragments compete for attention in the foreground. The clear hierarchy of landscape is misplaced in the tropics.

The entangled parts of Bukit Brown—architectural, vegetal, infrastructural, historical, cultural, natural, built—float in a field of subjects. The concept of 'field' holds a common element that inscribes a particular order governed by a specific pattern of growth. In the case of the cemetery's tropical overgrowth, order inscribed through the vigour of under- and over-growth, constitute a different kind of field. While a grass field can be journeyed *across*, a tropical jungle distinctively requires one to walk *into* it. The overgrown cemetery is an interior, one that is to be inhabited as much as walked through. The interiority of the field is geometrically and psychically defined—one enters into bounded space, and in that space, one is already occupying a differentiated temporality—partaking in the distinct spacing of things, specific patterns of movement, and particular configurations of light, air and wind.

A
B
C
D
1
2
3
4
5
6
7
8

Essayist and cultural historian John Berger describes 'field' as *both* a self-bounded space which contains a series of actions and events, the latter being particularly relevant for Bukit Brown. For Berger, a field must have a 'maximum possibility for exits and entrances' so that it possesses an 'attendant openness to events'.[25] Berger describes how our encounter with a first in a series of events (connected by the field) draws our attention to the space we are in, and consequently, to our own positions in that space:

> Having noticed the dog, you notice a butterfly. Having noticed the horses, you hear a woodpecker, and then see it fly across a corner of the field. You watch a child walking and when he has left the field deserted and eventless, you notice a cat jump down into it from the top of a wall. [B]y this time you are within the experience ... The visible extension of the field in space displaces awareness of your own lived time.[26]

Berger shifts from the interiority of the field to the interiority of the observer. Disinterested observation transforms into self-awareness, 'you are within the experience'. 'The field that you are standing before', concludes Berger 'appears to have the same proportions as your own life'.[27] Here is the suggestion that the field with its heightened sense of something about to happen—the possibility of several events unfolding—sensitises the observer not just to his or her surroundings but to his or her own presence within the same field. The field becomes a temporary but affective frame from which to observe, grasp and reflect upon a fleeting event. Conceptually perceived as a field, Bukit Brown enables a recognition of different fragments that make up the cemetery, and how their overlapping processes are made possible, equally by the space in which they are contained, as well as being the very means through which this space comes into focus. In other words, the equivalence of fragments held in the field of the cemetery necessitates an alternative kind of recording—one in which knowledge, such as the architecture of the tombs, becomes inconceivable without the tangle of the jungle and the various myths and superstitions intertwined with the burial ground. Histories, whether architectural, social or natural, are necessarily juxtapositions between fact with story, myth and/or superstition; between what took place and what is make-believe.

The simultaneity of fragments enabled by the conceptual frame of a field aligns with the *spectral* character of the Chinese burial ground, whose root word 'spectre' points to the ghostly, or to an apparition particularly when evoked in relation to a cemetery. Yet, in broader terms, the spectral refers to an alternative kind of vision whether it is the ability to see ghosts or something else. Most of

[Opposite] **7.5** An informal online photographic archive spontaneously grew from individual blogs and websites.

all, the spectral gestures to a different mode of seeing which does not trivialise what we cannot yet apprehend:

> In a series of more or less equivalent words that accurately designate haunting, specter, as distinct from ghost (revenant), speaks of the spectacle. The specter is first and foremost something visible.[28]

Spectrality is the coincident sense of absence and presence, as if these two opposing qualities were perfectly juxtaposed within a single encounter. The spectral not only displaces place by alluding to 'the just perceptible, the barely there, the nagging presence of an absence',[29] it destabilises the very certainty of what we know to be present. 'The spectral', geographer John Wylie attests 'is thus the very conjuration and unsettling of presence, place, the present, and the past'.[30] The spectral makes it difficult to assert that something in the present is not always already entangled with what is hidden away in the past.

The Chinese cemetery is founded on spectral principles of landscape—following the rules of geomancy which read the lie of the land as having direct implications on physical factors such as light, water and wind, and also, metaphysically influencing family fortunes and relationships. To see the Chinese cemetery as spectral is to recognise the impossibility of separating the seen and the factual, from the invisible and the imagined. It demands that we take seriously what we intuit as much as what we see.

The conceptualisation of Bukit Brown as a spectral field demands a mode of recording that can account for what is there as well as what is barely there; to imply both presence and absence. The recording here assumes the form of 'notes' taken 'in the field' at a time when Bukit Brown is altered by the simultaneous wills of development and preservation. The 'field notes' revolve around an inventory of navigational tools, which are divided into two types—constructed markers (diagrams as maps, wooden stakes, ribbons, names and numbers) and natural markers (earth pits and jute mats, paths). These two markers operate physically and symbolically. The constructed markers are the outcome of exhumation works and highway building. Returning to Mitchell's triad, these align with the marking of territory, and with the installation of a proper 'place'. The constructed markers signal the demise of the old cemetery, while the natural markers gesture to a continuity of cultural and familial practices.

In addition to these constructed and natural markers, the cemetery is pictured in two modes—in maps and in photographs—posing two modalities of knowledge. The map is definitive and aspires to authoritative wayfinding; the photograph records individual experience and is often atmospheric and evocative. The map and the photograph may be read as parallels to the constructed and natural markers respectively. The former locates and fixes the new-found configuration of the cemetery, the latter preserves a rapidly eroding historical space.

The field notes weave together affective evidence, and their attendant modes of knowing. Braiding together histories, theories, moods, customs, taboos and experiences, they construct a discourse based on the burial ground's spectral character. The field notes multiply the evidence and stories of the cemetery; and the ways one might see, understand and write about it. They attempt to challenge the sites, agencies, resources and narratives of architectural history which emphasise distance and objectivity, complementing this trajectory with intimate and often, amateur perspectives. The conception of architectural history, in this sense, moves restlessly between the past and the contemporary, the collective and the intimate, the cognitive and the affective.

Field notes 1: *constructed markers*

(i) Diagram becoming map | Diagramming vs. Mapping | Blindsighted journeying

The *Bukit Brown Wayfinder* is a 117-page digital publication commissioned by the Singapore Heritage Society.[31] It features photographs of flora, fauna, statues, tomb engravings and layouts; annotated diagrams of the Chinese tomb anatomy; a colour map of two hills with prescribed routes towards 25 tombs, and four other detour paths. The uniqueness of selected tombs are enumerated. Their particularity derives from the materials used; specific engravings or statues relevant to cultural beliefs and practices, or simply because they are proximate to natural features such as a rare tree. The tombs may also be prominent because of the deceased's fame.[32] As an online resource, the *Wayfinder* configures the vast and complex cemetery into more or less, distinct categories: 'personalities (pioneers), tangible cultures (material culture), intangible histories (stories of World War Two), and landscape elements (nature)'.[33]

Prior to the *Wayfinder*, the most commonly used map of the cemetery was a black-and white A4-sized drawing of Bukit Brown. This diagram was made by Asia Paranormal Investigators (API)[34] in 2011. Then, API were one of very few who had any systematic knowledge of the neglected cemetery, and the only group who had made a cursory sketch of the hills. Their 2013 version 'DIY Walking Trail Map Bukit Brown Chinese Cemetery' shows 63 named and numbered tombs with several arrows denoting possible directions from the gate.[35] The diagram was an archive-in-the-making with the numbers indicating an evolving burial registry. It required the passage into Bukit Brown be made by instinct and memory. Four years later, this diagram became the *Wayfinder's* primary source of information.

The source document of 2013 shows Bukit Brown bounded by four roads, the main gates, and a thick, bold line branching from the gates into a series of curling routes, outlining the cemetery's four hillocks. As a diagram with few details, it worked through and relied largely on, a user's imagination of the cemetery. Its purpose was not wayfinding. Without measurement, landmarks, or contour lines, it was not a proper map. Still, a diagram delineates a spatial organisation of what is yet undefined, and makes visible what is to come, 'a map of possible worlds'.[36] Although 'graphically reductive', diagrams are instructive of how a space is structured, its organisational and relational possibilities.[37]

The 2013 API diagram represented the tombs in their literal state—buried deep within the jungle. This diagram, rather than lacking detail, was a creative act that enacted the physical conditions of the cemetery—one in which intuition, familiarity, or the willingness to stay lost, were central to successfully locating tombs and other funerary artefacts. The vagueness of the API diagram depicted the futility of imposing specific importance to particular cultural objects or natural features. The 'empty' and 'vague' API diagram was profound in its unwillingness to simplify the location of the tombs, or to enumerate any important landmarks. Indeed, to encounter Bukit Brown was not to see these landmarks, tombs, trees or birds as discrete but as completely enmeshed in the field.

This immersive quality was similarly demonstrated in the nascent observations made by two students of mine who studied the cemetery as part of their final year design thesis. The first was one of many volunteers involved in the early documentation work at the cemetery in 2012. Obfuscated by vines and vegetation, the tombs were illegible (Fig. 7.6). He was forced to negotiate the burial ground by instinct, paying perspicacious attention to the subtle changes underfoot. He had to remember transient markers as clues to the location of the tombs. He relied on the angle of light, birdsong and the direction of the wind. In time, he became acutely aware of what he had always taken for granted—the texture, and profile of the ground. When asked for a site drawing of Bukit Brown, he emphasised the provisionality of such a drawing. Paths painstakingly shorn by a scythe were as easily lost as these were made, rapidly buried in a matter of days by relentless undergrowth. Despite documentation efforts to measure and document individual graves, a proper site plan of the cemetery still does not exist today.

When the second student arrived in Bukit Brown in 2014, the cemetery's national narrative had coalesced. More background material on the site had been amassed and circulated. The cultural and natural histories provided plenty of evidence to support the case for preserving the cemetery. However, this articulate student declared he was overwhelmed by 'too many things' and chose to communicate instead through drawings. His finely made pencil sketches featured trees whose locations were obscure. In each tree portrait there was an accompanying detail, small yet notable: a tiny bird, a burial mound wrapped in vines, a broken tombstone. It became apparent that 'too many things' referred to the density of subjects he encountered on site, 'things' which in the recent categorisation of facts, had not been read in their singular adjacencies but rather identified and rapidly grouped as patterns of similarity—the architecture of tombstones, the species of birds, the types of trees. These objects fell too neatly into their proper new places, forced into recognised frames of knowledge.

The first student's experience of the cemetery was marked by the overlooked—an awareness of the terrain itself—confounding the ambitions to create a map for wayfinding and recording every tomb. The second student's account struggled with the enumeration of knowledge through a scopic regime—what political scientist and anthropologist James C. Scott calls 'seeing like the state'[38]—a forceful method which eclipsed the poignant adjacencies of intertwined subjects. Indeed, the utility of the map relies on its authoritative clarity, a quality which is at odds with the lived experiences at the cemetery. The map emphasises the quantitative and as such becomes placeless and timeless.[39] It is made, as Paul Carter argues, by being 'blind to the history of journeys',[40] and ultimately, at the cost of 'myth, poetry and literature'.[41]

(ii) White stakes, red-and-white ribbons | Drawing | Cleansing | Imagining | Occupying

By 2013, 3,901 timber stakes, painted white, stood a metre-high from within a field of knee-high *lalang* grass.[42] The stakes were visible from the main gravel path. Each denoted a tomb obscured by undergrowth. They also marked the individual presence of the 2012 documentation team—researchers, students, conservationists, historians, archaeologists and volunteer guides—who hammered these numbered stakes by hand into the ground. From the main paths, they inscribed a dotted landscape of white points and lines which made a wave-like pattern on the hilly ground.[43] With the completion of the highway in October 2018, these white points were slowly submerged beneath the elevated roadbed.

The act of staking can be likened to what architectural historian Paul Emmons describes as 'drawing in dirt', an ancient ritual of making marks on the ground to inaugurate future construction.[44] Alberti also described how these markings preceded construction, 'to mark out the line of the intended wall with a trail of powdered white earth', and became the precursor of human intervention on an empty site.[45] *Ichnographia* or ground marking included 'laying out the site by using stakes and ropes, drawing on site in dirt and plaster and walking a

[Overleaf] **7.6** Tombs are buried deep within the rainforest interior with its ancient trees.

snowy site to mark out the future building's plan'.[46] 'Drawings in the dirt' were used to introduce religious structures on a site—the coming of a holy building, or a divinity-inspired city. Dust, plaster, whitewash, gypsum, flour, ash and cotton seeds—materials used historically to draw 'white' plans in the dirt—were soaked in naphtha, and then, set alight in flames.[47] The city of Faro first drawn in flour, 'allowed diviners to interpret the future of the city based upon if birds were attracted by the edible plan … "foreshadow[ing] that the city would abound in provisions."'[48] Bukit Brown's white markers heralded the coming of a highway; one might say that, in developmental terms, transport infrastructure is but another kind of 'sacred' structure.

While the white stakes emphasised the transformation of the fallow cemetery into land for future infrastructural development, they were simultaneously 'stained' with ethnic, spiritual and metaphysical meanings, associated with a Chinese burial ground. As a site of paranormal investigation, visitors to Bukit Brown would be familiar with the use of stakes to control spirits—a 'stake [driven] through the heart' was used to exorcise the unnatural.[49] At the cemetery where the ground is laden with the metaphysical residue of deceased bodies, staking is an act of violence. It recalls the brutality of tomb exhumation where hoes, axes and mechanical diggers are employed to hack into and topple over gravestones to retrieve coffins, which would be forced opened using chainsaws.

As the white stakes were inserted into the re-consecrated burial ground, red and white ribbons festooned the upper branches of trees around the surveyed graves. Each ribbon had a small plastic card stapled to its end, containing details of the deceased's name and directional arrows pointing to the grave.[50] The cards dangled obtrusively at eye-level. The ribbons were fashioned from red-and-white hazard tape, the type used to cordon sites where there had been an accident, an emergency, or ongoing construction works. Yet, not unlike the white stakes, the hazard tape became something other in the cemetery. The red and white ribbons moved gently in the breeze. In their animated state, they resembled wounded aerial tree roots, particularly the banyan species which are found scattered all around Bukit Brown. In the twilight hours, the flapping white bits had an uncanny resemblance to pieces of white cloth, which in turn, alluded to unearthly spirits.

Finally, the ribbons denoted the bodily reaches of the cemetery's volunteer community. It was evidence of where they had walked, what they had seen, and which graves they had chosen to mark. They marked the ephemeral traces of human occupation twice over—of bodies long buried in the ground, and of the recent activities of dedicated volunteers who had returned repeatedly to the cemetery to record the gradual passing of this space. Here, the objective instruments of survey and mapping gained viscerality and imagination. Each instrument was already multiply invested; the evidence of the volunteers' discoveries, processes, experiences and passage through a largely forgotten, though still, hazardous terrain.

(iii) Burial Register | Numbering | Fixing | Surveying | Prospecting

At the National Archives of Singapore, the cemetery's burial registers are organised in yearly sequence. Originally recorded by hand, the register was subsequently digitised for public reference. It contains details such as the sections of the cemetery, grave numbers (1102…1131…1160), causes of death (most entries are left blank), date of internment, name of the deceased, gender, age at time of death, residence and former profession of the deceased (again, most entries are left blank), and the registrar's signature.[51] These records date between April 1922 and December 1972, coinciding with the cemetery's active tenure. During the mass documentation of tombs in 2011, numbers were again designated for each newly discovered tomb, and spray-painted red onto white wooden stakes. The numbers gave each newfound grave a specific location, which meant that they were also accorded a unique identity.

The numbering ascertained not just tombs for exhumation but located prominent ones which would subsequently be traceable through satellite navigation systems.[52] The revival of the burial register has significant spatial implications. The numbers accorded hierarchy, discerning 63 elaborate tombs which belonged to the upper classes.[53] With these tombs highlighted as destination points, the unwieldy space of the cemetery was rapidly compressed and controlled, with the named tombs consumed as heritage. This perceptual shrinking of space concurrently signals a diminishing complexity in the cemetery's history, and the persistent justification and circulation of numbers (not just of graves but also natural species of birds and plants) meant that surveillance, control and measurement were pressed on this previously unkempt place.[54] Numbers create legibility. Through the placement of landmark tombs, an exact delineation of the cemetery grounds to be preserved becomes probable. At the same time, this demarcation gives distinct shape to Bukit Brown as a site. A 'site' as architect and theorist Robin Dripps argues, appeals because it has a 'reassuring degree of certainty' and 'can be owned', eventually taking on qualities of the institution to which it belongs.[55]

Carter writes that the legibility of space on a map, results in the contraction of that space encouraged by 'selection, omission, simplification, classification, [and] the creation of hierarchies'.[56] The passage through the old cemetery is embedded with a 'mythic-symbolic' aspect that when denied 'destroy[s] the very domain that we [seek] to understand'.[57] This sense of passage haunts the work of surveying Bukit Brown. Despite the sophisticated methods of construction for the elevated highway, measuring devices and survey methods in the cemetery were largely analogue. The act of numbering required individual volunteers to physically venture into the field to gather, record, estimate and evaluate tombs which existed largely in fragments; semi-buried in the undergrowth. Indeed, what became significant for each volunteer was the adventure of discovering these tombs, and the passage taken through the cemetery. In fact, because of the dense jungle, the act of locating these tombs was highly predicated on chance and individual perspicacity.

The numbers themselves hold significance. Numbers found amongst graves bear an otherworldly presence, designating certain fortunes or misfortunes in Chinese culture. Lucky 'red' numbers are hunted down by regular punters in search of hidden earth-deity shrines or tombs thought to be imbued with particular generosities—enabling the applicant to secure a promotion, a job, a spouse, a family, wealth and good health.[58] Indeed, the presence of numbers in a Chinese tomb landscape is believed to mediate between the unseen and the everyday. Punters bring paper strips written with numbers, or a small die, and offerings of 'crabs, sweets and biscuits'; they implore the landscape to provide specific numbers that might promise a windfall in the lottery.[59] Red numbers are thus, informal contracts made with the netherworld.

The red spray-painted numbers of newly found tombs were not exempted from this otherworldly status. They took on the dual status of being precise (giving a unique identity to a tomb) and excessive (portending luck or misfortune). At Bukit Brown, numbers do not merely function as instruments of order but harbour the spectral. They are embellished with cadastral and geomantic aspects of land, thus complicating the neutral physical dimensions of property with the mythic-symbolic connotations of a burial ground.

Field Notes 2: *natural markers*

(i) Earth Pits + Jute Mats | Burial Grounds | Holes | Exhumations | Presence and Absence

Post-exhumation, the complexion of Bukit Brown changed dramatically (Fig. 7.7). Lush greenery was transformed into greys and browns of broken stone tablets, dismantled tomb ornaments and jute mats. Coarsely woven jute mats were thrown over emptied burial pits. At the end of 2013, there were some 304 jute-covered pits.[60] In the tropics, jute mesh is temporarily laid over exposed

[Opposite] **7.7** Post-exhumation prayers.

earthworks in construction sites to control soil erosion during the rainy season. The presence of jute mats often points to a transition stage for the ground they overlay. At a construction site, jute mats are laid when the ground has been disrupted by building activity; they are only removed when a new architectural platform is put in place. In a garden, the jute mat used for weed control disintegrates into the ground when new plants thrive from beneath. Used to prevent erosion at Bukit Brown, these jute coverings were also inevitably markers of an in-between time-space, making visible the interval when the graves were forcefully emptied.

In traditional Confucian funerary practice, mourners put on primitive-looking jute garments, usually a vest with shroud. The mourners' jute garments constitute a symbolic shedding of identity, in reverence of the deceased. The shrouded entourage make themselves visible to all by purposefully walking a distance with the cortege at the start of the funerary procession. Upon arrival at the cemetery gates, the mourners again dismount from their vehicles to walk towards the burial plot. The marking of the deceased's final passage, here by walking mourners shrouded in jute garments, imbibes the commonly used construction and landscape mat with the human dimensions of death and mourning. In the weeks and months of exhumations leading up to the Bukit Brown's actual demolition, the jute-covered pits presented this duality of future construction and absent bodies. The jute-covered pits further reinstated, if only momentarily, the significance and materiality of the burial ground to Chinese burial rites. Being culturally ingrained, critical performances of burial grounds are exemplified in two defining artistic pieces—*Moving House* (1991), a documentary about exhumation and national development by Singaporean filmmaker Tan Pin Pin, and *The coffin is too big for the hole* (1984), a satirical play about a Chinese burial ceremony gone awry by Singaporean playwright and author Kuo Pao Kun. Both works spatialise the Chinese burial ground as having exceptional vitality, performativity and cultural significance.

In her award-winning debut *Moving House* (1991), Tan revisits the questions of making space for national development in an exhumation ritual undertaken by the Chew family. It follows the family's reactions in the aftermath of a visually violent exhumation segment. The latent significance of the burial ground is enacted in the first ten minutes of the film's opening sequence, where rites are performed calmly and reverently upon the grave. Almost without notice, the gravediggers brutally hack into the unyielding coffins with noisy electrical chainsaws. At this point, the stench of the ground and its contents are visually manifest. Subsequently, bits of human skulls, bones and disintegrated graveclothes are nonchalantly tossed out of the burial plot. The narrator reminds: 'The Singapore government requires this land for further national development. The Chews' parents have to move'. Tan's film juxtaposes the very visceral moment of exhumation with the developmental narrative, supporting this with archival footage of the nation's earliest public housing estates. The most powerful scene resides in this first sequence, albeit in the making of a hole for the purposes of exhumation.

Kuo's allegorical monologue *The Coffin is too big for the hole* (1984) also revolves around a hole in the burial ground.[61] A grandson to the deceased recalls a recurring nightmare. His wealthy grandfather reserved for his own burial a hefty coffin which requires sixteen men to hoist. The young man realises that this custom-made coffin is too large for the standard burial plot, a metaphor for the collision between Chinese burial rites and standard provisions made for modern burial. The grandson is asked to consider several choices including changing the coffin for an approved smaller version, moving the corpse to another cemetery, sawing away the excess parts of the ornate coffin, or burying the deceased in bamboo sheets. Through its deadpan dialogue, *The Coffin* reveals the contradictions between family and nation state. This separation is manifested in the clash between age-old customs and developmental progress. Each side is differentiated by their reading of the burial ground. For the young man and his family, the hole in the ground is an extension of their ancestral space. For the state, the hole in the ground is a set of measurements for a standard coffin.

The understanding of a Chinese burial ground is antithetical to land use principles. A carefully chosen burial site is guided by ancient and auspicious geomantic, or *feng-shui*, principles and signifies a ritualised space or what geographer Robert Sack calls 'mythical-magical space'[62] wherein space and substance are understood as conceptually one in a landscape.[63] This means that it is not possible to separate 'space' ('the location, orientation and physical space occupied by an ancestral grave') from the notion of 'burial' ('the "contents" or meanings that such a space held for its descendants').[64] Intricately bound with ancestral worship, a peaceful and propitious location for the dead ensures that the living will be rewarded with prosperity, longevity and progeny who will successfully sustain the family line. The burial site is not just a symbolic resting space for the dead. In Chinese custom, it is venerated and maintained by descendants as part of a contract between the dead and the living.

Feng-shui (literally meaning 'wind and water') follows a system of Chinese cosmological thought which invests the landscape with a mythical force (Fig. 7.8). The Chinese believe that '[w]hen the winds blow harmoniously and the rains come down regularly, the realm shall flourish and the people live in peace and comfort'.[65] Man's physical and psychological well-being are held in direct correlation with his immediate physical environment since '*feng-shui* unifies the geological, atmospheric, aesthetic and psychological qualities of the environment in one theory and code of practice'.[66] Drawing upon a kinship between the ancestral dead and the living descendants, *feng-shui* is engaged to seek specific landscape configurations for a peaceful burial. The most ideal sites enable unobstructed flow of winds, rains and watercourses. Further, the geomantic

7.8 *Feng Shui* follows a system of Chinese cosmological thought, investing the landscape with a mythical force.

landscape may be read as a series of dynamic animal structures, understood not only for their outer forms but also alluding to their active qualities. The four sides of a grave would be flanked by a tiger's steep sloping ground with a 'male' quality on its left, a dragon's gentle, undulating 'female' ground on its right, a tortoise's slow serpentine stream at its rear, and a bird's vantage point open to breezes at the front.[67] The Chinese migrants held this burialscape sacrosanct. Indeed, the Chinese, as historian Oswald Spengler argues, saw the landscape itself as the very substance of any architecture—whether of grave or building—and so, 'it is the architecture of the landscape itself, and only that' which mattered.[68]

Aligned with these geomantic principles, the Confucian death rites[69] dramatised by Kuo and Tan are deeply ceremonial and sophisticated. In comparison to the Buddhists who cremate their dead in the belief that the latter might reach paradise more quickly, Confucian funeral rites use burial as its primary mode of body disposal. Confucians believe that the dead body has to be returned, in its intact state, to their ancestors. A well-appointed burial plot and its well-preserved grave are expressions of filial piety. These attributes safeguard 'the well-being of both the dead and the living'.[70] The overlapping significance of jute cloths and burial pits—simultaneously present in traditional mourning rites as well as in the processes of exhumation and construction sites—complicate their perceived uses and meanings. The jute cloths and pits speak concurrently of absence and presence; rites and efficiency. They are affective residues which linger in an unsettled landscape.

7.9 Yellow joss strips and burnt paper offerings strewn on a footpath.

(ii) Paths | *Qingming* and Ritual Landscapes | Paths of Consensual Making | Path-Making as Social-Spatial Practice

Bukit Brown is a space of remnants (Fig. 7.9). Footfall is detected in an assortment of prints: in trodden, dried grass; in the slight but discernible parting of long grass trails, and in burnt incense sticks planted carelessly at the edges of walking trails where at times, the scent of sandalwood still lingers. There are also patches of charred earth and small piles of white ash, the remains of burnt paper offerings or 'spirit money' to appease ancestors and ghosts.[71] Rotting fruit and grey mouldy peel from food offerings in turn attract flies and rodents. Here and there, a crushed plastic water bottle, and balls of dehydrated wet wipes divulge a previous gathering. A scattering of yellow joss paper strips, one of the 'protean' forms of 'ghost bills' used as offerings, frequently litter footpaths.[72] These remnants are especially augmented during *Qingming*, a time when paths both designed and spontaneous gain visibility around the cemetery. Traditionally, a large congregation at the ancestral tomb is a testament to generational well-being, abundant progeny, reunion and celebration. Rising before dawn, relatives bringing food and paper offerings, are also armed with torch lights and scythes to hack through the undergrowth. Tomb sweeping demands a bodily engagement with the cemetery's undulating and overgrown terrain. This ritual is intimately shaped by the lie of the land, such that the acts of walking through the cemetery and creating these evanescent paths are central to the cemetery's continued occupation, as they are to the sustenance of the annual pilgrimage.

Celebrated in the month of April, the tomb sweeping festival or *Qingming* coincides with the monsoonal wet season. Wet earth, wet grass and wet paper offerings are invoked in a Chinese poem by the Tang poet Du Mu, familiar to a generation of Singaporean elementary school students in the 1990s. Du Mu's poem '*Qingming*' starts with the line '清明时节雨纷纷—a drizzling rain falls like tears during the *Qingming* festival'.[73] The poet speaks of melancholic wet weather traditionally expected during the monsoon season. In the poem, a grandmother carefully makes her way up the dirt tracks to her ancestor's tombs during *Qingming*. She warns her grandchildren to be mindful of where they tread: 'Don't step onto the paper offerings, don't slip on the wet grass'. After the rainy season, groundcover grows quickly, concealing these spontaneously made paths.

Landscape writer Robert Macfarlane describes informal footpaths, which must be distinguished from planned networks, to be in fact not all 'plain', or 'mundane'.[74] Rather, he extols these paths as 'a labyrinth of liberty, a slender network of common land that still threads through our aggressively privatised world of barbed wire and gates, CCTV cameras and "No Trespassing" signs'.[75] Such at-the-moment paths signal a freedom beyond ordered urban mobility. Paths are not merely instrumental for travel; they are collectively and opportunistically made, re-made and unmade by individuals or communities— they denote a spatial-social practice. The *Qingming* festival compels the urban dweller to make their own paths and obliges them to retrace previous journeys which they and others have taken over the same terrain. Walking across such a landscape is a deeply embodied and social process, as is the making of such transient paths:

> Paths are the habits of a landscape. They are acts of consensual making. It's hard to create a path on your own. … Paths connect. This is their first duty and their chief reason for being. They relate places in a literal sense, and by extension they relate people. Paths are consensual, too, because without common care and common practice they disappear: overgrown by vegetation, ploughed up or built over (though they may persist in the memorious substance of land law). Like sea channels that require regular dredging to stay open, paths *need* walking.[76]

Bukit Brown gained meaning through and for those who walked, searched, worshipped, cleaned and navigated its terrain. The repeated erasure of routes and the necessary clearing and making of annual paths, renews and sustains the cemetery whose natural but equally collaborative rhythms of opening and closing depict a cycle where landscape is 'made and noticed, acted or lived or thought with some degree of consciousness'.[77] The transient paths to the graves are fragile. They exist only through 'common care and common practice', without which they disappear.[78] Writer and activist Rebecca Solnit vividly describes implications of walking in the vestiges of other walkers, describing a path as a '*shul*', the Tibetan word for track, that brings an 'impression of something that used to be there'.[79] The *shul* is a 'mark that remains after that which it has passed by—a footprint, for example. … [it] is used to describe the scarred hollow in the ground where a house once stood, the channel worn through rock where a river runs its flood, the indentation in the grass where an animal slept last night'.[80] Family grave visits depend on a sustained generational memory of these momentary impressions on the ground:

> [An] uncle who died nearly five years ago was the only guy who knew how to navigate through the whole labyrinth of graves … It was like, north of the tallest tree if you go straight on the road, and then travel 30 steps till you come to a big trunk and then turn left. But, of course, it never works out because there was like, 'Oh no! There are 50 trees in this place!' [The graves] were not numbered and it was impossible to find unless you really remembered that place. So there were years when we lost it completely. So we went there and couldn't find it, and go back quite dejected.[81]

Solnit remarks that paths, and the landscapes they traverse, are also constantly made in the collective imagination. Within the family, the *Qingming* journey is a rewriting of the *shul* taken by each generation, with the help of sticks, scythes, torches, landmark trees and by mirroring the learnt footsteps of an elder, grandparent, or parent. The route is learnt by rote—paths in the landscape are practiced yearly, with the route lodged in familial and social memory until the practice dies with the last family member returning to sweep the tombs. The paths created during each *Qingming* festival align with Macfarlane's dual definition of paths as a space to traverse as well as a channel to communicate—'a route for communion as well as means of motion'—old paths connect with historical time and space.[82]

As the *Qingming* tradition fades away in Bukit Brown, another journey emerges in its place. Since 2011, the civil society interest group *All Things Bukit Brown* created a series of guided walks around the cemetery to raise awareness of its existence (Fig. 7.10). As of September 2020, over 862 guided walks have been conducted, involving more than 22,000 participants.[83] A guided walk typically lasts two hours. Participants are asked to wear sensible shoes and suitable attire, carry a water bottle and apply insect repellent. The ground is naturally uneven and frequently slippery. Beyond the history and trivia that the volunteer guides enthusiastically dispense, it is the act of walking *together*, through the rugged landscape, which matters. Walking in unison across Bukit Brown enacts 'the rare and magical possibility of a kind of populist communion', Solnit describes, encouraged by 'moving through the streets with people who share one's beliefs'.[84] The popularity of the Bukit Brown walks signal a first in Singapore

7.10 As of September 2020, over 862 guided walks have been conducted.

especially where walking a cemetery—taboo for visitations apart from specific occasions where the dead are ritually venerated—has become a social, cultural, and for some, a politicised act.

Much has been written about how walking across unfamiliar terrain might trigger a heightened consciousness of time and space, rich in juxtaposition, incongruence and difference.[85] This sensual engagement is largely alien to the city dweller whose body is over-conditioned by the urban grid. Many know only unimpeded and purposeful movements, smoothened by 'regulatory procedures, planning, cultural conventions and values, and spatial divisions'.[86] The urban dweller living in the grid identifies spatial engagement with efficient blocks of work-or-rest. The 'unchoreographed space' of the cemetery, however, offers a different degree of 'affordance', exerting distinctive potentials—physical, material, psychical and imaginative—on its users.[87]

To walk Bukit Brown is essentially to track an unknown terrain. Tracking involves taking 'steps in the footsteps of the past, [and] ... deriving information from the depth and direction of ... marks' on the trail.[88] The practice of tracking—looking for old marks as well as leaving new ones[89]—is an act of making space. 'Tracking' is also etymologically related to 'tract', a term denoting a measurement of land, particularly a large, bounded area with a designated purpose. Carter expands the meanings to 'contract',[90] linking the active tracing and remaking (walk/track) of space (tract), to arrive at a text (contract) which binds action and space. Indeed, the etymological relationship between these three terms connect authoritative textual knowledge in a contract to the uncertain and open-ended tracking across space. Yet, the cleaving apart of authoritative knowledge from open-ended practised knowledge persists.

In John Wylie's research of landscapes performed through walking, he argues that the methodological impasse surrounding such scholarship comes from separating practised, interpretive knowledge gathered through walking, against knowledge derived from 'archival, archaeological or ethnographic fieldwork'.[91] A walk through the century-old cemetery reveals an understanding of Bukit Brown that has the dual status of an object and an experience; the former allowing analysis at a distance, from the exterior; the latter holding us in close proximity, deep inside its interiors. The fascination and difficulty of thinking/writing/conceptualising Bukit Brown is to decide where and how to situate practised knowledge within the conceptual debate. Yet, the topographic uncertainty and physical effort of the Bukit Brown walk lift it from the status of a mundane activity to a self-conscious performance that deserves interrogation.[92] Indeed, it is this performative body—moving across the cemetery and making marks on the landscape—who exceeds the outcomes of academic knowledge, archival and fieldwork. Walking the cemetery produces visceral knowledge attuned to a disappearing cultural landmark. Philosopher Paul Virilio differentiates the living body as 'what is living, present, conscious, here, ... only because there is an infinity of little deaths, little accidents, little breaks, little cuts'.[93] 'Consciousness' unfolds at points in the walk where uncertainties are encountered—the negotiation of the cemetery's tough physical terrain is symmetrical with an acute awareness of the site's vulnerability. Walking the cemetery sensitises, and politicises, the walker.

A space of public care

Photographs of Bukit Brown in numerous private blogs and websites are *memento mori*, pre-emptively lamenting the cemetery's extinction. By drawing out an emergent location at the intersections of the shared and the personal, these photographs speak of an 'inter-relational'[94] space that hold both individual and collective meanings. They also describe the site's ambulant epistemology, where knowledge formation is organic, instinctive and transient—gathered primarily through walking the cemetery—and as such, deeply subjective and agent-dependent. These photographs delineate an 'unintentional landscape' whose aesthetic values spontaneously derive from interest in the cemetery's cultural and natural qualities (Fig. 7.11).[95]

The spontaneously constructed online-Bukit Brown is the cemetery's spectral double, its virtual twin. Its existence in various blogs and websites mark not just the cemetery's significance but denote a far-reaching and relational network of guardians and stakeholders, located in Singapore and beyond. The network's

7.11 A wild cemetery in the nation's cultural consciousness.

relationship to the cemetery resembles what architectural philosopher Peg Rawes calls an 'architectural ecology of care'. Rawes advances a reframing of the term 'ecology', moving this beyond architecture's object-based utopian thinking of technocratic sustainability, which has in recent years, been bolstered by large-scale, material and infrastructural technologies.[96] Instead, she proposes a gendered and politicised understanding of nature's alterity enabled through longer-lasting modes of social, material and ethical relations to a space. Rawes emphasises relationality for ecological balance, '... relational architectural ecologies are the poetic, political, social and psychic relationships through which modern subjectivity, and our habits, habitats and modes of inhabitations, are co-constituted'.[97] Describing artist Agnes Denes' project *Wheatfield* (1982)—four persons planting and harvesting a field of wheat on Manhattan's landfill site of Battery Park—as exemplary of the 'complex relationship between environmental resources, urbanism, architecture and spatial practice',[98] Rawes argues for a relational constellation of on-the-ground actions caring for a space. The landfill site, the field of wheat, the participants and Denes are held together by the actions of clearing, planting, tending, growing and harvesting. This process-based relationship, or what Rawes calls 'natural geometry', prioritises 'human and non-human relations of care'.[99]

Walking Bukit Brown approximates an architectural ecology of care. At the cemetery, walking is relevant as a persistent act of discovery, a mode of everyday occupation, a religious ritual, and/or a form of silent protest. These walkers—the Brownies, the descendants, the newly inducted cemetery walkers—are its informal custodians, forging intimate knowledge and memory work of the landscape. For many, it became clear that the act of walking a relatively untouched site, coalesced affective experience, cultural and ecological stewardship, and a ground-up politics of belonging. The walker has a tactile relation to the ground—puddles from the previous night's rain deep enough to sprain an ankle, the risk of a tomb buried underfoot, the whirring sound of mosquitoes feeding at earshot, and the imprint of another's foot on freshly trodden grass. Walking confers an ambulant perspective of space and constitutes 'both story gathering and storytelling'.[100] Each Bukit Brown walker potentially takes on the dual roles of agent and activist, having the capacities to discover their own relationship to this landscape, and to invigorate its multiple meanings.

By the time I arrived at Bukit Brown, the cemetery had already anticipated its end (Fig. 7.12). The bulldozers had moved in; one of its hillocks had disappeared; the raintrees lining the road to the cemetery's gates had vanished; its grand iron gates were already replaced with construction hoarding. Every walk, photograph, ritual, exhumation, prayer and offering had been taken in homage of its eventual destruction. These actions were nostalgic. They mourned the loss of a cemetery whose aura had already been destroyed the moment the highway was announced. This field guide is thus, a document of situated affective knowledge. It reflects the multiple forces that shape the cemetery, and reflects my position as entangled with my own encounter of such a space, attempting to catch the webs of emotion, imagination and movement, specifically the sensations of being lost and being displaced, of precarity and belatedness. The messy coming together of different and sometimes mismatched threads is central to an architectural scholarship that embraces affect. The cemetery's sensual but architecturally inconsequential details were compelling but also opaque to architectural discourse in a reversal of the burial 'field' and architectural 'figure'. Throughout these uncertainties, people kept walking Bukit Brown. Walking and gathering knowledge; telling stories and making new spaces for friendship; cultivating care and alliance from these stories. The knowledge offered here derives from these ambulant and affective configurations.•

[Overleaf] **7.12** The cemetery's 1920s cast-iron gates, and the highway beyond them.

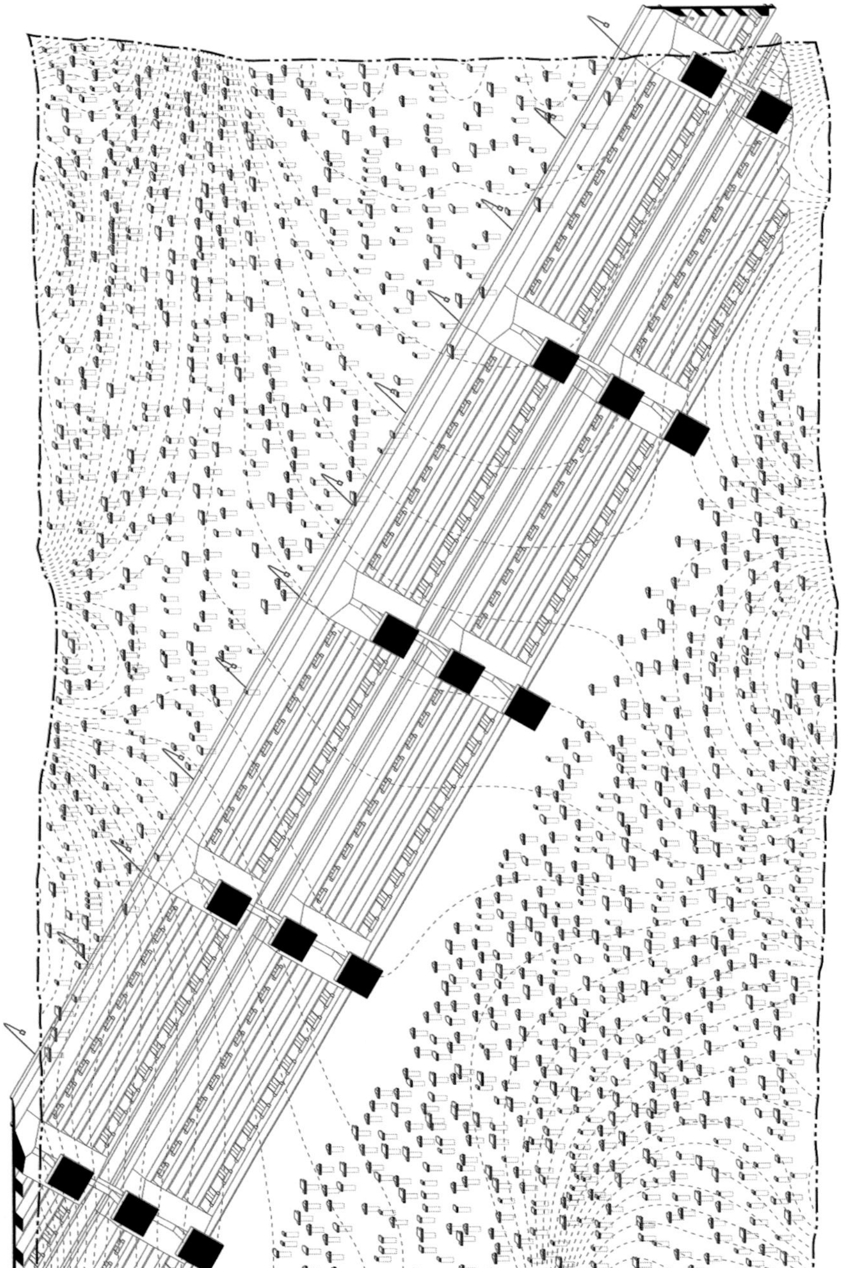

VAULT
DEAD
FLOOD

8 • Holes in the Ground

Lilian Chee and Wong Zi Hao, 2020.

Holes in the Ground examines physical and metaphysical elements in the exhumation of Chinese graves. Two-and-a-half dimensional drawing-models replicate the Chinese burial ground's habitual inaccessibility and the cemetery's undulating terrain. There is significant violence during exhumation. So here too, architectural plans are burrowed into; topographical sections are pulled apart.

The textual fragments layered onto the drawing-models are excerpts recounting taboos, superstitions and spiritual encounters. These were extracted from documentaries, tabloid newspapers, online cemetery-trekking hobbyist blogs, and paranormal adventure websites. The text-image intersection narrates a double landscape in tension: a cemetery passively awaiting demolition, and an active spectral landscape where reparations must be made in the face of such damage. The spectral landscape is mired in what cannot be seen but sensed: the forced opening of a timber coffin is followed by the sound of thunder; a plastic pail may save one's misplaced fortunes when used to bail water from a flooded pit.

The six drawing-models focus on key material objects and mishaps during exhumation. They are to be read in threes—*Tarpaulin, Rubble* and *Wood stack* from top to bottom as plans; *Vault, Dead End* and *Flood* are vertical sectional cuts into the burial ground. The sequential reading of these drawing-models mirrors the progression of cuts. Their thickness—individually each image was assembled by layering together more than 20 sheets of drawings—gestures to the myths and mysteries buried deep in a jungle cemetery.

Holes in the Ground is an affective architecture of a burial ground undone.•

TARPAULIN 425x302x145mm, watercolour, inkjet print and pencil on sketch paper, cardboard, writing ink and pencil on note paper, cotton string, steel wire, bamboo sticks, found twigs.

The tarpaulin sheet is installed for the dead. Its shadow obscures the view of the excavated grave. Taoists believe that light should not shine on the remains of one's deceased ancestors. Over the years, trees have grown around, and onto the graves. Later, trees become anchors upon which to fasten the tarpaulin sheet.[1]

[1] Rojak Librarian, 'Exhumations in Bukit Brown', *Rojak Librarian* (blog), July 8, 2014, https://www.mymindisrojak.blogspot.com/2014/07/exhumations-in-bukit-brown.html.

RUBBLE 425x302x123mm, watercolour, inkjet print and pencil on sketch paper, cardboard, writing ink and pencil on note paper, cotton string, steel wire, cement and plaster fragments, found twigs.

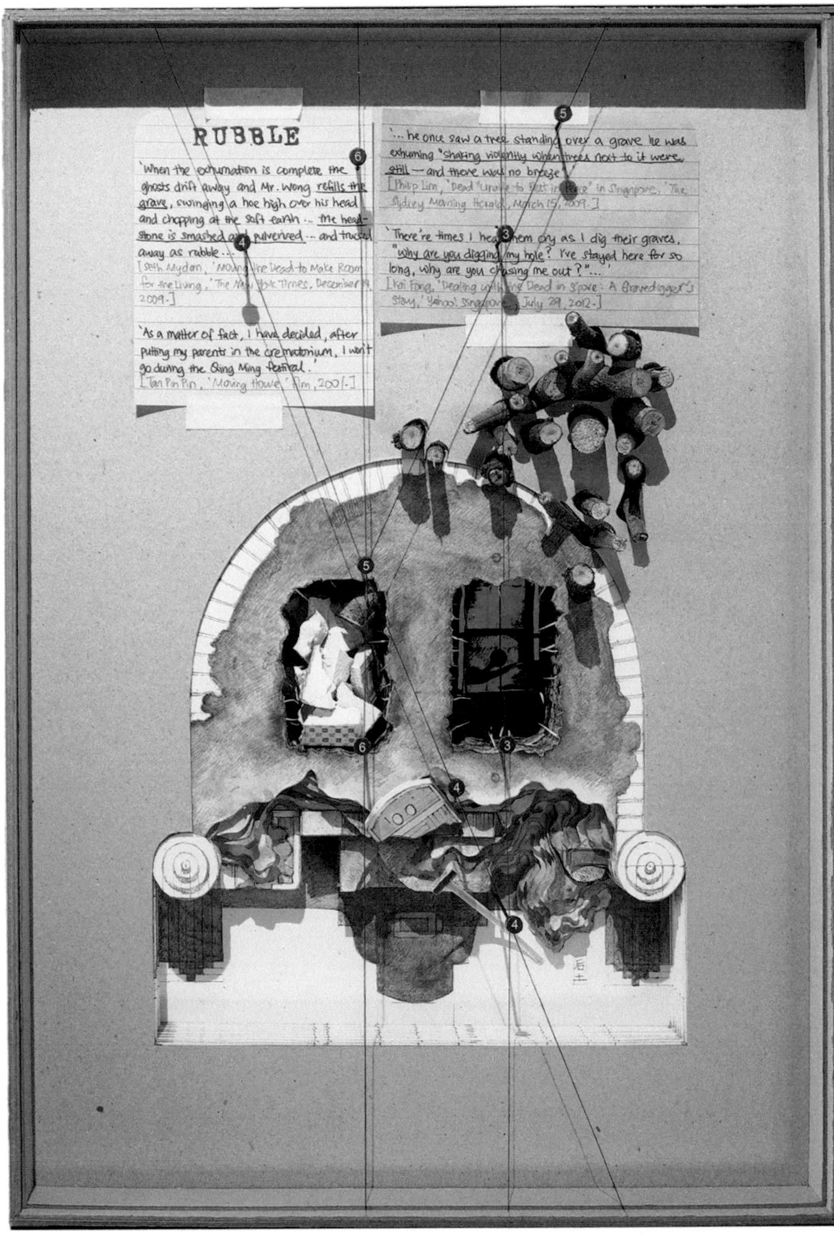

Hammers, hoes, plowshares, crowbars and chainsaws are used to open the tombstone. The coffin is forcibly undone. The aftermath of exhumation resembles a demolition site. Stone, plaster and concrete debris are mixed with soil and human remains. For the Taoists who believe that the tomb is the physical link between the living and the dead, such violence is sacrilegious.[2]

[2] Philip Lim, 'Dead "Unable to Rest in Peace" in Singapore', *The Sydney Morning Herald*, March 15, 2009, https://www.smh.com.au/world/dead-unable-to-rest-in-peace-in-singapore-20090315-8yuc.html.

WOOD

425x302x45mm, watercolour, inkjet print and pencil on sketch paper, cardboard, writing ink and pencil on note paper, cotton string, steel wire.

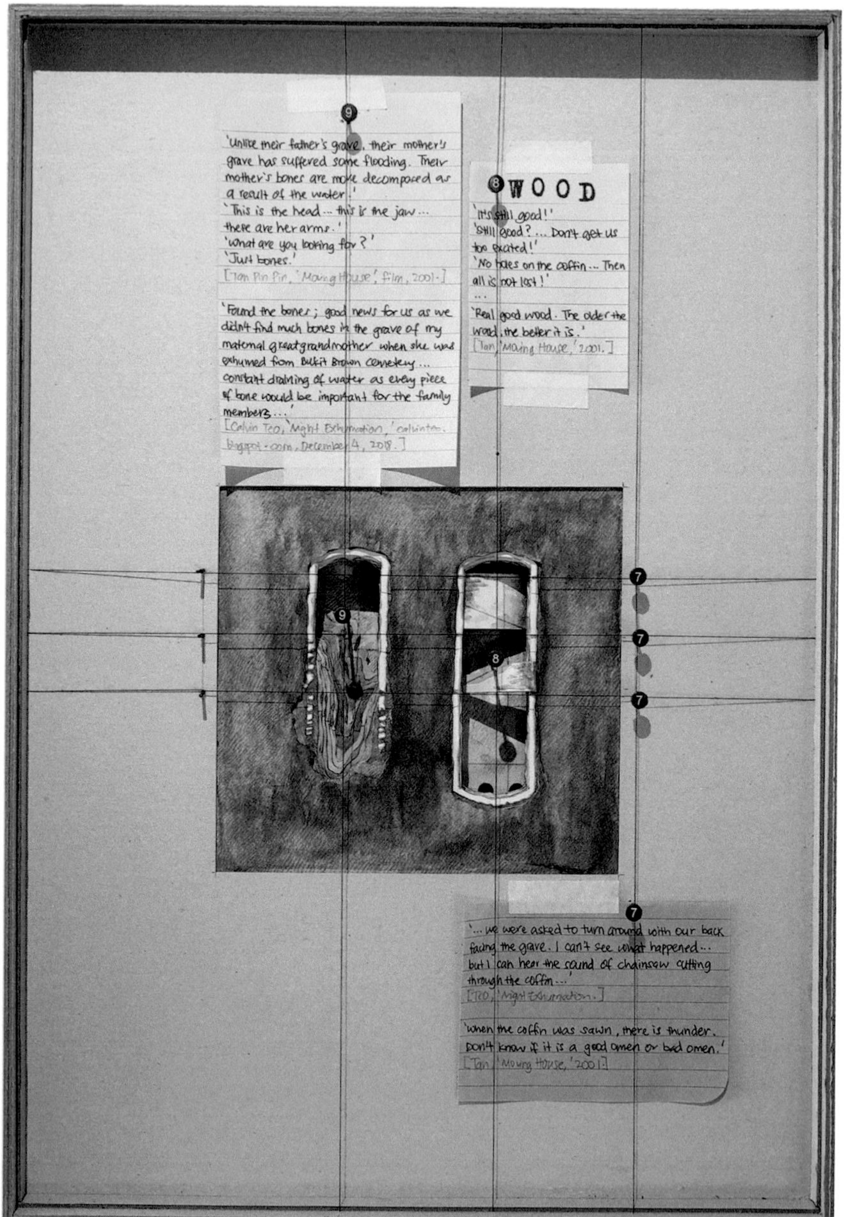

The gravediggers' tools are primitive, often using only their bare hands. This is recognized as the best way to feel around for delicate bones. Their task is near impossible. It is also filthy especially when the wooden coffin is waterlogged and filled with mud. Ultimately, the quality of the coffin matters. 'No holes on the coffin… Then all is not lost!'[3]

[3] Tan Pin Pin, *Moving House* (2001), 16mm, Documentary, 22:26 (Discovery Channel International, 2001), https://www.tanpinpin.com/about/moving-house/.

VAULT

425x302x90mm, watercolour, inkjet print and pencil on sketch paper, cardboard, writing ink and pencil on note paper, cotton string, steel wire, cement and plaster fragments, bamboo stick.

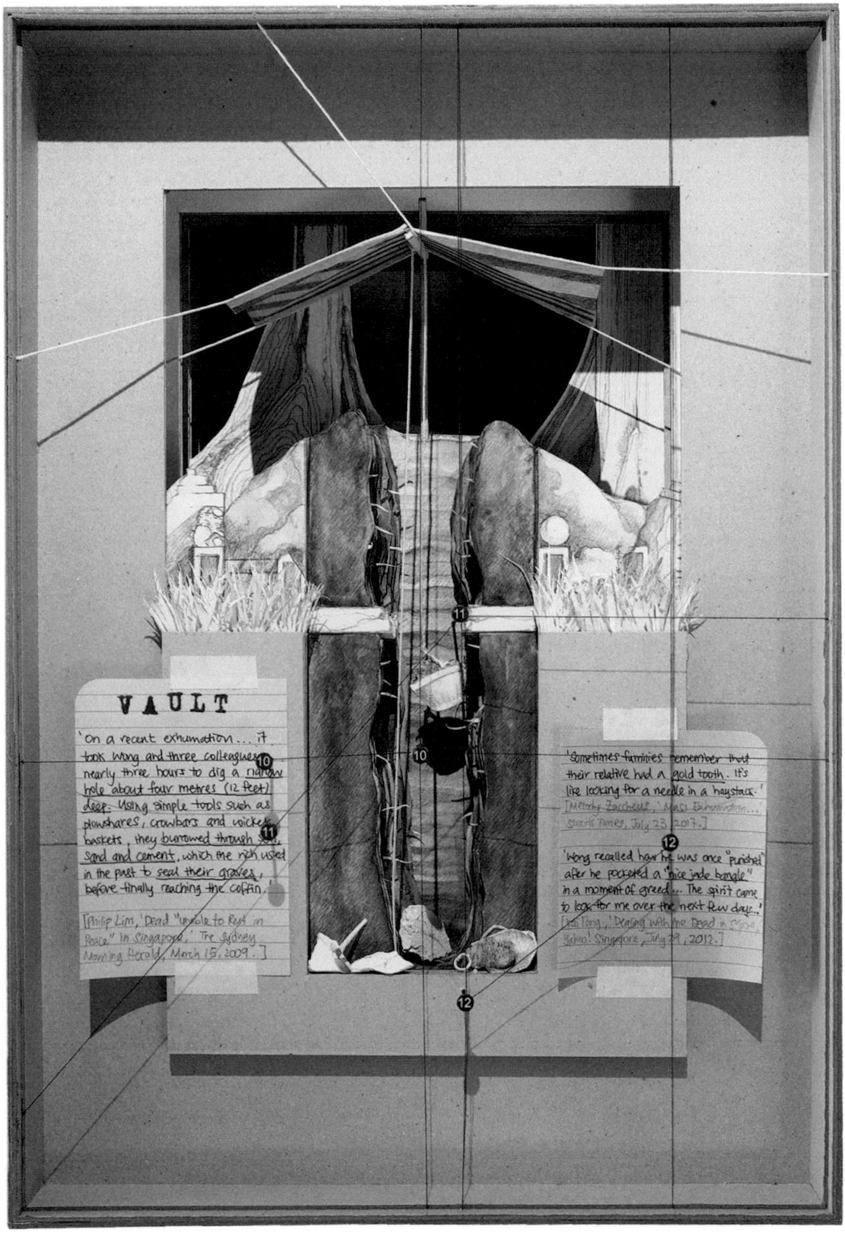

The model-drawing shows a four-metre-deep shaft. The depth of the pit signals a wealthy or influential occupant. The tomb was securely sealed and buried deep to keep its owner and his/her earthly treasures safe to prevent looting. A gravedigger attempted to extract several gold teeth from a skull but failed. Later, his infant daughter had a bad fall and her skull 'cracked in an eerily similar manner'.[4]

[4] Fong Kai, 'Dealing with the Dead in S'pore: A Gravedigger's Story', *Yahoo! News*, July 29, 2012, https://www.sg.news.yahoo.com/dealing-with-the-dead-in-s%E2%80%99pore--a-gravedigger%E2%80%99s-story-20120729.html.html.

DEAD END 425x302x92mm, watercolour, inkjet print and pencil on sketch paper, cardboard, writing ink and pencil on note paper, cotton string, steel wire, found twigs.

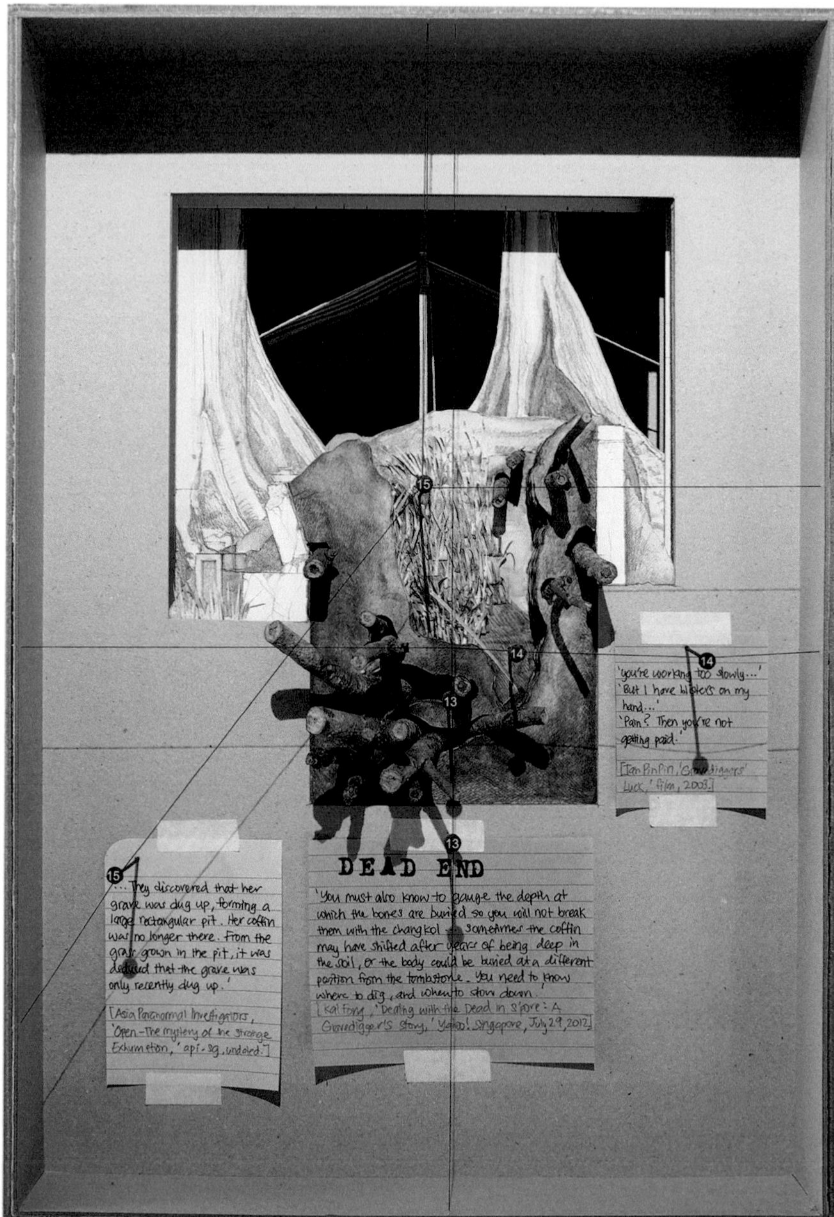

The burial ground is teeming with life; the coffin is caught in a tangle of tree roots, insects' nests, and organic rotting matter. After many years, the coffin may have shifted from under the tombstone. The gravedigger must gauge the depth of the pit so as not to cause more damage. Sometimes, after hours of back-breaking work, an empty pit is discovered.[5]

[5] Asia Paranormal Investigators, 'Open – The Mystery of the Strange Exhumation', accessed May 6, 2020, http://www.api.sg/open-the-mystery-of-the-strange-exhumation/.

FLOOD

425x302x50mm, watercolour, inkjet prints and pencil on sketch paper, cardboard, writing ink and pencil on note paper, cotton string, steel wire.

A flooded grave is the outcome of monsoon rains in Singapore. Stagnant water is taboo in Chinese geomancy. Floating debris consisting of wood and bone, plant matter and putrid water must be bailed out promptly. Done manually with the help of a pail and a sieve, the stench lingers in the air long after the job is completed.[6]

[6] Cavin Teo, 'Night Exhumation For My Maternal Grandfather's Grave (Private Arrangement) @ Lim Chu Kang Cemetery [Singapore]', *Entree Kibbles: Cavinteo.blogspot.com* (blog), December 4, 2018, https://www.cavinteo.blogspot.com/2018/12/exhumation-chinese-grave-night-lim-chu-kang-cemetery-private-arrangement.html.

SAJAHAT
SAJAHAT

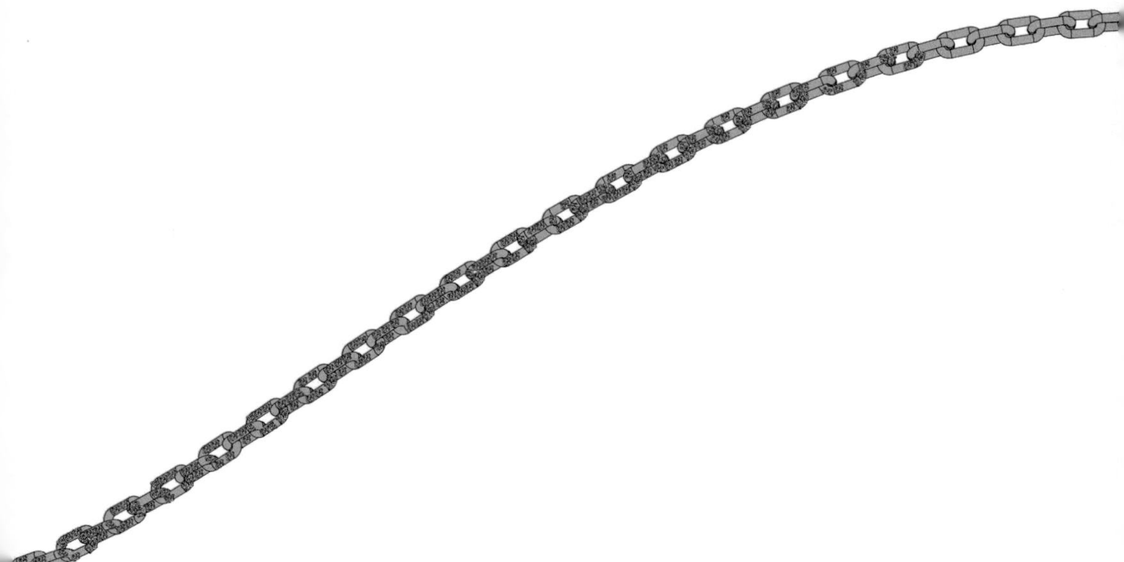

9 • The Sea, and the Sea

Infrastructure and the dialectical image

SINGAPORE'S COASTAL PERIMETER TEEMS WITH PRODUCTIVE, TERRITORIAL AND LUCRATIVE INFRASTRUCTURE: concrete sea walls, waterfront highways, container ports and terminals, industrial refineries and petrochemical tanks, garbage plants, military forests and life-firing reserves, waterfront gated properties, private yacht clubs and marinas, country clubs and golf courses.[1] Consequently, the sea recedes further from common reach (Fig. 9.1).[2] Meanwhile, the burgeoning sea/coast/island-as-infrastructure constitutes a thriving network of cutting-edge maritime and water technologies populated by canals, barrages, tunnels, monsoon storm drains, and underground caverns. Such infrastructure 'sustain the flows, connections and metabolisms'[3] of the island to create a techno-natural edge which blurs the distinctions between nature and artifice. The islanders' imagined sea—local, common, free, limitless—with its seafaring history and working-class fishing communities long extinct, is crowded with global, profitable and adaptive maritime technologies and real estate. Yet, the sea is still perceived as 'arcane'—there are definitive prehistories of sea and coast,[4] or the sea is seen as a specialist entity 'circumscribed by technical expertise, applied science and positivist impulses'.[5]

9.1 Sunseekers in an artificial beach lagoon against a backdrop of private yachts, container ships, and offshore rigs.

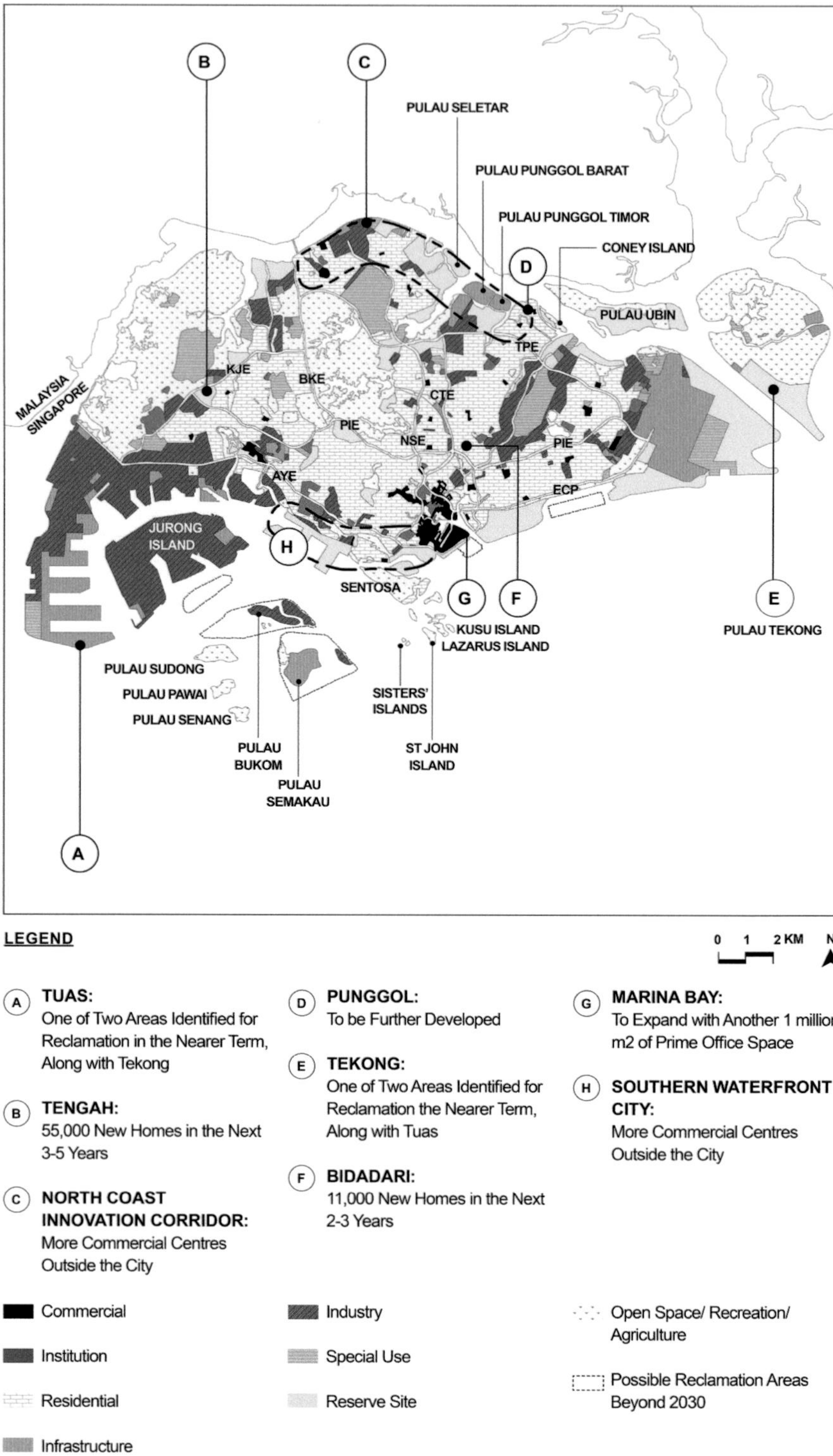

9.2 Future land use, Singapore, 2016.

The sea is polemical; it divides perception and opinion. Art historian Rosalind Krauss argues that the sea, by virtue of its isolation, self-enclosure and limitless plenitude, becomes 'a special kind of medium for modernism' that might be flattened into, 'the no-space of sensory deprivation'.[6] In a world without oceans, Michel Foucault warns that 'dreams dry up, espionage takes the place of adventure, and the police take the place of pirates'.[7] Krauss' cool, detached and isolated modernist sea is antithetical to Foucault's romantic universe of seas, boats, adventures and pirates. The physically demanding and mortally precarious realm of piracy versus the optically dominated modes of surveillance and counter-surveillance. Contradictions abound here: proximity and distance; intimacy and aloofness; the modernist sea and the swashbuckling sea. The sea encourages reinvention.

These contradictions reappear in the persistent memory of the coast. In anthropologist Michael Taussig's allegorical essay about beaches, he reminisces about his childhood suburban Sydney. Taussig draws on two other experiences recorded in literature—poet Goethe's first visit to Venice, and American poet Sylvia Plath's early childhood memories living by the sea. Goethe wrote how, when he finally saw the sea and walked on the beach, he felt like a child greedily picking up 'a good many' seashells on the beach that had a 'beautiful threshing floor of the sand' left by the ebbing waves.[8] 'My childhood landscape was not land,' so begins Plath's memoir, 'but the end of the land—the cold, salt, running hills of the Atlantic'.[9] Learning to swim by instinct, feasting on oceanic bounty, beachcombing for 'treasures' washed ashore, and watching a hurricane arrive at midnight, Plath's sea was mythic and palpable; unknown and substantial. After her father's death when Plath turned nine, the family moved away from the seaside house. Her seaside childhood quickly became ossified, it 'stiffen(ed) … (and was) sealed … off like a ship in a bottle—beautiful, inaccessible, obsolete, a fine, white flying myth'.[10] Taussig's working class beach was a little scruffy on its edges, housing a mostly poor, migrant population, who lived 'close to the oysters clinging to the orange-faced sandstone', right by the water's edge.[11] They fished, built boats and their own houses, swam in the sea they lived by, and made use of the water. Upon his return to Sydney decades later, Taussig observed that the seaside suburbs had become unaffordable. Wealthy residents who were disengaged from the coast do not notice the Vietnamese migrant families who fish through the night at their jetties. The residents and the migrants occupy two opposing ends of the sea's economic spectrum, as the sea's history becomes inextricable from an economic one. This is exemplified in the contemporary co-option of the sea as a lucrative urban edge to be occupied, extracted and branded.

Surrounded by sea, Singapore is pelagic. It is the biggest of sixty-odd islands which are mostly located on its southern coast. Approximately ten of these islands have geologically vanished; either merged to become a larger land mass

or sacrificed for the greater good of land reclamation on the main island (Fig 9.2). Its economic ambitions have justified continual land reclamation; the island has grown from 581.5km^2 (1960) to 724.2km^2 (2018). In turn, the sea has been partitioned to take on infrastructural capacities on top of becoming real estate. Parts of the sea are managed by the Singapore Land Authority, a state-controlled agency with land-based investments.[12]

While it is commonplace to wrest control of the sea for resource, territory, security, conservation and entertainment, the Singaporean anxiety of diminutive scale yields a particularly architectural perspective of this frontier. On the basis of land scarcity, the sea is increasingly commoditised as Singapore's hinterland. A modernist architectural project, its coastal components are persistently cleansed, re-arranged and maintained. Beaches, islands, islets and rocks are approached as interchangeable, urbanised and temporary instruments, organised in a never-ending modernist city puzzle. Extensions of the urban masterplan, these natural configurations have been absorbed into technologies of expansion—islands are merged, coastlines are extended, rocks are intensively built upon, beaches are moved and repositioned. Ascribed ephemeral functions and tracked in terms of their performance, over time, these acquire techno-natural attributes. The Singapore sea is deeply infrastructural.

Yet, proximity to water, as writer Amitav Ghosh argues, is recent; a 'colonial vision of the world' wherein 'power and security, mastery and conquest' have been normalised in a globalised pattern of middle-class occupation along coastlines all over the world.[13] Ghosh points out that conversely, the working class whose livelihoods once depended on the sea were, in fact, wary about living so close to the shore: 'They generally did not build large settlements on the water's edge … It is as if, before the early modern era, there had existed a general acceptance that provision had to be made for the unpredictable furies of the ocean—tsunamis, storm surges, and the like'.[14] The water-edge condition in Singapore, a city also mentioned in Ghosh's writing, is more complex. Almost three-quarters of its waterfront is given over to infrastructure owned by military, industry, maritime and reclamation works. The popular public beach at East Coast is reclaimed land, but it is already naturalised such that this beach is synonymous with 'the beach'. On a typical weekend, sunseekers and young families play, swim, sunbathe and picnic under coastal casuarina and *Ketapang* trees, watched in the horizon by oil tankers and cargo ships parked nose-to-tail (Fig. 9.3). How does this infrastructural sea and its hard economic value figure in the islanders' re-imagination?

[Opposite] **9.3** Sand, sea, salt, sky, oil tankers and barges on Singapore's East Coast Beach.

Situated at the contradictory junction of an imagined sea and its commoditised presence, this chapter examines how three water-based infrastructure—buoys, lighthouses and urban water channels (drains, canals and pools)—co-produce new images and imaginations of the sea. Referencing architect Keller Easterling's argument of infrastructure's potential to transform spatial usage and perception, the imagination of the infrastructural sea is reconsidered through Walter Benjamin's dialectical image, vis-à-vis Taussig's theorisation of the commoditised sea returning as 'second nature', at the moment of its extinction. What is the dialectical image that moves beyond the polemics of the sea as a site of loss and commoditised territory?

This argument adopts as its archive, the seascapes and water bodies portrayed in *SEA STATE* (2004–present)—a multimedia, ongoing visual art project by Singapore-based artist and sailor Charles Lim. *SEA STATE* makes infrastructure accessible and vulnerable. These are climbed over, sailed past, walked on, crawled through; things grow on them, they decay. Yet, where hard technologies become 'leaky, partial and heterogeneous entities',[15] the dispositions and potentials of the infrastructural sea gain visibility.[16] In *SEA STATE*, infrastructure in states of transition—failure, naturalisation, monumentalisation—produce knowledge of the sea they inhabit, and increasingly, define. The in-transition infrastructural sea is the sea's dialectical image.

The infrastructural sea.

SEA STATE has a double meaning: a meteorological maritime measure used to rate the height of waves in different sea conditions (Sea State 0 denoting a calm sea, and Sea State 9 signalling calamity), and a description of an island nation denuded of its natural coast. The making of *SEA STATE* stems from Lim's childhood in a small fishing village on Changi beach at the easternmost tip of the island, where he witnessed coastlines changing. Later, training as a competitive sailor, Lim's survival relied on his tacit knowledge of currents, tides and coastal variations:

> Wind is generally invisible to the human eye, so you need to develop ways to try to locate it by looking at the effect it has on the water, the feeling it has on your body. Quite a lot of sailors cut their hair quite short, so their skin becomes more sensitive to the wind. And even smell … when there is a storm coming, you can smell the increased humidity or you can even taste it'.[17]

Such knowledge is visceral and life-depending. The sea is both enveloping and distant—to be gauged by the sailor's body and requires perspicacity.

In that sense, *SEA STATE* is shaped by an aesthetic of distance. The sea is situated outside the visible frame of the artwork although it continually haunts our reading, akin to how the sailor must detect what is invisible, piecing together fragments that may destroy him/her in the aftermath of what is imminent.[18] Contrasting the sublime emptiness of artists Tacita Dean's and Roni Horn's waterscapes, Lim's work demonstrates a controlled and mechanised landscape hard at work. The sea is rendered quantifiable and finite, but also elegiac, expansive and lush by proximity to the infrastructure it inspires. Through a supporting cast of instruments—lighthouses, canals, drains, jetties, ports, swimming pools, underwater caverns, buoys, boats, barges, tractors, diggers, lorries, construction sites, piles of sand—the sea comes into view and gains a work-like sensibility. Art curator Shabbir Hussain Mustafa who worked with Lim on *SEA STATE* describes the work as 'informatic naturalism', 'informational'[19] and 'technoshamanistic'.[20] The sea is *not* indexical but 'concrete and geometric, a solid, material force of unyielding gravity. It does not narrate; it informs. It is not so much subject matter, but a medium, in the biological, economic and aesthetic senses of the word'.[21]

SEA STATE neither romanticises the sea nor vilifies the technologies that domesticate it. As such, the work moves closer towards Krauss' reading of the modernist sea as a politicised zone of deprivation. It explores modalities of the 'industrialised, concretised, politicised and militarised'[22] sea, pointing to what we see and know, but also, to what we do not see but know. The work gives expression and unexpected urgency to a sea that no longer holds eternal form or meaning. At the same time, the secret power of the sea erupts when the instruments used to tame this vast resource are momentarily overwhelmed by the sea's force, rendering them useless. A boat capsizes when tides are disrupted by encroaching landforms; a buoy submerged in warmer-than-usual tropical waters grows a thick skin of barnacles; an abandoned swimming pool turns green with algae; a dry network of drains and canals dramatically fill and overflow in a thunderstorm.

The artwork's aloofness and instrumentalism reflect an industrialised and mercantile sea, something to be controlled, owned and built upon. During British colonial rule, Singapore perceived its islands in the same way. The colonial gaze started from a high point, frequently a fort on a hilly outpost, and often settled on an infrastructural focus in the horizon—a harbour, a lighthouse, a buoy, a jetty, an oil refinery. Infrastructure, rather than the sea, was prized. When the colonial government laid the foundations in 1850 for the Horsburgh Lighthouse on one of the smaller eastern islands called Pedra Branca, recognisable for its white rocky outcrop, the inscription for this lighthouse read: 'Our island, of which this Rock is a dependency'.[23] The declaration on the foundation stone points to the recognition of the white rock, given its characteristic whiteness by bird excrement, as an asset of the crown colony through the installation of

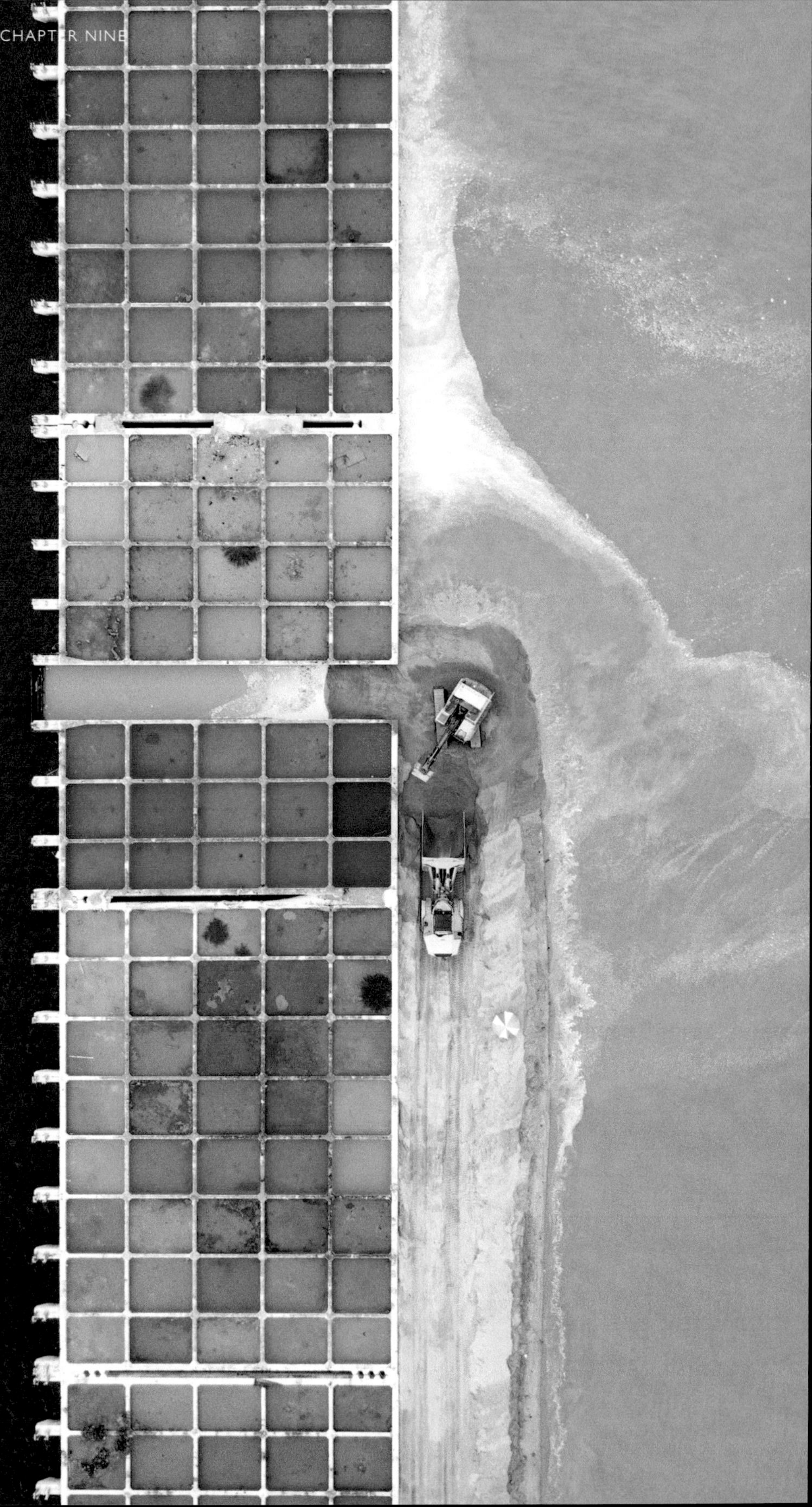

its lighthouse. Furthermore, forts such as those found on Singapore's smaller southern islands including Pulau Blakang Mati and Pulau Brani were out-of-bounds by law; anyone found sketching or photographing within 3000 yards could be arrested and charged.[24] Archipelagic infrastructure necessitated occupation, order and dominion.

Infrastructure, while often perceived as instruments of the military or used in nation-building, are equally treated as evidence of 'economic liberalism' or facilitating 'global exchange'.[25] Easterling describes how infrastructure space is in perpetual action and collaborative with both free-market and institutionalised power, though this is often unremarked:

> As a site of multiple, overlapping, or nested forms of sovereignty, where domestic and transnational jurisdictions collide, infrastructure *space* becomes a medium of what might be called *extrastatecraft*—a portmanteau describing the often undisclosed activities outside of, in addition to, and sometimes even in partnership with statecraft.[26]

Easterling argues that the importance of infrastructure space—such as the infrastructural sea in *SEA STATE*—is the degree to which it challenges the monolithic conception of an 'object form', a format which architecture and urban planning favour. As an 'active form', infrastructural space may be read as a composition of multiple and complex parts—some physical, others informational—not unlike 'each ripple on the water or each bit of code in the software', whose in-between relations are consequential to the operative whole.[27] The 'object form' privileges certitude while the 'active form' evolves and adapts in relation to changing circumstances and situations.[28]

Infrastructure is powerful because of its specific and immanent 'dispositions'—the potential capacities of the infrastructure's relational parts to adjust according to need and situation (Fig. 9.4). These parts are further constituted by physical (instruments, spaces) and abstract (laws, policies, moralities) entities.[29] The term 'disposition' may thus be associated with Michel Foucault's *dispositive* wherein the *dispositif* is defined as a 'system of relations' holding together different components including discourses, institutions, architectural forms, scientific statements, regulations and laws, as well as moral and philosophical positions.[30] As such, the *dispositif*'s influence is both far reaching and entangled.

Further, the different parts of the *dispositif* may be redesigned to enable particular desired outcomes: 'To access disposition is to assess how an organization deals with the variables over time—how it absorbs or deflects the active forms moving within it'.[31] Thus, an analysis of an active form's disposition might

[Opposite] **9.4** Chi Yin Sim, *Shifting Sands*, 2017. Tuas Port land reclamation project.

'uncover accidental, covert, or stubborn forms of power ... hiding in the folds of infrastructure space'.[32] Easterling emphasises that infrastructure is powerful because it has the capacity to act in various modes, and in their combinations as multipliers, remotes, interplays of networks, and carriers of scripts, infrastructure is triply enabled as technological, organisational and social forms (Fig. 9.5).

A 'multiplier' sets processes and agents into motion, and into relation with each other. Illustrating how the car galvanised the need for roads which in turn influenced the way in which suburban housing was conceptualised as a packaged commodity 'from their frames and roofs to their TVs and washing machines', Easterling argues that the city's morphology, use and experiences are transformed because of the circulation of active 'multipliers' such as 'cars, elevators, mobile phones, laws, real estate formulas, structural innovations, and security technologies'.[33] As a remote device, infrastructure acts like a switch to reroute circulation, such as in electrical, internet, highway or maritime networks.[34] Apart from controlling the circuit, these nodal switches have the potential to monopolise traffic and markets. The design of the golf course is a strategic interplay of active form networks. To maximise profit, the saleable surface area

of the golf course allocated to golf villas follows a formula which subsequently governs the shape, and thus design, of the course. In addition, endorsement from celebrity golfing champions further raises the price of the villas. Thus, even while the outward form of the golf course is important, this 'object form' is less significant than its 'software—the powerful bits of code underlying millions of acres of development all around the world'.[35] Finally, infrastructure gains social weight when it manifests a script or story. Easterling leans on Bruno Latour's actor-network-theory when she advocates that infrastructure can be a productive non-human actant to 'channel a flow of meanings'.[36] In another example, she evokes how Levittown's postwar suburban housing was successful in its association with 'familial and patriotic narratives',[37] which then infiltrated, formulated and influenced the social networks that proliferated the suburb. Opportunities are multiplied even further when there are deviations from the assigned narrative, revealing possibilities for 'both the naturally occurring dislocations of meaning as well as the duplicitous politics of extrastatecraft'.[38] In this technique of 'extrastatecraft', infrastructure's discrepancies and fictions are maximised for agility, compatibility and relevance to the speed and spread of global power.

9.5 Chi Yin Sim, *Shifting Sands*, 2017. Marine infrastructure.

SEA STATE depicts the sea through marine infrastructure's active forms. Buoys, subterranean grottos, lighthouses and water channels feature as the sea's proxies. Infrastructure is internalised as visual language and as subject matter. The artwork follows the awe-inspiring monumentality, enviable efficiency and undisputable autonomy of maritime instruments. Yet it also introduces discrepancies and ambiguity by amplifying infrastructure's dispositions, offering in turn unplanned infrastructural glitches even within a tightly controlled system. The technique employed here is akin to 'extrastatecraft', particularly in its non-binary, doubling and non-oppositional modes of operation, or what Easterling terms 'auxiliary activism'.[39] The auxiliary activist is concerned with partial inversions, with undeclared activities, and with manifesting dissensus without destroying what is certain and declared.[40] Auxiliary activism in *SEA STATE* relies on practical knowledge of the sea, and improvisational survival skills. The criticality of the artwork lies in its complex rendering of the sea as we encounter it today; clearly depicting this 'sea state' not as the result of a single power or agency, but instead reflecting 'countless mirrorings of power in a world where no one is innocent'.[41]

Further, the multimodal approach in *SEA STATE*—using cameras, 3D modelling, maritime charts, archival artefacts and documents, interviews, podcasts, object-based interventions and surveillance—offer multiple and contradictory optics. These ways of looking cultivate an oblique critique of a landscape that is both ubiquitous and distant for Singapore.[42] As anthropologist Michael Fischer argues, these modes of looking are fostered out of a 'nomadic curiosity' to project a poetic aesthetic, which while remaining enigmatic, also allow us to co-produce meanings through our identification with the sea, and our embodiment of it as a working landscape, or what Fischer terms 'our increasingly reworked ecologies, our forms of advanced re-manufacturing with the earth itself'.[43]

SEA STATE's aesthetics are poignant and beautiful: giant machines, colourful buoys, vast underground caverns, cruising sand barges, even the mundane monsoon drains, canals and swimming pools are given stunning filmic presence. The artwork embraces these technologies as the stuff of science fiction but also pulls the viewer back to an archaic past when the instruments stutter and falter, when these grow barnacles and algae, and when they gradually become naturalised. Indeed, the affective quality of SEA STATE lies in this sense of momentary wonder and magic, a sense of 'what if' when the sea is evoked through technology, all the while suspending judgement whether these instruments might ultimately destroy or save it. Lim's technonatural imagery constructs a dialectical image of the contemporary sea. It glimpses into the sea's, and technology's, unyielding power, their relationship, and reciprocal vulnerability.

The dialectical sea.

The dialectical image, Benjamin argues, challenges the sequential continuum from past to present. Rather, it works like a montage, bringing the past and present together 'in a flash' at a time of a commodity's crisis:

> It's not that what is past casts its light on what is present, or what is present its light on the past; rather, image is that wherein what has been comes together in a flash with the now to form a constellation. In other words, image is dialectics at a standstill. For while the relation of the present to the past is a purely temporal, continuous one, the relation of what-has-been to the now is dialectical: is not progression but image, suddenly emergent.[44]

What Benjamin means by 'image' here is a momentary but influential impression encountered, one in which colliding perceptions of past and present merge. The dialogical relationship set up by the dialectical image connects the viewer directly with the object being viewed. Thus, the dialectical image signals 'a method of seeing images', or of reading texts, rather than pointing to a particular type of image.[45]

The radicality of the dialectical image is how it sees history as something embodied and constructed out of multiple temporalities. Benjamin's proposition challenges the hegemony of ideological frameworks and narratives which predetermine the detached ways in which history is interpreted. The dialectical image instead reconsiders history from the inside out, insisting that it is ultimately complicit with the viewer's/reader's perspective, and that our own time of the present must also be folded into our perception of history. In other words, the dialectical image privileges an embodied encounter juxtaposing temporalities of past and present, compressing in a single instance personal memories and shared events, bringing together an individual's and a collective's consciousness. Significantly, Benjamin's dialectical image is 'recovered' from the material of dreams, outmoded populist objects, and 'half-forgotten experiences of childhood' such that these images transport our imagination to occupy spaces between 'the realm of things and the realms of allegory'.[46]

The dialectical image is when the past is recognised in the present as a ruin—something that has failed having lost its use value. The dialectic raises a contradiction; wherein the utopian desires for this commodity have proven delusional.[47] The image shocks and disrupts thinking, 'the usual patterns of thinking and everyday living stop and new ones are given a chance to emerge'.[48] The dialectical image shows the said commodity to be vulnerable while also threatening to forcibly remove it from its familiar surroundings. In this case, the dialectical image constitutes the sea which is no longer inhabited. It is something to be contemplated—sea-fronting real estate replacing working class fishermen shacks, and thus, because this sea is displaced at a distance from us, there is an intense urgency to grasp what it meant before, and what it means to us now.

In *SEA STATE 7: Sandwich* (2015), Lim challenges the narrative conventions of the moving image. The 'sandwich' is made from side-turned strips of moving images, each strip documenting a coastal scene (Fig. 9.6). The layering forces a reading of disjunctive scenes in proximity—migrant construction workers sitting on boulders by the beach, makeshift waterside houses on stilts, millionaire bungalows on Sentosa Island, the busy container port, piles of sand transported on barges, a sand hill in the background, sand slowly dropped into the sea, rocks and fortification prepared for the land reclamation process. Its formal composition mimics the layering of materials in the reclamation of land at the coast. The result is a conceptual space for unpacking the abstract activity and understated implications of coastal reconfiguration by industry and territorialisation.

> The idea of sandwich was the moving image. We don't normally see it as spatial. We tend to throw in a narrative into the moving image. But through sandwich I was trying to devise a new way of looking at the sea, in a sense. I am cutting out the water, but in a way you are looking at the sea.[49]

For Benjamin, the detritus of history is key to unravelling the delusions of a utopian narrative. In a dialectical image, the detritus—constituting 'small pieces of historical experience otherwise dismissed as insignificant, beneath attention, unassimilable'[50]—is emplaced through the Surrealist technique of montage. The montage is chosen because it interrupts thought by deferring 'synthesis' or 'unity of meaning' in an image.[51] An interpretation of the dialectical image can thus never be authoritative. Instead, such an image communicates allegorically by associative, disjunctive and non-linear connections:

> ... texts read through other texts; fragmentary, partial hieroglyphic, ambiguous; paradigm of palimpsest; critical in the involvement of misreading; disregard for aesthetic boundaries,

9.6 Charles Lim, *SEA STATE 7: Sandwich*, 2015, single-channel HD digital video, c. 5 mins.

> especially between the visual and the verbal; 'piling up the fragments ceaselessly'; emerging from an 'appreciation of the transience of things, and the concern to rescue them for eternity'.[52]

In *Sandwich*, the sea is evoked through its hardworking economic edges. At the same time, the increasingly antiquated notion of the sea—as a naturally productive landscape sustaining livelihood and leisure—are slipped into the mix. The disappearing fishing families and their thinning catch, their tenuous coastal dwellings, and the country's largely invisible foreign labour taking a weekend off on the beach, are strategically sandwiched into this montage of a productive Singapore sea. In this amalgamation of maritime futures and presents-becoming-extinct, a familiar yet also uneasy dialectical image of the sea unfolds. The dialectical image facilitates a consciousness—apprehended in a single glance—of the contradictions, fictions, violence, motives and fantasies which pile ceaselessly, determining our 'common' perception of 'what is the sea'. The sea as a dialectical image is thus unsettling and precarious, offering its beholder a vantage point from which to remember, doubt and question.

Taussig proposes that the dialectical image offers more than a shock to thinking; it presents the commodity or object of its focus as archaic, or pre-historical, and in turn 'abruptly release their significance'.[53] The archaic, introduced by Benjamin in his examination of commodity as fetish, aims to recuperate not just what has passed but also to give this past a space and significance in the contemporary present. Benjamin argues that it is at the moment of loss, or in the face of impending extinction, that commodity gains immense value, and is accorded an aura. The juxtaposition of the past in the present challenges the conception of sequential time and traditional ideas of historical progression. As Taussig explains, the surfacing of the archaic in the present, makes apparent 'that different orders of time may coexist with a past precisely because the past is both real and fictional, nature and "second nature"'.[54] The dialectical sea is such an archaic space and a fetishised commodity. The sea is where time is not progressive but overlapping; residues of lost childhoods and evidence of competitive economies pile upon each other. The residual sea is as embellished and constructed as the infrastructural sea. Both versions have come to reconstitute the sea's 'second nature', such that how we now see, talk,

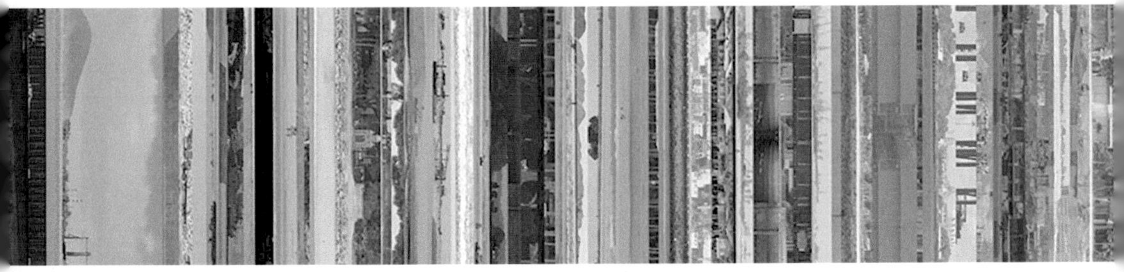

remember and fantasise about the sea is taken as inevitable and natural. Taussig argues that lure of the sea today is because of this dialectic—the sea as an archaic vanishing commodity overlaid on the sea as lucrative commodity.

Indeed our 'common sense' knowledge of the sea is revealing because it connects contemporary personal histories with the economic histories of oceans, and the colonial histories of territorialisation. The sea and coast are fetishised as a Dionysian space of wildness, freedom and childhood play. This fetish is a precondition for how the sea becomes an intractable form of capital; its nostalgic value expands beyond its use value. Yet what *SEA STATE* in turn reveals is an efficient infrastructural sea, a working landscape similarly fetishised for its discipline. The artwork possesses what Taussig calls an 'aesthetic of revelation'—it does not point to a hidden secret but surfaces the sea's ineluctable contradictions.[55] Where 'knowledge comes only in flashes',[56] the dialectical image strategically mines the unseen, unsaid and unreachable. At the core of each *SEA STATE* instalment, subtle fragments reprise the dialectic—barnacles proliferating on a buoy; a dying fish writhing to a popular Malay song at the end of a triumphant video segment on the technologically complex construction of underwater caverns; guppies caught from monsoon drains filling a plastic bottle after a sudden downpour. Threats of climate change, toxic pollution, ecological imbalance, capitalist greed and transnational rivalry crowd out a placid childhood sea. The penultimate section of this chapter traverses Singapore's watery surrounds following three infrastructural routes taken in the wake of *SEA STATE*. Tracing the contrasting dispositions of the infrastructural sea, I argue that the artwork evokes yet another surprising Dionysian twist—of maritime machines and monuments commanding the sea but ultimately slowly absorbed into its depths.

Buoys.

In July 2015, at the Reed Bank in the South China Sea, hundreds of bright yellow buoys formed a horizontal barrier 'stretching as far as the eye could see'.[57] On pulling close to one, a Filipina fishing boat observed that the buoy's surface was covered with Chinese inscriptions. When the fishing crew tried to cut the line free, a large Chinese patrol ship promptly appeared on the horizon. On one side of the yellow wave, Manila had established their presence through oil and gas expeditions, and on the other side, was the Spratly Islands, China's 'seven man-made islands [built] on top of coral reefs'.[58] In January 2019, at the mouth of the Straits of Johor, a white Malaysian vessel, the *Polaris*, designed for laying buoys in the sea was identified as arguably trespassing Singaporean territorial waters. It had been anchored at that spot for a month.[59] The buoy-laying

Polaris was itself construed as territorial infrastructure, docked on a strip of water whose territorial edges were constantly redrawn in view of boundary contentions and misalignments.[60] In these two instances, the buoy approximates a mobile territorial body. It persistently encroaches, defends and makes visible the boundaries enforced in an amorphous sea space. The buoys bobbing rhythmically along stretches of territorial waters signal what philosophers Gilles Deleuze and Felix Guattari call 'refrains'—reproductions, or repetitive images, that reinforce and re-present the sea as a delineated commodity with tangible edges.[61]

Unlike the solitary and monumental structure of a lighthouse, instrumented buoys are ubiquitous in numbers, sizes (ranging from small, 10 cm to big, 10m in diameter) and types, as well as serving a multitude of functions crucial to the operational status of oceanic research, commerce and industry. They are best known as territorial instruments; used for mooring especially for congested waters, and as navigational markers to demarcate safe waters (coloured in red or green depending on directional location in relation to the harbour, or stripes of red and white/black to mark different water depths and safety of passage), and to delineate prohibited zones, pipelines, limits for fishing and wreckages (coloured in yellow). Buoys are equally employed for oceanographic and meteorological research and forecasts (air and water temperature, wind, ocean currents and ocean wave motion), military exercises (naval warfare and simulation practice) and for rescue in emergencies. In anticipation of poor visibility in bad weather, they are brightly coloured. Fixed to the seabed with long chains (usually three times the depth of the water), larger buoys are made from steel and can weigh as much as eight tonnes, while smaller buoys deployed in shallower and sheltered waters are often fabricated in lighter weight fibreglass-reinforced plastic. Thus, the buoy-wave-wind-water-coast assemblage constitutes a spatial organisation of highly sensitive movements and flows consequently affecting territory, economy, safety and security, and as such, demand incessant tracking, monitoring and policing.

In Singapore's maritime history, buoys were used by the colonial government to stake territorial claim. Archipelagic sea routes governed by the British Straits Settlements in the nineteenth century which controlled three major trading ports lining the Straits of Malacca—Singapore, Penang and Malacca—were made visible with beacon buoys. Buoys were preferred as a maritime frontier technology because of their relative affordability, compared to lighthouses and watch towers, and for their ease of installation.[62] Together, the gaslit buoys and lighthouses delineated an infrastructure of territory, with the English in competition with the Dutch; the former commanding the Malay archipelago, the latter controlling the Indonesian one. Alongside the colonial enterprise of mapmaking, these governments saw the lighting of their sea routes as viable means to consolidate, 'make rigid' and 'concretize' a series of 'imaginary lines' in the ocean. The floating lines of light demarcated colonial power, stability and influence within an ever-shifting landscape.[63]

The imaginary lines of control secured by innocuous-looking buoys are depicted in *SEA STATE 1*. Subtitled *Inside/Outside* (2005), 110 brightly coloured buoys are photographed from both sides of territories they demarcate. Each photograph shows either Singapore or its neighbouring nation in the background. Singapore's extensive buoy network is a mode of surveillance. The widespread installation of buoys in 1988 was attributed to the radar and computer systems used to 'monitor ship movements within the port area and in the Singapore Straits'.[64] Visually, these photographs give finite form to an easily overlooked maritime infrastructural network of isolated floating markers and geopolitical anchors. Lim highlights that buoys are site-specific. Each buoy is custom-made for its own location, with no two being alike. Deployed as alarm systems or to collect maritime data, they are markers of maritime temporary autonomous zones:

> They project interior discontents and anxieties onto outside spaces: fortified islands, territorial water markers, beacons of safety, zones of exception. These temporary and often shifting zones, are in the physical world, analogues of 'out of bound markers' ... They are also analogues of contested and shifting sovereign demarcations and national security concerns in the maritime environment.[65]

In *SEA STATE 2: As Evil Disappears* (2014), Lim gestures towards the fate of an island—Pulau Sajahat (*Sajahat* is Malay for evil, cursed, bad)—which seemingly disappeared from Singapore's maritime maps between 2000 and 2012.[66] The buoy, together with the island, located off the southern coast which it marked, was perhaps consumed by the onslaught of land reclamation, receding as the sea slowly filled up with sand. As much as new infrastructure enables the founding of new space, the failure of that infrastructure might destabilise this space. Pulau Sajahat has a chequered history of mishaps. Sailors were thought to be cursed when approaching this island because of its treacherous rocks. However, as folklore would have it, the island was said to be where the waterbound Achenese marines committed mass suicide in the eighteenth century when they were overcome by their Portuguese colonisers. It is believed that the island was haunted thereafter, a belief anecdotally confirmed by Singaporean national servicemen sent to this island for training in recent years. During the second World War, Pulau Sajahat was one of 51 gun batteries and pillboxes along Singapore's southern coast anticipating the Japanese invasion, which in the end manifested in Japanese soldiers unexpectedly mounting their attack from the northern part of the island.

As a rejoinder to the missing Sajahat buoy, Lim commissioned a new buoy as its replacement. The replacement buoy was sunk, for a month, into the waters around Changi beach, where it sustained a prolific growth of barnacles. Hauled up and placed on land, the barnacled buoy looked nothing like a sea instrument. This barnacle-encrusted object was intricate, beautiful, decorative; an oversized artefact, having no other purpose than to be enjoyed and admired. Behaving more like an anchor than a buoy, the calcified specimen also started to oxidise; the steel took on a reddish-brown colour (Fig. 9.7). Thriving barnacles in seawater environments result in biofouling, a form of contamination on a ship's hull or anchor line, reducing their performance. The growth of barnacles on ships can increase its weight and drag by as much as sixty percent, resulting in excessive fuel consumption. Furthermore, the calcium glue secreted by the barnacle is so strong that chemicals are required to dislodge the plate-like surface formation of these crustaceans. The manipulated Sajahat replacement depicts infrastructure-becoming-maritime-wreckage-becoming-crustacean. The transformation of 'invisible', albeit unremarkable infrastructure, acts out what art historian Kevin Chua calls 'infrapolitics'—a low profile form of resistance which, when sustained, becomes troublesome.[67] Having lost its use value, the now remarkable Sajahat replacement recalls Benjamin's archaic commodity—reminding one of a sea monster conjured by childhood imagination, but here conjoined with myths of a vanished island and blighted maritime technology.

Lighthouses.

The sea is heavily marked with legal boundaries. The term 'territorial waters' is defined by a distance of three nautical miles from the coastline, or historically, by the firing distance of a cannonball.[68] This original legal limit was later extended to 12 nautical miles from land, plus another 12 nautical miles where 'hot pursuit' is legalised. The naval bases extend the shore to the edges of territorial water—geologically and lawfully termed the Continental Shelf, which denotes the outermost margins of a coastal state's territorial waters designated arbitrarily at 200 nautical miles.[69] These abstract lines of ownership are measurements taken at the seabed, itself a topography recognised geologically to be perpetually changing and thus, unstable. Further, marking of territory with maritime markers floating at sea, such as buoys, are equally susceptible to misalignment. Thus, when circumstances allow, the use of more permanent infrastructure, such as a lighthouse, is preferred.

Since the late nineteenth-century, lighthouses have been used as surveillance devices to monitor maritime movement in the archipelagos around South and Southeast Asia.[70] Aiming to quell indigenous piracy and unpoliced informal trading activities, and thus, to dispel European fears of traversing these waters, the Southeast Asian British-Dutch hydrographic space-race reached its peak in the 1870s:

9.7 Detail of barnacles growing on the replica Sajahat Buoy.

> Dozens of new lighthouses, lightships, and beacons went into operation throughout the western half of the archipelago. … [B]y the 1880s and 1890s, this expansion on both sides of the Straits of Malacca was underway in earnest.[71]

Singapore's nineteenth-century straits lighting campaigns constitute a motley assembly of lighting structures. Lighthouses were installed on high vantage points including hills and roofs of taller buildings; they were sited on the mainland as well as on a scattering of islands. These lighthouses were mobile, in that the beacon could be moved elsewhere once the coastal outlines called for such a shift. In 1903, a lighthouse built on an existing 1855 light-station on top of Fort Canning on Government hill was subsequently shifted in 1958 to the General Post Office Building (now the Fullerton Hotel), and then to the top of a 26-storey apartment tower in the southeastern suburb of Bedok in 1978.[72] In 2015, this beacon was shifted once again to a 25-storey block at Marine Terrace, one of the oldest housing blocks built on reclaimed land on Singapore's east coast, facing the sea. The shifting light and its lighthouse thus respond to, and register, an outward moving shoreline. The lighthouse functions not only to determine the safety of those using the sea, but equally, to signify political and economic autonomy, and in the Singapore example, to anticipate changing shifting territorial limits through land reclamation.

Preceding *SEA STATE*, Lim's artwork *In Search of Raffles Light* (2013-14) began with Lim and Mustafa attempting to gain entry to a lighthouse of the same name on Pulau Satumu (One Tree Island). Completed in 1855, Raffles Light was built by Indian convict labour. The lighthouse, named after Sir Stamford Raffles, colonial founder of Singapore, functioned as the southernmost marker of Singapore's territorial waters. The video shows an approaching boat which continually loops around the island. Lim and Mustafa never gained access to the Raffles Lighthouse. Instead, their search brought them closer to a panoply of maritime infrastructure, protocols and agencies working in service of the lighthouse. Thus, *In Search of Raffles Light* sets out to debunk the romanticised perception of the desolate lighthouse. The multi-part film- and data-based artwork refers to the said lighthouse but also, to the search for finite information key to understanding the sea. *In Search of Raffles Light* makes visible 'the sea as a process of knowledge'.[73] The intentionality of this artwork illuminates the outcome of a more controversial lighthouse, one in which the unfolding knowledge of the sea, gathered through maintenance of the lighthouse itself, became central to ownership rights.

On 21 December 1979, neighbouring Malaysia published a map (*Territorial Waters and Continental Shelf Boundaries*) based on the seabed extension of the continental shelf boundaries, which subsumed Pulau Batu Puteh (Malay for 'Island of the

White Rock') or Pedra Branca (the same name in Portuguese), a tiny island off the eastern coast of Singapore, under Malaysia's territorial boundaries.[74] As far as Singapore was concerned, Pedra Branca had always been under its jurisdiction. The island had been significant to Singapore's trade routes, particularly since it commanded the 'entire eastern approach to the Straits'. Situated in the middle of the Straits, Pedra Branca witnessed approximately 900 ships per day, with 'more than 80 percent of these ships arriving and departing from the port of Singapore'.[75] In 2008, Malaysia's case of Continental Shelf limits was overthrown in an International Court of Justice ruling, which accorded Pedra Branca to Singapore. Amongst the most compelling evidence on Singapore's side were its historical investments in the building and maintenance of the Horsburgh Lighthouse.

The Horsburgh Lighthouse, built in 1851, is constructed of granite masonry, and painted in alternating white and black bands, with the topmost lantern dome funnel rising to a height of over 34 metres above the white rock on which it stands (Fig. 9.8).[76] Granite was the only suitable material owing to the harsh tropical seas and treacherous rocks.[77] During the construction of the lighthouse, the government surveyor, J. T. Thomson, recorded the severity of the weather saying that Pedra Branca 'experiences the full effects of the waves created by the prevailing winds; a swell, more or less heavy, breaks on the rocks almost without intermission … a broken sea gets up such as no boat can live in'.[78] The island, no larger than the size of a football field,[79] was identifiable predominantly by its lighthouse. Without natural resources of food or water, Pedra Branca was unfit for inhabitation. Occupants of the Horsburgh Lighthouse relied on regular boat trips that shuttled provisions and stores to its rotating crew of seven persons who were 'relieved every month'.[80] Over a century and a half, multiple supporting structures and spaces were constructed on the rock. Around the base of the tower, a series of infrastructural installations were added: a docking jetty, a helipad, a cookhouse, a general building, living quarters for up to seven men, a water desalination plant and an electrical room. Altogether, these structures expanded to occupy and overwhelm the girth of the original rock.[81]

The infrastructure at Pedra Branca required regular maintenance and repair. Crew and contractors visited for building projects, boats mooring and leaving, materials and stores exchanging hands, and the beacon was replaced numerous times.[82] The heavy traffic even necessitated a helipad in 1991.[83] Technicians and engineers, planners and governmental officials came to inspect, conduct tests, troubleshoot and repair equipment on this eastern-most bit of offshore terrain, and maritime patrols conducted periodic checks on boats and other shipping vessels.[84] Singapore's sovereign claim over the 'white rock' was premised around the amount of work and resources needed to upkeep the lighthouse over one and a half centuries. This argument was remotely bolstered by global media coverage which persistently represented the island through

9.8 Pedra Branca: Not quite an island; a landform identified primarily by its lighthouse.

the iconic Horsburgh Lighthouse. The stoic black-and-white-striped lighthouse contrasted against a background littered with a cumulative cast of modernising infrastructure. This ensemble of lighthouse-with-modern-peripherals had a further 'constituting power' of the refrain.[85] It asserted territorial entitlement by continually visualising infrastructural ownership and maintenance. In other words, the repeating image of the lighthouse with peripherals became the expressive basis for future ownership. The Horsburgh Lighthouse and its compound, in becoming the aesthetic and affective representation of Pedra Branca, had the constituting power to shift the perception of sovereign entitlement from Malaysia to Singapore.

Channels.

Jurong Island, located on Singapore's southwestern industrial edge, is an amalgamation of seven smaller offshore islets. The outcome of a 30-year land reclamation project completed in 2009, the uninhabited island connected to 'mainland' Singapore by a single causeway, is an infrastructural landscape of tanks and pipes, passing and docking ships, and petrochemical plants.[86] The parts of this island-as-machine is largely invisible. Since 2007, tunnelled to a depth of 130 metres and blasting through 1.8 million cubic metres of undersea rock, the Jurong Rock Caverns is an extraordinary feat of vision and engineering. Located beneath the Banyan Basin, it provides 1.47 million cubic metres of oil and liquid-gas underground storage, and is a state-of-the-art exploration into the vertical dimension of future territorial expansion.[87] In the video for *SEA STATE 6: phase 1* (2014) a nostalgic Malay melody resounds through a still-empty cavern, when an explosive charge reverberates through a subterranean tunnel (Fig. 9.9). Reinforced by thick concrete walls, the cavern is water-tight and shock-proof. There is little impact within as two workers carry a small boat unperturbed, and above them, a sand barge languidly passes, oblivious to the disruptions beneath the water's surface.[88] The end of the video shows a casualty of the blast—a writhing fish whose jerky movements jar against the musical backdrop of the Cascades' song with the tell-tale refrain, 'There must be a reason, I like to know why, oh why?'

Beyond the spoils of territorial expansion, *phase 1* reveals something more sinister. Political geographer Rachael Squire argues that the human body may be considered a three-dimensional volume intimately immersed in the 'depth' and 'volumes' of spaces they occupy, such that 'objects, bodies, atmospheres, and lived experiences are all of the "things" that fill volume but remain somewhat incalculable'.[89] The volumetric infrastructural voids underneath the surface of sea and land, as with underwater and subterranean tunnels, challenge the perceived horizontal limits of land, air and sea as merely

'flat, bounded entit[ies]'.[90] The impact of such new conceptual and physical depths on the human body is disorientating since we 'act, inhabit, shape, and are immersed in the three-dimensional whilst also having [our] own three-dimensional geographies and volumes that interact intimately and minutely with surrounding space'.[91] At 130 metres below the Banyan Basin, the depth of the Jurong Caverns surpasses all previous infrastructural depths including shopping malls (15m), Mass Rapid Transit tunnels (35m) and sewerage systems (up to 60m).[92] In a deeply subterranean, sub-oceanic, quasi-natural cavern, the common notions of thresholds, boundaries and directionality become muddled. This space is simultaneously land-and-sea-in-one. As a corollary to the cavernous excavated space, Lim shows a reciprocal video titled *Capsize* where he capsizes his sailboat. In the video, the viewer catches the bearings of sea and sky by following Lim's body as he moves from underwater to above, and back under again several times in an effort to turn his boat the right way up. The intimate scale of sailboat-human-body-undulating-sea is wholly different from the circular elevator shaft that brings the viewer deep down into the bowels of the Jurong Caverns. While both views challenge our bearings, the capsized sailboat reinstates the viewer's anthropocentric control while the descend into the cavern erodes all notions of size and scale, above and below the sea.

SEA STATE indicates other ways of seeing; prompting the viewer to be attentive to traces, dissonances and gaps that point elsewhere. Lim recounts: 'I devised a strategy for my work where instead of ... a political position, what I did was execute a gesture. A very neutral gesture, through which these situations expand'.[93] There are many such traces of containment and expansion which Lim allows us to glimpse through his wide output.

In *SEA STATE 4: Line in the Chart* (2008), we see a long-rusted metal hoarding, the same type as those found around construction sites (Fig. 9.10). The hoarding is in the middle of the sea. On it there is a sign which reads 'No Entry | Restricted Zone | HDB'. A man swims towards this line, and upon reaching it looks backwards before climbing over the rusted hoarding in an act of trespass. In the first iteration of this project, Lim held a lottery where winners travelled with him to the actual site, to inspect that line on the chart. The hoarding was erected by Singapore's Housing and Development Board, a statutory body in charge of managing and commissioning public housing architecture. The hoarding conceives of the sea as future real estate. In *SEA STATE 0: It's not that I forgot but I chose not to mention* (2008), a man is seen swimming up and down a lap pool (Fig. 9.11). He moves very slowly, trying to clear away floating algae with each swim stroke. The pool is a standard facility offered by every gated condominium on the island, making it the most recognisable and 'natural' infrastructure containing a very large volume of water. The presence of algae,

[Opposite] **9.9** Collage from Charles Lim, *SEA STATE 6: phase 1*, 2015, single-channel HD digital video, c. 7 mins.

9.10 Charles Lim, *SEA STATE 4: Line in the Chart*, 2008.

like that of barnacles, is naturally encouraged by Singapore's hot and humid climate. Swimming pools and buildings are fastidiously cleaned, filtered, treated, scrubbed and repainted to repress weathering and algae invasion. The swimming pool is a sign of luxury, and like the sea, it is sometimes perceived as a status symbol. The less a pool is used, like the abandoned one in Lim's video, the more susceptible it is to algae.

Lim's obsession with how Singaporeans understand their islander status led him to observe how often people peered into canals. In *SEA STATE 0: All The Lines Flow Out* (2011), water is traced and contained in makeshift plastic bottles, and in a sophisticated system of monsoon drains and canals (Fig. 9.12). As a storm begins, water fills up shallow drains, is channelled towards larger storm drains, and gushes into the extensive network of canals running all over the island. There is water everywhere. The efficient drainage prevents flooding. A man garbed in plastic overalls from head to toe, stoops over a section of the

canal seemingly engrossed by the sight of guppy fish trapped in a plastic bottle. As the canal fills, more men in plastic laboratory coats appear. The scene cuts to another part of the canal. We watch from the vantage of a moving boat. The canal crosses many suburban areas until, at one point, all the lines (of water) flow out into the open sea. The sight, sound, smell and viscosity of water is tactile and immediate. The taming of water, and by extension, the sea, is so normalised in the concrete containers (which behave like expanded plastic bottles holding guppy fish in the drain) that when strong currents are encountered towards the end of the film, the sea is all the more menacing and unexpected. This is heightened by a woman's voice audible towards the end of the video, when a swimmer in an orange life jacket is viciously thrashed by the open sea. The camera takes us across a dilapidated wooden pier towards a pavilion on stilts. The woman speaks of the perils of living so close to the sea, of the pavilion's vulnerability to the monsoon seasons and storms. When the film ends, we are literally at the edge of the island, cut adrift, and facing the Straits of Johor.

[Above] **9.11** Charles Lim, *SEA STATE 0: It's Not That I Forgot Rather I Chose Not To Mention*, 2008, single-channel HD digital video, c. 44 mins.

[Overleaf] **9.12** Collage from Charles Lim, *SEA STATE 0: All The Lines Flow Out*, 2011, single-channel HD digital video, c. 21 mins.

Water is politicised in Singapore. Fresh, drinkable water is supplied by neighbouring Malaysia.[94] Singapore began to produce its own desalinated water when talks with Malaysia over water prices reached breaking point on several occasions. In *All The Lines Flow Out,* water is visceral, even as it is carefully controlled by infrastructure. There is no hint of a crisis except when this water flows out into the sea. There is no spatial threshold between the water in the canal and the open sea. It is the suddenness of the leap between the two opposing situations—one placid, the other tumultuous—which shocks. *All The Lines Flow Out* 'allegorizes subconscious knowledge, the surprise of recognising things suppressed, forgotten, classified, or screened out lest they disturb civilian life'.[95]

This video presents several gaps into which the viewer might burrow. One inspiration for tracing the flow of water through drains and canals, and into the sea was given by Singapore's most extensive manhunt for alleged terrorist Mas Salamat, who in February 2008, escaped from the high security prison of Whitney Detention Barracks. Salamat, who was eventually captured in Skudai (in Johor, Malaysia) about a year later, reputedly used this watery route to make his getaway. Another backstory is of the house which comes into our view at the end of the film. The Cashin House or the Pier House was built in 1920, and belonged to the Cashins, one of Singapore's oldest Jewish families. Located in Lim Chu Kang, on the northern reaches of Singapore, it is connected to the straits by a pier built in 1906 by Henry Cashin for moving supplies to and from his rubber plantation. This pier was also one of the northern landing sites used by the Japanese to invade Singapore in 1942. The Cashins used the house as a weekend resort until lawyer Howard Cashin's death in 2009.[96] It was known that the Cashins were friendly with the Sultan of Johor, who would visit the family on his own private boat, coming into Singapore directly through the Cashin Pier.[97] One might speculate that the Pier House was a space of informal diplomacy between two nations, and that the continued presence of both house and pier today make visible this special relationship. In this sense, the Pier House and the pier act much like Easterling's infrastructural 'switch'—they orchestrate or set in motion events which influence the flow and meaning of other things and processes around them. In *SEA STATE,* Lim gestures to the dispositions of these channels, lighthouses and buoys, opening our eyes to the complex processes of the sea as a politicised space.

The return of a whale.

In 1892, on a beach in Sa'batu, eighteen miles to the south of Malacca in neighbouring Malaysia, a carcass of a 'fish of monstrous size *(ikan besar skali)* [that] had been stranded upon the shore ...' was salvaged.[98] The Indian Fin whale was 'offensive' and bloated with 'gases from decomposition'.[99] Releasing this gas, it made a great deal of noise for three days.[100] A sizeable portion of the whale lay 'embedded in the soft mud' of considerable depth making it difficult to go up close to the carcass even in low tide.[101] The stench emanating from the leakage of bodily fluids into the surrounding mud, was made worse by flies gathering around the great animal. Unable to get close, curious onlookers went aboard 'a steam launch' to discover 'this monster of the deep'.[102] Some observers mistakenly thought the whale still alive, taking a whole week to die,[103] due to its 'groaning'.[104] It was presented to the Natural History Museum in Singapore, and for many years served as its key exhibit until it was given away in 1974.

On the morning of 10 July 2015, the Lee Kong Chian Natural History Museum was alerted to 'a large indistinguishable mass of grey and red' that had surfaced near the footings of a concrete jetty at Jurong Island.[105] Viewed from the beach, it was a 'heart-breaking mass of blood and stretched skin', with only a single flipper raised out of the water.[106] The marine debris tinted the surrounding waters red. Compared to the Sa'batu whale which was washed up gently onto the open Malaysian coast, the Jurong Island whale was beached on a rocky outcrop with little room for manoeuvre.[107] The zoologists' original plan to extract the skeleton by natural decomposition through a 'bury and wait' method on the nearby ash landfill site of Pulau Semakau was rejected on the grounds that the whale was unprocessed trash, which Semakau did not accept.[108]

The carcass was subsequently towed in open waters for six hours to a reclaimed embankment on Tuas, where it was hauled up to a rocky outcrop.[109] The second 'landing' of the whale on the Tuas coast was particularly laborious. A jagged coastline had to be smoothened with a mechanical excavator which extracted large rocks and accumulated debris before the fragile carcass could be hauled up.[110] Extra care was taken to prevent further breakage. This manoeuvre was especially challenging at low tide and in the dark, when the submerged rocks, heavily enmeshed with barnacles, sea glass, crushed plastic, fishing lines and hooks, became entangled with the ropes used to lift the whale.[111] Amidst an industrial backdrop of trash barges, dumpster skips, compactors and incinerators, the natural history crew spent the next sixty-nine days working on this salvaged carcass.[112]

A vinyl museum banner and a disused roll of carpet were laid over the rocks and the whale respectively, to mark a temporary zoological laboratory. When water was thrown onto the banner, it 'act[ed] as a slide, "lubricating" the ascent'[113] of the carcass up to the outcrop, cushioning the specimen as it was slowly taken apart. The ad hoc carpet and banner surfaces made a soft, smooth and pliable surface, uncannily replicating the sand and mud which cushioned the first whale on Sa'batu beach.

As the carcass rotted further, gas began to build up quickly beneath the whale's blubber (Fig. 9.13). Fearing the build-up would lead to an explosion of the viscera, the zoologists began making incisions into the whale's skin, inserting straw-like metal pipes to release the accumulated gas.[114] In an uncanny repetition of the Sa'batu encounter, '[f]rom the open holes came the gurgling and spurting of some curious odours and sounds, and cautiously [they] took a few steps back, just in case'.[115] The discovery of the Jurong Island/Tuas whale stirred the imagination. The scientists admitted feeling 'uneasy' because they seemed to be 'mimicking ... historic whalers [by] using their 19th century methods of hooks and knives ... [with which they] began slicing and tugging to strip the thick layer of blubber from the carcass'.[116] They thought their experience resembled that of 'Moby Dick'; it amazed them that their contemporary discovery could evoke in them 'childhood wonder', with the whale being doubly 'magical' given it was found in Singapore.[117] The scientists' instinctive wonder of the Jurong Island-Tuas whale was, in fact, in keeping with regional beliefs. According to ancient folkloric beliefs of island and coastal-dwelling Southeast Asia, the sighting of whales represent kindred ancestral spirits 'willing to assist deserving communities', or the safe return of a lost kin out at sea.[118] In Indonesia and Vietnam, whale skeletons and whale-shaped deities are still venerated and worshipped.[119] Historian Barbara Andaya highlights that 'the porous nature of human/non-human boundaries' in such human-whale relations focuses on the body of the whale as a metaphor for the vast ocean, its complex coastlines and their coastal communities.

The Jurong Island-Tuas whale is an uncanny repetition of the Sa'batu one. Both articulate the sea's propensity for the archaic, where the 'past is both real and fictional, nature and "second nature"'.[120] The second whale was remarkable in how it encapsulated a moment where the vanishing sea of Moby Dick coincided with the infrastructural sea of machines, compactors and mechanically made embankments. The image of a whale with steel straws in its body, laid out on a vinyl banner over a reclaimed embankment surrounded by machines, poses this dialectic. It is staged around a sea mediated by infrastructure to create, secure, and maintain territory. At the same time, it calls to mind another kind of sea; one that is wild and mired in folklore. The Jurong Island-Tuas whale embodies our material *and* imagined sea where technology and science reside next to, and overlap with, myths and stories.

The whale and the sea are affective constructs which evolve in-relation-to and reconfigure in-the-midst-of. 'Affect', anthropologist Kathleen Stewart emphasises, 'is the commonplace, labour-intensive process of sensing modes of living as they come into being'.[121] The zoologists who laboured tirelessly to dissect the beached whale grasped the coming into being of a precarious configuration in their midst. The whale-embankment-sea-trash was a formation, or a worlding which 'link[ed] some kind of everything'.[122] Human and nonhuman alignments momentarily crossing. 'World' in worlding does not refer to a fixed entity but a set of fluid relations in the making.[123] Thinking the sea through the whale on the embankment changes us. Climate change and sea

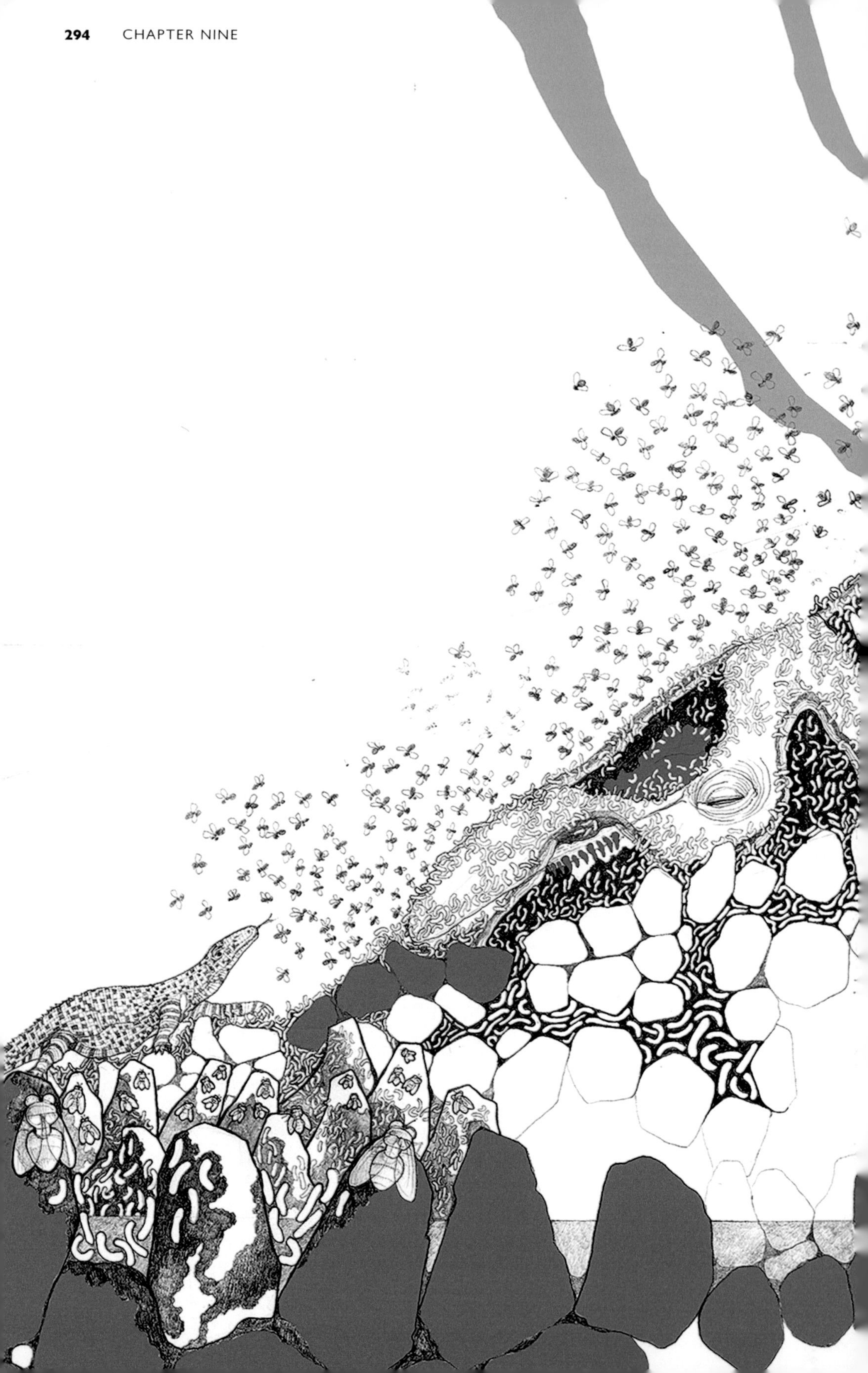

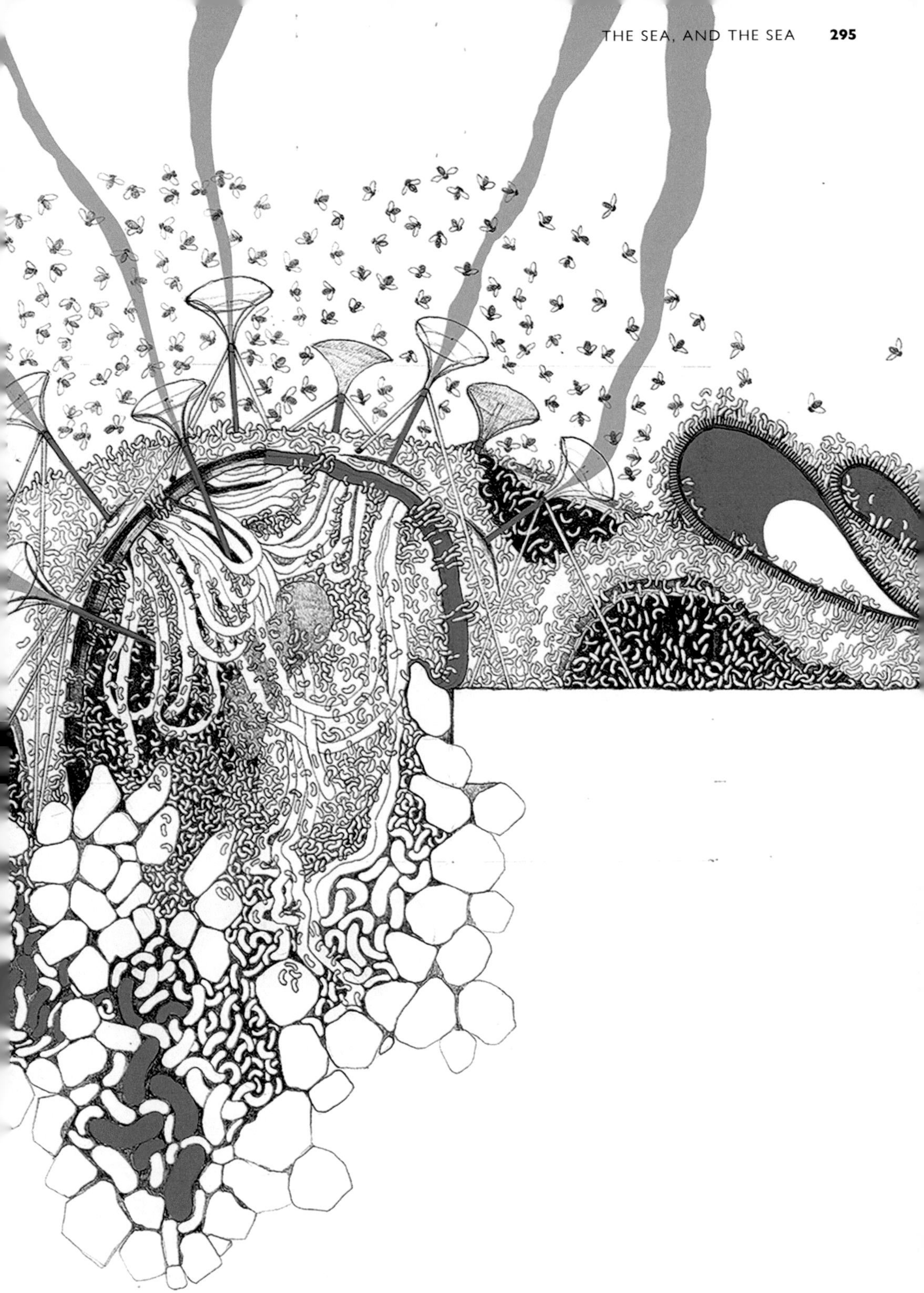

9.13 A rotting whale, maggots and flies.

level rises matter on a bodily level. The beached whale matters to our personal stories and childhood memories. This is a process of worlding, of how we form (or break from) relations, how we articulate what we see and sense through images, descriptions, stories and theories. As feminist theorist of science Donna Haraway reminds us, 'It matters what matters we use to think other matters with; it matters what stories we tell to tell other stories with…'.[124]

The whale encounter beckons attunement to, and care for, possible worlds which intersect the one we inhabit. In our curiosity towards and investment in such encounters, we actively enact an embodied relationship with our subjects, attending fully to what, how, and where we find ourselves. We may even be relieved, momentarily, of our blind spots. Produced through affective encounters, the processes of worlding itself—the construction of this architectural discourse, your present reading and subsequent re-articulation of it—produces its own affects. Architecture after affect is the contingent outcome of the discipline's unfolding, in the midst of others.•

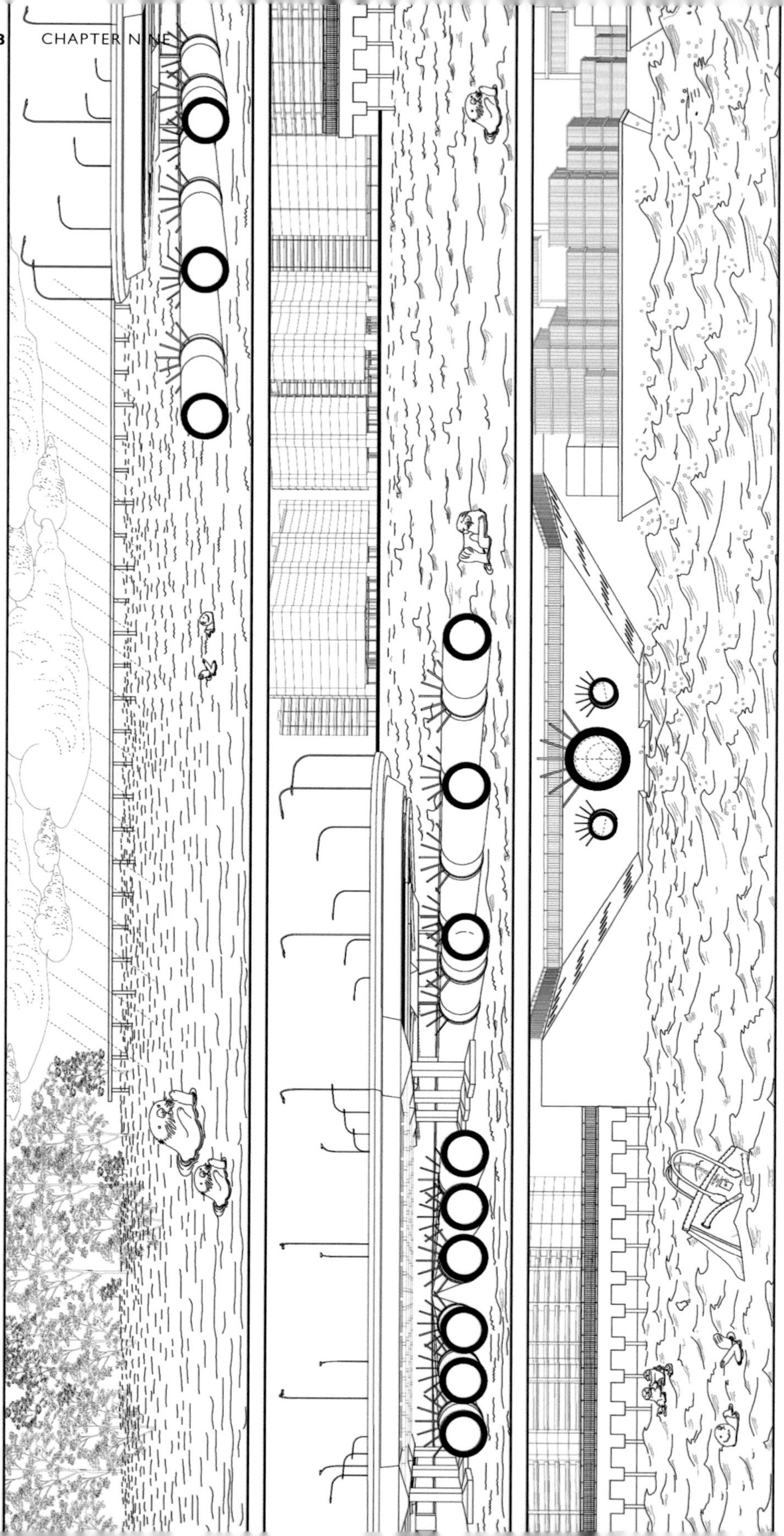

NO ENTRY
RESTRICTED ZONE
TEMPAT LARANGAN
DI LARANG MASUK

NOTES

PREFACE: SQUINTING FROM A BLIND SPOT

1. Lilian Chee, 'Materializing the Tiger in the Archive: Creative Research and Architectural History', in *Feminist Practices: Interdisciplinary Approaches to Women in Architecture*, ed. Lori Brown (Surrey: Ashgate, 2011), 155–65.

2. Lisa Blackman and Couze Venn, 'Affect', *Body & Society* 16, no. 1 (1 March 2010): 7, https://doi.org/10.1177/1357034X09354769.

3. Brian Massumi, *Politics of Affect* (Cambridge: Polity Press, 2015), 47.

4. Massumi, 10.

5. Yael Navaro-Yashin, *The Make-Believe Space: Affective Geography in a Postwar Polity* (Durham, NC: Duke University Press Books, 2012), 172.

6. Hélène Frichot, 'Infrastructural Affects: Challenging the Autonomy of Architecture', in *Architectural Affects after Deleuze and Guattari*, ed. Marko Jobst and Hélène Frichot (Oxon: Routledge, 2021), 33–55.

7. I am grateful to Barbara Penner for this insightful reading of my intentions.

8. Lisa Blackman, *Immaterial Bodies: Affect, Embodiment, Mediation* (London: SAGE Publications, 2012), 1, https://doi.org/10.4135/9781446288153.

9. Anna Gibbs, 'Writing and Danger: The Intercorporeality of Affect', in *Creative Writing: Theory beyond Practice*, ed. Nigel Krauth and Tess Brady (Teneriffe: Post Pressed, 2016), 159.

10. Gibbs, 159.

11. Eugenie Brinkema, *The Forms of the Affects* (Durham, London: Duke University Press, 2014), xii.

12. Brinkema, xv.

13. Brinkema, xvi.

14. Massumi, *Politics of Affect*, 151.

15. Massumi, 11.

16. See discussion on Raymond Williams' 'structures of feeling' in the Introduction.

INTRODUCTION: KNOWING OTHERWISE: ARCHITECTURE AFTER AFFECT

1. Eugenie Brinkema, *The Forms of the Affects* (Durham: Duke University Press, 2014), 39.

2. Michael Taussig, *I Swear I Saw This: Drawings in Fieldwork Notebooks, Namely My Own* (Chicago: Chicago University Press, 2011), 2.

3. Brian Massumi, *Architectures of the Unforeseen: Essays in the Occurrent Arts* (Minneapolis: University of Minnesota Press, 2019), 157.

4. Michael Taussig, *The Corn Wolf* (Chicago: University of Chicago Press, 2015), 5.

5. Taussig, 6.

6. Taussig, 6.

7. Taussig, *I Swear I Saw This*, 2.

8. Taussig, 9.

9. Taussig, xii.

10. Taussig, *The Corn Wolf*, 76.

11. Taussig, *I Swear I Saw This*, 9.

12. Roland Barthes, *A Roland Barthes Reader*, ed. Susan Sontag (London: Vintage, 2000), 318.

13. Barthes, 320. Barthes' multiple definitions of the term 'obtuse' is worth reiterating here. 'I even accept for the obtuse the meaning the word's pejorative connotation: the obtuse meaning appears to extend outside culture, knowledge, information; analytically, it has something derisory about it: opening out into the infinity of language, it can come through as limited in the eyes of analytic reasoning; it belongs to the family of pun, buffoonery, useless expenditure. Indifferent to moral or aesthetic categories (the trivial, the futile, the false, the pastiche), it is on the side of the carnival. *Obtuse* is thus very suitable'.

14. Barthes, 326.

15. Taussig, *I Swear I Saw This*, 6.

16. Taussig, 22.

17. Taussig, 6.

18. Barthes, 332.

19. Kathleen Stewart, 'Atmospheric Attunements', *Society and Space*, D, 29 (2011): 445, https://doi.org/10.1068/d9109.

20. Massumi, *Architectures of the Unforeseen*, 156.

21. Massumi, 174.

22. Massumi, 175.

23. Massumi, 175.

24. Massumi, 176.

25. Massumi, 183.

26. Hélène Frichot, *Creative Ecologies: Theorizing the Practice of Architecture* (New York: Bloomsbury Publishing, 2018), 10.

27. Stewart, 'Atmospheric Attunements', 445.

28. Stewart, 445.

29. Taussig, *I Swear I Saw This*, 144.

30. Kathleen Stewart, 'Precarity's Forms', *Cultural Anthropology* 27, no. 3 (2012): 519.

31. Barbara Penner and Charles Rice, 'Detecting Architecture: Questions of Evidence in Architectural History', in *Detecting Architecture* (College Art Association Annual Conference, New York, 2009). Unpublished paper.

32. Joan W. Scott, 'The Evidence of Experience', *Critical Inquiry* 17, no. 4 (1991): 774. http://www.jstor.org/stable/1343743.

33. Samuel R. Delany, *The Motion of Light in Water: Sex and Science Fiction Writing in the East Village, 1957-1965* (New York: Arbor House Pub Co, 1988), 173. Cited by Scott, 'The Evidence of Experience', 364.

34. Scott, 'The Evidence of Experience', 775–6. Emphasis mine.

35. Scott, 777.

36. Scott, 783.

37. Scott, 780.

38. Scott, 777.

39. Scott, 792. Emphasis mine.

40. Brian Massumi, *Semblance and Event: Activist Philosophy and the Occurrent Arts* (Cambridge, MA: MIT Press, 2011), 1.

41. Lauren Berlant, *Cruel Optimism* (Durham: Duke University Press Books, 2011), 5.

42. Berlant, 5.

43. Berlant, 52.

44. Berlant, 53.

45. Berlant, 79.

46. Berlant, 71.

47. Brian Massumi, *Politics of Affect* (Cambridge: Polity Press, 2015), 147.

48. Massumi, 70–82.

49. Massumi, 94.

50. Massumi, 147.

51. Massumi, 148–49.

52. Berlant, 82.

53. Brian Massumi et al., 'Affect and Immediation: An Interview with Brian Massumi', 2019, 114, https://doi.org/10.13023/DISCLOSURE.28.09. Massumi argues that affect occupies an 'infralinguistic' space in language.

54. Keller Easterling, *Extrastatecraft: The Power of Infrastructure Space* (New York: Verso, 2014), 215.

55. Massumi, *Politics of Affect*, 92–93.

56. Saidiya Hartman, 'Intimate History, Radical Narrative', AAIHS, May 22, 2020, https://www.aaihs.org/intimate-history-radical-narrative/. Hartman is citing Sarah Haley.

57. Roland Barthes, *Camera Lucida*, trans. Richard Howard (London: Vintage, 2000), 19.

58. Davide Panagia, *Ten Theses for an Aesthetics of Politics*, Forerunners: Ideas First 16 (Minneapolis: Minnesota University Press, 2016), https://manifold.umn.edu/read/ten-theses-for-an-aesthetics-of-politics/section/efa1c1e5-8b4f-4264-947c-06e4361137f6#ch01.

59. Panagia.

60. Davide Panagia, *The Political Life of Sensation* (Durham: Duke University Press, 2009), 154.

61. Barthes, 21.

62. Barthes, 21.

63. Barthes, 26, 28.

64. Barthes, 40-1. Schor says that the 'studium' 'participates in the economy of meaning', that is, it 'works in the service of the message'. See Naomi Schor, *Reading in Detail: Aesthetics and the Feminine* (London: Methuen, 1987), 91.

65. Barthes, 38.

66. Barthes, 47. See also Schor, 90. The 'punctum' may be compared to Walter Benjamin's 'fragment'. See Graeme Gilloch, *Myth and Metropolis: Walter Benjamin and the City* (Cambridge: Polity Press, 1996), 8-9.

67. Literary critics such as Naomi Schor and Trinh T. Minh-ha have registered Barthes's emphasis on the reader's desire as a 'feminine' method of interpretation. Schor, 96-7; Trinh T. Minh-Ha, 'The Plural Void: Barthes and Asia', in Diana Knight (ed.), *Critical Essays on Roland Barthes* (New York: G. K. Hall & Co., 2000), 209-18.

68. Barthes, 55.

69. Jacques Derrida, 'From "The Deaths of Roland Barthes"', in *Critical Essays on Roland Barthes*, ed. Diana Knight (New York: G. K. Hall & Co., 2000), 131, 132.

70. Barthes, 57.

71. Barthes, 41.

72. Barthes, *Roland Barthes Reader*, 326.

73. Barthes, 328.

74. Barthes, *Camera Lucida*, 21.

75. Sara Ahmed, 'Happy Objects', in *The Affect Theory Reader*, ed. Melissa Gregg and Gregory J. Seigworth (Durham: Duke University Press, 2010), 29, 30.

76. Ahmed, 33.

77. See also Shanti Sumartojo and Sarah Pink, *Atmospheres and the Experiential World: Theory and Methods* (Oxon: Routledge, 2019), 3.

78. Sumartojo and Pink, 6.

79. Sumartojo and Pink, 6.

80. Williams' work saw 'culture' as a contested field of class struggle and "articulation of working class consciousness and experience" in the 1950s. Sean Matthews, 'Change and Theory in Raymond Williams's Structure of Feeling', *Pretexts: Literary and Cultural Studies* 10, no. 2 (1 November 2001): 191, https://doi.org/10.1080/10155490120106032.

81. Roland Barthes, *Politics and Letters: Interviews with the New Left Review* (London: New Left Books, 1979), 159. Cited in Matthews, 179.

82. Raymond Williams, *Marxism and Literature* (New York: Oxford University Press, 1977), 130.

83. David Simpson, 'Raymond Williams: Feeling for Structures, Voicing "History"', *Social Text*, no. 30 (1992): 21, https://doi.org/10.2307/466464.

84. Ben Highmore, 'Formations of Feeling, Constellation of Things', *Cultural Studies Review* 22, no. 1 (4 April 2016): 157, https://doi.org/10.5130/csr.v22i1.4413.

85. Matthews, 191.

86. Isobel Armstrong, *The Radical Aesthetic* (Oxford: Blackwell Publishing, 2000), 115. Armstrong reads Freud and Kristeva's definitions of affect through structures of repression and melancholia respectively.

87. Armstrong, 109.

88. Armstrong, 123.

89. Armstrong, 123.

90. Armstrong, 124.

91. Kathleen Stewart, *Ordinary Affects* (Durham: Duke University Press, 2007), 127, 128.

92. Hartman, 'Intimate History, Radical Narrative'.

93. Stewart, *Ordinary Affects*, 3.

94. Kathleen Stewart, 'In the World That Affect Proposed', *Cultural Anthropology* 32, no. 2 (n.d.): 192–98, https://doi.org/10.14506/ca32.2.03.

95. Stewart, *Ordinary Affects*, 4–5.

96. Stewart, 4.

97. Stewart, 4–5.

98. Sara Ahmed, 'Travelling with Strangers', *Journal of Intercultural Studies* 42, no. 1 (2 January 2021): 9, https://doi.org/10.1080/07256868.2020.1859204.

99. Massumi, *Architectures of the Unforeseen*, 156. Here, Massumi cites the artist Simryn Gill. But 'drawing' recalls Taussig's fascination with the spontaneity and instinctiveness of his sketch, and a reminder that the word 'drawing' holds many meanings including 'a depicting, a hauling, an unravelling, and being impelled towards something or somebody.' See Taussig, *I swear I saw this*, xii.

100. Massumi, 163.

101. Massumi, 163.

102. Jane Rendell and Pamela Wells, 'The Place of Prepositions: A Space Inhabited by Angels', in *Architecture: The Subject Is Matter* (London: Routledge, 2001), 137.

103. Rendell and Wells, 141.

104. Jane Rendell, *Site-Writing: The Architecture of Art Criticism* (London: I.B. Tauris, 2010).

105. Jane Rendell, 'Prelude: The Ways in Which We Write', in *Writing Architectures: Ficto-Critical Approaches*, ed. Hélène Frichot and Naomi Stead (London: Bloomsbury, 2020), 1–10.

106. Rendell, 8.

107. Lesley Naa Norle Lokko, 'Black Matter(s): Such as? Does It?', in *Architecture: The Subject Is Matter*, ed. Jonathan Hill (London: Routledge, 2001), 183. See also: Lesley Naa Norle Lokko, *White Papers, Black Marks: Architecture, Race, Culture* (Minneapolis: University of Minnesota Press, 2000).

108. Lokko, 'Black Matter(s)', 189.

109. Massumi, *Architectures of the Unforeseen*, 182–83.

110. Massumi, *Semblance and Event*, 12–13.

111. Erin Manning, *The Minor Gesture* (Durham: Duke University Press, 2016), 32–33. Manning cites Whitehead (1929), 4, 10.

112. Manning, 33.

113. Jennifer Bloomer, *Architecture and the Text: The (S)Crypts of Joyce and Piranesi* (New Haven: Yale University Press, 1995); Katja Grillner and Klaske Havik, 'Between Sites and Stories, Between Text and Times: A Dialogue about Fiction and Narrative in Architectural Inquiries', in *Writing Place: Investigations in Architecture and Literature*, ed. Klaske Havik et al. (Rotterdam: nai010, 2016), 158–72; Karen Bermann, 'The House Behind', in *Places Through the Body*, ed. Heidi J. Nast and Steve Pile (London, New York: Routledge, 1998), 165–80.

114. Hélène Frichot, 'Following Hélène Cixous' Steps Towards A Writing Architecture', *Architectural Theory Review* 15, no. 3 (1 December 2010): 312–23, https://doi.org/10.1080/13264826.2010.524310; Naomi Stead, 'Architecture and Memory in W. G. Sebald's Austerlitz', *Arq: Architectural Research Quarterly* 19, no. 1 (March 2015): 41–48, https://doi.org/10.1017/S1359135515000263; Anoma Pieris, 'Dwelling in Ruins: Affective Materialities of the Sri Lankan Civil War', *The Journal of Architecture* 22, no. 6 (18 August 2017): 1001–20, https://doi.org/10.1080/13602365.2017.1363265.

115. Hélène Frichot and Naomi Stead, 'Waking Ideas from Their Sleep: An Introduction to Ficto-Critical Writing in and of Architecture', in *Writing Architectures: Ficto-Critical Approaches*, ed. Hélène Frichot and Naomi Stead (London: Bloomsbury, 2020), 12.

116. Frichot and Stead, 15.

117. Massumi, 'Interview', 115.

118. Kathleen Stewart, 'Pockets', *Communication and Critical Cultural Studies* 9, no. 4 (2012): 366.

119. Roland Barthes, *The Neutral: Lecture Course at the College de France* (1977-1978), trans. Rosalind E. Krauss and Denis Holier (New York: Columbia University Press, 2005), 10. Cited in Michael Taussig, *Mastery of Non-Mastery in the Age of Meltdown* (Chicago: Chicago University Press, 2020), 10.

120. Michel Foucault, *Ethics: Subjectivity and Truth*, ed. Paul Rainbow (New York: The New Press, 1997), 323.

121. Stewart, 'Atmospheric Attunements', 445.

122. Stewart, 'Precarity's Forms', 518.

123. Karen Burns, 'Architecture/Discipline/Bondage', in *Desiring Practices: Architecture, Gender, and the Interdisciplinary*, ed. Katerina Rüedi, Sarah Wigglesworth, and Duncan McCorquodale (London: Black Dog Publishing, 1996), 73–83.

124. I am grateful to Charles Rice for this succinct and accurate description of my work.

125. See Jane Gallop, 'Anecdotal Theory', in *Anecdotal Theory*, 1-11, 85.

126. Nancy K. Miller, 'Feminist Confessions: The Last Degrees Are the Hardest', in *Getting Personal: Feminist Occasions and Other Autobiographical Acts* (London: Routledge, 1991), xi-xii.

127. Lauren Berlant and Kathleen Stewart, *The Hundreds* (Durham: Duke University Press Books, 2019), 42.

128. Jennifer Bloomer, 'Hold It (Meditations upon a Gorgonzola Cheese)', *Journal of Architectural Education (1984-)* 40, no. 2 (1987): 8–9.

129. Bloomer, 8.

130. Jennifer Bloomer, 'Abodes of Theory and Flesh: Tabbles of Bower', *Assemblage*, no. 17 (April 1992): 18, https://doi.org/10.2307/3171221.

131. Bloomer, 18.

132. Lauren Berlant, *The Queen of America Goes to Washington City: Essays on Sex and Citizenship* (Durham: Duke University Press, 1997), 12.

133. Jonathan Hill, *Occupying Architecture: Between the Architect and the User* (Routledge, 1998).

134. Gregory J. Seigworth and Melissa Gregg, 'An Inventory of Shimmers', in *The Affect Theory Reader*, ed. Melissa Gregg and Gregory J. Seigworth (Durham: Duke University Press, 2010), 3.

1. THE RULED AND THE UNRULY: ANIMALITY, ANECDOTES AND STORYTELLING

1. Paul Carter, *Material Thinking: The Theory and Practice of Creative Research* (Carlton, Victoria: Melbourne University Publishing, 2004), 76.

2. Felix Driver and Brenda S.A Yeoh, 'Constructing the Tropics: Introduction', *Singapore Journal of Tropical Geography* 21, no. 1 (2000): 1–4, doi: 10.1111/1467-9493.00059.

3. Anne Collett, *Tracking the Literature of Tropical Weather: Typhoons, Hurricanes, and Cyclones* (New York: Springer, 2016), 10.

4. David N. Livingstone, 'Tropical Hermeneutics: Fragments for a Historical Narrative; An Afterword', *Singapore Journal of Tropical Geography* 21, no.1 (2000): 96, doi:10.1111/1467-9493.00066.

5. Livingstone, 96.

6. Jacques Derrida, *Dissemination*, trans. Barbara Johnson (London: Athlone Press, 1981), 355.

7. Nancy K. Miller, *Getting Personal: Feminist Occasions and Other Autobiographical Acts* (New York: Routledge, 1991), xii.

8. Karen Burns, 'Architecture/Discipline/Bondage', in *Desiring Practices: Architecture, Gender and The Interdisciplinary*, ed. K. Ruedi, S. Wigglesworth and D. McCorquodale (London: Black Dog Publishing, 1996), 77. For a feminist interdisciplinary critique of an architectural history focused on 'the architect', see Barbara Penner, 'Researching Female Public Toilets: Gendered Spaces, Disciplinary Limits', *Journal of International Women's Studies* 6, no. 2 (2005): 81–98. Examples of 'architect-centred' discourses, which form a large part of the architectural canon, include: Nikolaus Pevsner, *An Outline of European Architecture* (Harmondsworth: Penguin, 1977); Nikolaus Pevsner, *Pioneers of the Modern Movement: From William Morris to Walter Gropius* (London: Faber, 1936); Kenneth Frampton, *Modern Architecture: A Critical History* (London: Thames and Hudson, 1980); Spiro Kostof, ed., *The Architect: Chapters in the History of a Profession* (New York: Oxford University Press, 1977).

9. Jacques Derrida, *Archive Fever: A Freudian Impression*, trans. Eric Prenowitz (Chicago: University of Chicago Press, 1996), 1–3.

10. Arundhati Roy, *The Ministry of Utmost Happiness* (London: Hamish Hamilton, 2017), 3.

11. Roy, 58.

12. Roy, 4.

13. Roy, 67.

14. Roy, unpaginated.

15. Roy, 98.

16. Roy, 98–100; 136.

17. Roy, 98, 100.

18. Roy, 98.

19. Roy, 68.

20. Cherian George, *Singapore: The Air-Conditioned Nation; Essays on the Politics of Comfort and Control, 1990–2000* (Singapore: Landmark Books, 2000), 16.

21. Roy, 21.

22. Anthony Burgess, *The Malayan Trilogy* (London: Vintage Books, 2000), 25.

23. Thomas Mann, 'Death in Venice', in *Death in Venice: Complete, Authoritative Text with Biographical and Historical Contexts, Critical History, and Essays from Five Contemporary Critical Perspectives*, ed. Naomi Ritter (Boston: Bedford Books, 1998), 25.

24. Mann, 24

25. Mann, 25.

26. Mann, 26.

27. Amitav Ghosh, *The Great Derangement: Climate Change and the Unthinkable* (Chicago: University of Chicago Press, 2016), 30.

28. Amitav Ghosh, *Wild Fictions* (Essay), accessed February 3, 2022, 12, https://www.amitavghosh.com/docs/Wild%20Fictions.pdf.

29. Ghosh, *Wild Fictions*, 12.

30. Monique Allewaert, *Ariel's Ecology: Plantations, Personhood, and Colonialism in the American Tropics* (Minneapolis: University of Minnesota Press, 2013), 36.

31. Lesley Wylie, *Colonial Tropes and Postcolonial Tricks: Rewriting the Tropics in the Novela de La Selva* (Liverpool: Liverpool University Press, 2011), 50, doi:10.5949/UPO9781846315220.

32. Davide Torri, 'Her Majesty's Servants: The Tamed and the Wild Under the British Raj', in *Charming Beauties and Frightful Beasts: Non-Human Animals in South Asian Myth, Ritual and Folklore*, ed. Fabrizio M. Ferrari and Thomas Dähnhardt (Sheffield, Bristol: Equinox, 2013), 32–3.

33. Torri, 33.

34. Roy, 122.

35. John Berger, *About Looking* (London: Bloomsbury, 2009), 81.

36. Parts of the following sections on the Billiard Room at the Raffles Hotel were previously published as: Lilian Chee, 'Under the Billiard Table: Animality, Anecdote and the Tiger's Subversive Significance at the Raffles Hotel', *Singapore Journal of Tropical Geography* 32, no. 3 (2011): 350–64, https://doi.org/10.1111/j.1467-9493.2011.00437.x.

37. *Straits Times*, 'A Tiger in Town: Shot at Raffles Hotel under the Billiard Room', August 13, 1902.

38. Frank Buck, *Wild Cargo* (New York: Simon and Schuster, 1932), 203.

39. Boomgaard briefly discusses the history of Singapore's tiger menace and gives special mention to the tiger in Raffles Hotel, which he understands was under the billiard table. See: Peter Boomgaard, *Frontiers of Fear: Tigers and People in the Malay World, 1600-1950* (New Haven, CT: Yale University Press, 2001), 102–3, 140.

40. Meaghan Morris, 'Banality in Cultural Studies', *Discourse* 10, no. 2 (1988): 7.

41. Jane Gallop, *Anecdotal Theory* (Durham: Duke University Press, 2002), 8.

42. George Orwell, *George Orwell: A Life in Letters*, ed. Peter Davison (London: Harvill Secker, 2010), 3.

43. George Orwell, 'Shooting an Elephant', in *George Orwell: Essays* (London: Penguin Books, c.1968, 2000), 23.

44. Orwell, 22.

45. George Woodcock, *The Crystal Spirit: A Study of George Orwell* (London: Fourth Estate, 1967), 63.

46. Woodcock, 63.

47. Woodcock, 69.

48. Woodcock, 67. See also: George Orwell, 'A Hanging', in *George Orwell: Essays* (London: Penguin Books, c.1968, 2000), 14–18.

49. George Orwell, *Burmese Days* (London: Penguin Books, c.1934, 2014), 106–7.

50. Orwell, 14.

51. Woodcock, 67.

52. Jane Rendell, *The Pursuit of Pleasure: Gender, Space and Architecture in Regency London* (London: Athlone Press, 2002), 63.

53. Woodcock, 76.

54. Orwell, 17.

55. See for example: George Frederick Pardon [Captain Crawley], *The Billiard Book* (London: Longman, Green, and Co., 1866), 1; Sidney Gillett, *The Earlier History of Billiard Tables and Accessories as Seen from the Sales Journals of John Thurston, 1818–1843* (London: Thurston and Co., 1996), 7.

56. See for example: F. R. Cowell, *The Athenaeum: Club and Social Life in London, 1824-1973* (London: Heinemann, 1975), 36; Anthony Lejeune, *The Gentlemen's Clubs of London* (London: Parkgate Books, 1984). Club historian Anthony Lejeune states that at the turn of the nineteenth and twentieth centuries, billiard rooms were found in the Eccentric Club, the Junior Carlton Club, the Arts Club, the National Liberal Club, White's and the Savile Club, amongst others. For a gendered architectural history on London clubs, see: Rendell, *The Pursuit of Pleasure*, 86–103. For a contemporary perspective on the masculine setting of London clubs, see: Barbara Rogers, *Men Only: An Investigation into Men's Organizations* (London: Pandora, 1988), 165–99.

57. John Cameron, *Our Tropical Possessions in Malayan India* (London: Smith, Elder, 1865; Kuala Lumpur: Oxford University Press, 1965), 303. Citations refer to the Oxford University Press edition.

58. Charles Burton Buckley, *An Anecdotal History of Old Times in Singapore* (Singapore: Fraser & Neave, 1902; Singapore: Oxford University Press, 1984), 566, 622, 709–10. Citations refer to the Oxford University Press edition.

59. Kathy Creamer, *The Tiger Who Came to Tea* (Singapore: Oxford University Press, 1996); Kelly Chopard, *Terry's Raffles Adventures*, ill. Patrick Yee (Singapore: Landmark Books, 1996).

60. *Straits Times*, 'A Legend Roars Back to Life', February 11, 1986.

61. An early newspaper report announcing the opening of the new Raffles Hotel also promises a 'large and commodious billiard room containing four tables'. See: *Straits Times*, 'A New Hotel in Singapore', September 19, 1887. Other rival hotels in Singapore also advertised their billiard rooms as a major feature including L'Europe and Adephi, as well as the lesser-priced Sea View Hotel and Hotel van Wijk.

62. *Straits Times*, 'Raffles Hotel: New Billiard Room To Be Constructed and Other Improvements Made', October 9, 1906.

63. Ilsa Sharp, *There is Only One Raffles: The Story of a Grand Hotel* (London: Souvenir Press, 1981), 38–41, 96–115.

64. Gretchen Liu, *Raffles Hotel* (Singapore: Editions Didier Millet, 2006), 46.

65. Catherine Ingraham, *Architecture and the Burdens of Linearity* (New Haven, CT: Yale University Press, 1998), 80.

66. Rudyard Kipling, *The Jungle Book* (New York: Signet, 1961), 153–68, quoted in Robert Wessing, *The Soul of Ambiguity: The Tiger in Southeast Asia* (DeKalb, IL: Centre for Southeast Asian Studies, Northern Illinois University, 1986), 7.

67. Armand Denis, *Cats of the World* (Boston: Houghton Mifflin Co., 1954), 51, quoted in Wessing, *Ambiguity*, 8.

68. Kevin Chua, 'The Tiger and the Theodolite: George Coleman's Dream of Extinction', *FOCAS* 6 (2007): 138.

69. Buckley, 389.

70. Boomgaard, 5.

71. Cameron, 91.

72. Buckley, 220–1.

73. Richard Perry, *The World of the Tiger* (London: Cassell, 1964), 190.

74. The reward was halved in 1850 because the results were minimal, but increased again around 1860 when the additional 50 dollars was put up by the merchants' fund. See: Boomgaard, 103.

75. C. C. Brown, trans., *Sejarah Melayu or Malay Annals: An Annotated Translation* (Kuala Lumpur: Oxford University Press, 1970), 20, quoted in Ho Tzu Nyen, 'Every Cat in History Is I', *FOCAS* 6 (2007): 150.

76. Ho, 151.

77. Boomgaard, 6.

78. Thomas John Newbold, *Political and Statistical Account of the British Settlements in the Straits of Malacca... with a History of the Malayan States on the Peninsula of Malacca*, vol. 2 (London: Murray, 1839), 192, quoted in Chua, 'The Tiger and the Theodolite', 135.

79. Chua, 135.

80. Boomgaard, 187.

81. The first reference to weretigers in the Malay world is from an early fifteenth century Chinese source which recounts the story, told to visitors to Melaka at that time, of 'tigers which turned into men'. See: J. V. G. Mills, ed., *Ma Huan: Ying-yai sheng-lan; The Overall Survey of the Ocean's Shores; [1433]*, (Cambridge: Cambridge University Press, 1970), cited in Boomgaard, 187.

82. Chua, 138.

83. Chua, 136.

84. Michel De Certeau, *The Practice of Everyday Life*, trans. Steven Rendall (Berkeley: University of California Press, 1988), 124–9.

85. Buck, 203.

86. Buck, 203.

87. Penner, 81–98.

88. R. Allen, *The New Penguin English Dictionary* (London: Penguin Books, 2001), 48.

89. *Straits Times*, 'Raffles Hotel'. Emphases added.

90. Walter Benjamin, 'The Storyteller: Reflections on the Works of Nikolai Leskov', in *Illuminations: Essays and Reflections*, ed. Hannah Arendt (London: Pimlico, 1999), 91.

91. Benjamin, 87.

92. *Straits Times*, 'A Tiger Alarm', September 17, 1902.

93. Raymond Flower, *Year of the Tiger* (Singapore: Times Books International, 1986), 20.

94. Roberto Pregarz, *Memories of Raffles: 22 Years with a Grand Old Hotel* (Singapore: Treasury Publishing, 1990), 42–7.

95. Pardon, 12; Gillett, 5–6.

96. Peter Stallybrass and Allon White, 'Introduction', in *The Politics and Poetics of Transgression* (Ithaca, NY: Cornell University Press, 1986), 23.

97. Ingraham, *Burdens of Linearity*, 54.

98. Catherine Ingraham, *Architecture, Animal, Human: The Asymmetrical Condition* (London: Routledge, 2006), 237.

99. Ingraham, 238.

100. For the conservative role of the 'primitive' in architectural discourse, see: Adrian Forty, 'Primitive: The Word and Concept', in *Primitive: Original Matters in Architecture*, ed. Jo Ogders, Flora Samuel, and Adam Sharr (Abingdon, Oxon: Routledge, 2006), 3–14.

101. David Quammen, *Monster of God: The Man-Eating Predator in the Jungles of History and of the Mind* (New York: W.W. Norton, 2003), 253.

102. Chopard; Creamer.

103. Steve Baker, *Picturing the Beast: Animals, Identity and Representation* (Manchester: Manchester University Press, 1993), 159.

104. Baker, 123–4; Ursula K. Le Guin, *Buffalo Gals and Other Animal Presences* (London: Victor Gollancz, 1990), 10. For a philosophical discussion on the unthinkability of animals in Western philosophy, sexual difference and voice, see: Jacques Derrida, 'The Animal That Therefore I Am (More to Follow)', in *Animal Philosophy*, ed. Matthew Callarco and Peter Atterton (London: Continuum, 2004), 113–28.

105. De Certeau, *The Practice of Everyday Life*, 123; Jane Rendell, 'An Atlas of the Welsh dresser', in *Surface Tension: Problematics of Site*, ed. Ken Ehrlich and Brandon LaBelle (New York: Errant Bodies, 2003), 291.

3. AFTER THE LAST TRAIN: REMAINDERS AT THE TANJONG PAGAR STATION

1. Robert Segrest, 'The Perimeter Projects: Notes for Design', *Assemblage* 1 (1986): 25.

2. Ibrahim Iskandar, 'Why I Drove the Last Train out', *ST.Blogs*, *Straits Times*, July 1, 2011, http://blogs.straitstimes.com/2011/7/1/royal-on-the-rails.

3. *The Singapore Free Press and Mercantile Advertiser*, 'Opening of Singapore's New Station', May 3, 1932.

4. *The Singapore Free Press and Mercantile Advertiser*, 'Singapore's New Station'.

5. *The Singapore Free Press and Mercantile Advertiser*, 7; *The Malayan Saturday Post*, 'Opening of the new F.M.S.R. terminal station', May 7, 1932, 4, cited in Yong Chun Yuan, 'Tanjong Pagar Railway Station', *Singapore Infopedia*, accessed February 10, 2022, http://eresources.nlb.gov.sg/infopedia/articles/SIP_954_2005-01-10.html.

6. Preservations of Monuments Board and Urban Redevelopment Authority of Singapore, 'Historic Railway Stations to be kept for Future Generations', Media Release, April 8, 2011. https://www.nas.gov.sg/archivesonline/data/data/pdfdoc/20110415003/joint_pmb-ura_release_on_tprs_and_btrs__final_.pdf.

7. The Raffles Hotel is featured in the preceding chapter, 'The Ruled and the Unruly'.

8. Personal correspondence with the artist, 25 March 2016.

9. Caitlin DeSilvey, 'Observed Decay: Telling Stories with Mutable Things', *Journal of Material Culture* 11, no. 3 (2006): 328, doi:10.1177/1359183506068808.

10. Lilian Chee, 'The Domestic Residue: Feminist Mobility and Space in Simryn Gill's Art', *Gender, Place & Culture* 19, no. 6 (2012): 750, doi: 10.1080/0966369X.2012.674924.

11. There are numerous articles from both perspectives, see for example: *HistorySG*, 'Singapore Separates from Malaysia and Becomes Independent', accessed February 13, 2022, https://eresources.nlb.gov.sg/history/events/dc1efe7a-8159-40b2-9244-cdb078755013; Tim Huxley, 'Singapore and Malaysia: A Precarious Balance?' *The Pacific Review* 4, no. 3 (1991): 204–13, doi:10.1080/09512749108718919; Lily Zubaidah Rahim, 'Singapore-Malaysia Relations: Deep-Seated Tensions and Self-Fulfilling Prophecies', *Journal of Contemporary Asia* 29, no. 1 (1999): 38–55, doi:10.1080/00472339980000031; Chang Li Lin, 'Singapore's Troubled Relations with Malaysia: A Singapore Perspective', *Southeast Asian Affairs* 2003, no. 1 (2003): 259–74; Kadir Mohamad, *Malaysia-Singapore: Fifty Years of Contentions, 1965-2015* (Kuala Lumpur: The Other Press, 2015), 12–51.

12. Yao Souchou, *Singapore: The State and the Culture of Excess* (New York: Routledge, 2007), 19.

13. Yao, 19.

14. Lee Kuan Yew, *The Singapore Story: Memoirs of Lee Kuan Yew* (Singapore: Prentice Hall, 1998), 16.

15. Michael D. Barr, *Lee Kuan Yew: The Beliefs behind the Man* (Washington: Georgetown University Press, 2000), 75.

16. Barr, 75.

17. Lee Kuan Yew, 'Text of a Talk by the Prime Minister Lee Kuan Yew, Received as a Voicecast from London and Broadcast by Radio Singapore on 9th July 1963', (transcript, Radio Singapore, September 7, 1963), National Archives of Singapore, http://www.nas.gov.sg/archivesonline/speeches/record-details/73dd0b4c-115d-11e3-83d5-0050568939ad.

18. Lee, 'Text of a Talk'.

19. *Singapore Sunday Times*, 'Speech at UMNO General Assembly by Tunku Abdul Rahman', May 16, 1965, cited in Kadir, *Malaysia-Singapore*, 17.

20. Lee Kuan Yew, *From Third to First World: The Singapore Story: 1965-2000* (Singapore: Times Media, 2000), 759.

21. Kadir, 22.

22. Kadir, 23–24.

23. Tunku Abdul Rahman Putran, *Looking Back: The Historic Years of Malaya and Malaysia* (Kuala Lumpur: Pustaka Antara, 1977), 122, cited in Kadir, *Malaysia-Singapore*, 25.

24. Lee, *The Singapore Story*, 648.

25. Huxley, 'Singapore and Malaysia', 206.

26. For more information on Pedra Branca, see Chapter 9, 'The Sea, and the Sea'.

27. Kadir, 143.

28. Lee, *From Third World to First*, 282.

29. Kadir, 134–6.

30. In an agreement between Singapore and Malaysia signed in September 2010, Malaysia was given six land parcels in Marina South and the Ophir-Rochor area in exchange for Malaysia's six Singaporean railway sites in Tanjong Pagar, Kranji, Woodlands and Bukit Timah. See: Jessica Lim, 'Tanjong Pagar Station a National Monument', *Straits Times*, April 9, 2011, https://www.straitstimes.com/singapore/tanjong-pagar-station-a-national-monument.

31. Lim Yan Liang, 'Singapore and Malaysia Hold Ceremony to Resume Work on Woodlands-Johor Rail Link; Operations Expected to Begin at End-2026', *Straits Times*, July 30, 2020, https://www.straitstimes.com/singapore/singapore-and-malaysia-sign-deal-to-resume-work-on-woodlands-johor-rail-link.

32. Stephen Cairns and Jane M. Jacobs, *Buildings Must Die: A Perverse View of Architecture* (Cambridge, MA: MIT Press, 2014), 67.

33. Lee Hsien Loong, 'Prime Minister Lee Hsien Loong's National Day Rally 2011 (Speech in English), Sunday, 14 August 2011, at University Cultural Centre, National University of Singapore' (transcript, Prime Minister's Office Singapore, August 14, 2011), http://www.pmo.gov.sg/mediacentre/prime-minister-lee-hsien-loongs-national-day-rally-2011-speech-english-sunday-14-august.

34. Urban Redevelopment Authority of Singapore, 'Citation for Adaptive Reuse of the Former Tanjong Pagar Railway Station', *Rail Corridor*, accessed February 13, 2022, https://www.ura.gov.sg/MS/railcorridor/rfp/awards/Winners.aspx.

35. Urban Redevelopment Authority of Singapore, 'Overview', *Rail Corridor*, accessed February 13, 2022, https://www.ura.gov.sg/MS/railcorridor/rfp/overview.aspx.

36. Lily Kong and Brenda S. A. Yeoh, 'Urban Conservation in Singapore: A Survey of State Policies and Popular Attitudes', *Urban Studies* 31, no. 2 (1994): 247–65, doi.org:10.1080/00420989420080231; Brenda S. A. Yeoh and Lily Kong, 'Reading Landscape Meanings: State Constructions and Lived Experiences in Singapore's Chinatown', *Habitat International* 18, no. 4 (1994): 17–35, doi:10.1016/0197-3975(94)90015-9; Brenda S. A. Yeoh and Lily Kong, 'The Notion of Place in the Construction of History, Nostalgia and Heritage in Singapore', *Singapore Journal of Tropical Geography* 17, no. 1 (1996): 52–65, doi:10.1111/j.1467-9493.1996.tb00084.x; Brenda S. A. Yeoh and L. Y. Ho, 'Popular Cognition of National Monuments in Singapore', *Environmental Monitoring and Assessment* 44, no. 1 (1997): 21–31, doi:10.1023/A:1005755717141.

37. Yeoh and Kong, 'The Notion of Place', 60.

38. James S. Duncan, *The City as Text: The Politics of Landscape Interpretation in the Kandyan Kingdom* (Cambridge: Cambridge University Press, 2004), 19, cited in Yeoh and Kong, 'The Notion of Place', 60.

39. Cairns and Jacobs, 52.

40. Barbara Herrnstein Smith, *Contingencies of Value: Alternative Perspectives for Critical Theory* (Cambridge, MA: Harvard University Press, 1988), 125, cited in Cairns and Jacobs, *Buildings Must Die*, 49.

41. John A. Scanlan, *On Garbage* (London: Reaktion, 2005), 60, 116, cited in Cairns and Jacobs, *Buildings Must Die*, 51.

42. Cairns and Jacobs, 51.

43. Cairns and Jacobs, 51.

44. Cairns and Jacobs, 70–1.

45. Cairns and Jacobs, 70.

46. Edward Hollis, *The Secret Lives of Buildings: From the Parthenon to the Vegas Strip in Thirteen Stories* (London: Portobello Books Ltd, 2010), 8, cited in Cairns and Jacobs, *Buildings Must Die*, 64.

47. Aron Vinegar and Jorge Otero-Pailos, 'On Preserving the Openness of the Monument', *Future Anterior Journal of Historic Preservation, History, Theory, and Criticism* 9, no. 2 (2013): iii–vi.

48. Vinegar and Otero-Pailos, iii-vi.

49. Vinegar and Otero-Pailos, iii-vi.

50. Philip Brophy, 'Singapore Biennale 2006', *Philip Brophy*, October 2006, http://www.philipbrophy.com/projects/rstff/SingaporeBiennale2006_A.html.

51. Simryn Gill, *Guide to the Murals of the Tanjong Pagar Railway Station* (Singapore: Singapore Biennale 2006 and the Singapore National Arts Council, 2006). All citations about the station are taken from this edition.

52. Michael Fitzgerald, 'Against Blankness: The Inhabiting Spaces of Simryn Gill', *Art Asia Pacific*, accessed February 15, 2022, http://artasiapacific.com/Magazine/82/AgainstBlanknessTheInhabitingSpacesOfSimrynGill

53. Fitzgerald.

54. Fitzgerald.

55. Michel de Certeau, *The Practice of Everyday Life*, trans. Steven Rendall (Berkeley: University of California Press, 1984), 93.

56. Roland Barthes, *Camera Lucida*, trans. Richard Howard (London: Vintage, 2000), 25–60.

57. Murat Nemet-Nejat, *The Peripheral Space of Photography* (Kobenhavn: Green Integer, 2003), 37.

58. Segrest, 'The Perimeter Projects', 27.

59. Michel Serres, *Hermes*, ed. Josue V. Harari and David F. Bell (Baltimore: Johns Hopkins University Press, 1982), xxxv, cited in Segrest, 'The Perimeter Projects', 28.

60. Henri Bergson, *An Introduction to Metaphysics*, trans. T. E. Hulme (New York: G. P. Putnam's Sons, 1912), 65.

61. There were plans to reconstruct the pavilion in Australia. See: Debbie Cuthbertson, 'Australia's First Venice Biennale Pavilion Finds Second Life as New Building Progresses', *Sydney Morning Herald*, October 18, 2014, http://www.smh.com.au/entertainment/art-and-design/australias-first-venice-biennale-pavilion-finds-second-life-as-new-building-progresses-20141017-117imr.html.

62. Kitty Hauser, 'Artist Simryn Gill Becomes Australia's Accidental Ambassador at Venice Biennale', *The Australian*, May 25, 2013, http://www.theaustralian.com.au/arts/review/artists-simryn-gill-becomes-australias-accidental-ambassador-at-venice-biennale/story-fn9n8gph-1226649130185.

63. Antoine Picon and Karen Bates, 'Anxious Landscapes: From the Ruin to Rust', *Grey Room*, no. 1 (2000): 76.

64. Picon and Bates, 77.

65. Fitzgerald.

66. Walter Benjamin, 'The Work of Art in the Age of Mechanical Reproduction', in *Illuminations: Essays and Reflections*, ed. Hannah Arendt, trans. Harry Zohn (London: Pimlico, 1999), 232.

67. Gregory J. Seigworth and Melissa Gregg, 'An Inventory of Shimmers', in *The Affect Theory Reader*, ed. Melissa Gregg and Gregory J. Seigworth (Durham: Duke University Press, 2010), 1.

68. Kathleen Stewart, 'Afterword: Worlding Refrains', in *The Affect Theory Reader*, ed. Melissa Gregg and Gregory J. Seigworth (Durham: Duke University Press, 2010), 353.

69. Mary Douglas, *Purity and Danger* (London: Routledge, 1966), 160.

70. Hauser, 'Artist Simryn Gill'.

71. Tim Edensor, 'Waste Matter - The Debris of Industrial Ruins and the Disordering of the Material World', *Journal of Material Culture* 10, no. 3 (2005): 324, doi:10.1177/1359183505057346.

72. Edensor, 318–20.

73. Edensor, 324. See also: Tim Edensor, *Industrial Ruins: Space, Aesthetics and Materiality* (New York; Oxford: Berg, 2005), 139.

74. Edensor, 'Waste Matter', 324.

75. Edensor, *Industrial Ruins*, 162.

76. Edensor, 164.

77. Edensor, 164.

78. Svetlana Boym, 'Ruins of the Avant-Garde: From Tatlin's Tower to Paper Architecture', in *Ruins of Modernity*, eds. Julia Hell and Andreas Schönle (Durham: Duke University Press, 2010), 58.

79. Boym, 'Avant-Garde'.

80. Caitlin DeSilvey, 'Making Sense of Transience: An Anticipatory History', *Cultural Geographies* 19, no. 1 (2012): 34, doi:10.1177/1474474010397599.

81. DeSilvey, 35.

82. DeSilvey, 35.

83. Correspondence with artist, 25 March 2016.

84. Kevin Hetherington, 'Phantasmagoria/Phantasm Agora: Materialities, Spatialities and Ghosts', in 'Spatial Hauntings', ed. Kevin Hetherington and Monica Degan, special issue, *Space and Culture* 4, no. 11–12 (2001): 39.

4. KEEPING CATS, HOARDING THINGS: SITUATIONS IN THE HOUSING BLOCK

1. Fann Sim, 'Plight of "Homeless" Woman in Bedok Captures Netizens' Attention', *Yahoo! News*, January 8, 2013, https://sg.news.yahoo.com/plight-of--homeless--woman-in-bedok-captures-attention-of-netizens-153857867.html.

2. Jaipal Singh Gill, 'Commentary: Do We Need to Do More to Protect Our Community Cats?', *Channel News Asia*, June 2, 2021. https://www.channelnewsasia.com/commentary/community-cats-cruelty-ang-mo-kio-spca-cases-animal-abuse-1823451.; Yun Ting Choo, 'Cats Injured by Glue Traps Meant for Rats in Redhill', *Straits Times*, January 23, 2019, https://www.straitstimes.com/singapore/environment/cats-injured-by-glue-traps-meant-for-rats-in-redhill.

3. Ng Wei Kai, '"Every Few Steps There Was a Cat Lying on the Floor": How SCDF Officers Saved 13 Cats from Fire', *Straits Times*, April 30, 2022, https://www.straitstimes.com/singapore/every-few-steps-there-was-a-cat-lying-on-the-floor-how-scdf-officers-saved-13-cats-from-fire.

4. Kimberly E. Zarecor, 'Architecture in Eastern Europe and the Former Soviet Union', in *A Critical History of Contemporary Architecture, 1960–2010*, ed. Elie G. Haddad and David Rifkind (Surrey, England: Ashgate, 2014), 255–74.

5. Miles Glendinning, 'From European Welfare State to Asian Capitalism: The Transformation of "British Public Housing" in Hong Kong and Singapore', in *Architecture and the Welfare State*, ed. Mark Swenarton, Tom Avermaete, and Dirk van den Heuvel (London: Routledge, 2015), 299–320.

6. Ada Louise Huxtable, 'Russian Engineers Take on Problem of Mass Housing', *Chicago Tribune*, October 19, 1967, http://archives.chicagotribune.com/1967/10/19/page/10/article/russian-engineers-take-on-problem-of-mass-housing.

7. Lauren Berlant, *Cruel Optimism* (Durham: Duke University Press Books, 2011), 198.

8. Lauren Berlant, 'Austerity, Precarity, Awkwardness' (November 2011), https://supervalentthought.files.wordpress.com/2011/12/berlant-aaa-2011final.pdf.

9. Berlant, *Cruel Optimism*, 195.

10. Berlant, 195.

11. Berlant, 198.

12. Stephen Cairns et al., 'Void Decks – Singapore's Void Decks', in *Public Space in Urban Asia*, ed. William S. W. Lim (Singapore: World Scientific, 2014), 80–89.

13. The literary texts of the first two decades of Singapore's founding (1960s-1970s) describe the degree of standardization and the ensuing homogenization of the housing landscape. See Lilian Chee, 'Domestic Digressions: Interrogating Singaporean Public Housing through Its Literary Forms', in *The Routledge Companion on Architecture, Literature and The City*, ed. Jonathan Charley (London: Routledge, 2018), 37–57.

14. Lee Kuan Yew, *From Third World to First: The Singapore Story 1965-2000* (Singapore: Singapore Press Holdings, 2000), 117.

15. Georges Teyssot, *A Topology of Everyday Constellations* (Cambridge, MA: MIT Press, 2013), 5-6.

16. Jonathan Hill, *Occupying Architecture: Between the Architect and the User* (London; Routledge, 1998); Karen Burns, 'Architecture/Discipline/Bondage', in *Desiring Practices: Architecture, Gender, and the Interdisciplinary*, ed. Duncan McCorquodale, Katerina Rüedi and Sarah Wigglesworth (London: Black Dog, 1996), 73; Tim Anstey, Katja Grillner, and Rolf Hughes, eds., *Architecture and Authorship* (London: Black Dog, 2007).

17. Eunice Seng, 'Habitation and the Invention of a Nation, Singapore 1936-1979', (PhD Diss., Columbia University, 2015).

18. Chua Beng-Huat, 'Not Depoliticized but Ideologically Successful: The Public Housing Programme in Singapore', *International Journal of Urban and Regional Research* 15, no. 1 (1991): 24–41, doi:10.1111/j.1468-2427.1991.tb00681.x.

19. Lee, *From Third World to First*, 117.

20. Chua argues that housing stakeholding is a two-way situation which gives the state political legitimacy and the citizen a stakehold in the nation. See: Chua Beng-Huat, *Political Legitimacy and Housing: Stakeholding in Singapore* (London: Routledge, 1997).

21. Christopher Tremewan, 'Public Housing: The Working Class Barracks', in *The Political Economy of Social Control in Singapore* (London: St Martin's Press, 1994), 45–73.

22. Chua Beng-Huat, *Communitarian Ideology and Democracy in Singapore* (New York: Routledge, 1996), 136.

23. Housing and Development Board, 'Eligibility to Buy New HDB Flat', *HDB InfoWEB*, September 15, 2015, http://www.hdb.gov.sg/cs/infoweb/residential/buying-a-flat/new/hdb-flat.

24. Tai Ching Ling, *Housing and High-Rise Living: A Study of Singapore's Public Housing* (Singapore: Chopmen, 1988), 114, cited in Martin Perry, Lily Kong, and Brenda S. A. Yeoh, *Singapore: A Developmental City State* (New York: Wiley, 1997), 246.

25. Chua, *Communitarian Ideology*, 129.

26. Housing and Development Board, *Annual report 2020/2021*, 2021, https://www20.hdb.gov.sg/fi10/fi10221p.nsf/hdb/2021/assets/ebooks/annual-report-2021.pdf. The overall percentage has remained more or less constant, between 81-82%, from 2015-2019.

27. Robbie B. H. Goh, 'Things to a Void: Utopian Discourse, Communality and Constructed Interstices in Singapore Public Housing,' in *Theorizing the Southeast Asian City as Text*, ed. Brenda S. A. Yeoh and Robbie B. H. Goh (Singapore: World Scientific, 2003), 64.

28. Perry, Kong and Yeoh, *A Developmental City State*, 6.

29. Liu Thai Ker, 'Design for Better Living Conditions', in *Public Housing in Singapore: A Multi-Disciplinary Study*, ed. Stephen H. K. Yeh (Singapore: Singapore University Press and HDB, 1975), 118.

30. Liu, 117.

31. Le Corbusier, *Precisions. On the Present State of Architecture and City Planning*, trans. Edith Schreiber Aujame (Cambridge, MA: MIT Press, 1991), 36. Emphasis in original.

32. Liu, 117.

33. Jane M. Jacobs, 'The Only Way Is Up', panel discussion, National Library Board, Singapore, November 6, 2015.

34. Liu, 1–29.

35. Catherine Ingraham, *Architecture and the Burdens of Linearity* (New Haven, CT: Yale University Press, 1998), 30.

36. Ingraham, 46.

37. Ingraham, 51.

38. Ingraham, 51. Emphasis mine.

39. Ingraham, 30.

40. Lee Hsien Loong, 'Speech to Parliament on Reading of Penal Code (Amendment) Bill, 22 October' (speech, Parliament of Singapore, Singapore, October 22, 2007).

41. Ingraham, 42.

42. Ingraham, 52.

43. From Singaporean award-winning poet and author Alfian Sa'at's personal blog, which is no longer open to public viewing. Cited in Limin Hee, *Constructing Singapore Public Space* (Singapore: Springer Science+Business Media, 2017), 154. For observations on occupant driven public spaces in the HDB, specifically the 'persistent' and 'defiant', see: Hee, 144–57.

44. For a history of tropical architecture, see: Chang Jiat-Hwee, *A Genealogy of Tropical Architecture: Colonial Networks, Nature and Technoscience* (London: Routledge, 2016).

45. SIT 1947 census, cited in Housing and Development Board, *50,000 Up: Homes for the People* (Singapore: Housing Development Board, 1966), 20.

46. *Report of Survey: Colony of Singapore Masterplan* (Singapore: Government Print Office, 1958; 1955).

47. Housing and Development Board, *50,000 Up*, 24. All statistics for this period are obtained from this source.

48. For an argument on the politicised nature of open space in public housing, see Loh Kin-Kit, 'Domestic(ated) Landscapes: Contesting the Open Spaces of Public Housing in Singapore' (Master of Architecture Diss., National University of Singapore, 2014).

49. National Heritage Board, *Community Heritage Series III: Void Decks* (Singapore: National Heritage Board, 2013), 3.

50. Ooi Giok Ling and Thomas T. W. Tan, 'The Social Significance of Public Spaces in Public Housing Estates', in *Public Space: Design, Use and Management*, ed. Chua Beng-Huat and Norman Edwards (Singapore: Singapore University Press for Centre for Advanced Studies, Faculty of Arts and Social Sciences, 1992), 73.

51. Housing and Development Board, 'Annual Report 1977/1978' (Singapore: Housing and Development Board, 1978), 24.

52. Singapore Institute of Architects Magazine, 'Park Road Redevelopment', *Singapore Institute of Architects* 28–29 (October 1968): 9–19.

53. *Straits Times*, 'Multi-Use Buildings: First Goes up in Chinatown', April 21, 1967, http://eresources.nlb.gov.sg/newspapers/Digitised/Article/straitstimes19670421-1.2.24.aspx.

54. *Straits Times*, 'Multi-Use Buildings'.

55. Tony Tan Keng Joo et al., 'Physical Planning and Design', in *Housing a Nation: Twenty-Five Years of Public Housing in Singapore,* ed. Aline K. Wong and Stephen H. K. Yeh (Singapore: Maruzen Asia, 1985), 90.

56. Ooi and Tan, 80.

57. Cairns et al., 82.

58. The Singapore Improvement Trust (SIT) (1927-59) was the British government's colonial housing agency, and a precursor to the Housing Development Board. It built the Tiong Bahru housing estate which is now being conserved. It constructed 23,000 flats in its tenure, and initiated the first HDB satellite town Queenstown, which was eventually completed by the HDB.

59. Seng, 223–24.

60. Liu, 137.

61. Seng, 227.

62. Henri Lefebvre, *The Production of Space,* trans. Donald Nicholson-Smith (Malden, MA: Blackwell, 1991), 195.

63. Rizal Hassan, 'Some Sociological Implications of Public Housing in Singapore', *South-East Asian Journal of Sociology* 2 (1969): 23–26; Rizal Hassan, 'Social and Psychological Implications of High Population Density', *University of Singapore, Department of Sociology Working Paper*, no. 47 (1975).

64. Foreign workers using the Little India void decks caused concern by littering and fighting after being drunk. Eventually, gates were installed to privatise the void deck in this area. See: Gan Ling Kai, 'Little India Residents' Woes,' *New Paper*, June 20, 2011, http://news.asiaone.com/print/News/AsiaOne%2BNews/Singapore/Story/A1Story20110620-285089.html. Recent issues include allegations of possible ghettoism at the void deck when non-Singaporean residents posted notices for gatherings. This eventually led to an enforced quota of non-Singaporean residents in HDB blocks. See: Janice Heng, 'Quota for HDB Subletting to Foreigners: 8% Cent for Neighbourhoods, 11% for Blocks', *Straits Times,* January 16, 2014, http://www.straitstimes.com/singapore/quota-for-hdb-subletting-to-foreigners-8-cent-for-neighbourhoods-11-for-blocks.

65. Ingraham, 54.

66. *AsiaOne*, 'Man, 76, Found Dead in Trash-Filled Flat', August 15, 2012, https://www.asiaone.com/print/News/Latest%2BNews/Singapore/Story/A1Story20120815-365468.html.

67. Jean Lau, 'Woman's 20-Year Hoarding Habit Irritates Neighbours', *New Paper*, May 1, 2015, http://news.asiaone.com/news/singapore/womans-20-year-hoarding-habit-irritates-neighbours-0.

68. Ingraham, 31.

69. Sylvia Lavin, 'Architecture in Extremis', *Log*, no. 22 (2011): 51–61.

70. Susan Strasser, *Waste and Want: A Social History of Trash* (New York: Metropolitan Books, 1999), 6, cited in Neil Maycroft, 'Not Moving Things Along: Hoarding, Clutter and Other Ambiguous Matter', *Journal of Consumer Behaviour* 8, no. 6 (November 2009): 363, doi:10.1002/cb.298.

71. Maycroft, 358.

72. Maycroft, 358.

73. Lavin, 59.

74. Cat Welfare Society, 'Frequently Asked Questions | Cat Welfare Society', accessed March 23, 2022, http://www.catwelfare.org/faq.

75. Housing and Development Board, 'Keeping Pets', HDB InfoWEB, accessed March 22, 2022, http://www.hdb.gov.sg/cs/infoweb/community/neighbourliness/being-a-good-neighbour/keeping-pets.

76. Lucy Davis, 'Zones of Contagion: The Singapore Body Politic and the Body of the Street-Cat', in *Considering Animals: Contemporary Studies in Human-Animal Relations*, ed. Carol Freeman, Elizabeth Leane, and Yvette Watt (Abingdon, Oxon: Ashgate, 2011), 194.

77. For an urban account of the Singaporean street cat, see: Ying-kit Chan, 'No Room to Swing a Cat? Animal Treatment and Urban Space in Singapore', *Southeast Asian Studies* 5, no. 2 (2016): 305–29, https://doi.org/10.20495/seas.5.2_305.

78. Housing and Development Board, 'Keeping Pets'.

79. There have been recent calls to reconsider ownership of cats in public housing. At the time this book went to press, a public consultation exercise was being planned. Wallace Woon, 'ST Explains: Why Cats Are Banned from HDB Flats, and What Lifting the Ban Would Take', *The Straits Times*, September 5, 2022, https://www.straitstimes.com/singapore/environment/st-explains-why-cats-are-banned-from-hdb-flats-and-what-is-being-done-to-change-it.

80. Huw Griffiths, Ingrid Poulter, and David Sibley, 'Feral Cats in the City', in *Animal Spaces, Beastly Places: New Geographies of Human-Animal Relations*, ed. Chris Philo and Chris Wilbert (London: Routledge, 2000), 61.

81. Davis, 194.

82. Paul Rae, 'Cat's Entertainment: Feline Performance in the Lion City', *The Drama Review* 51, no. 1 (2007): 129.

83. Davis, 189.

84. Chew Hui Min, 'Another Cat Found Dead at Foot of Yishun Block; Eighth Cat Death in Six Weeks', *Straits Times*, October 30, 2015, http://www.straitstimes.com/singapore/another-cat-found-dead-at-foot-of-yishun-block-eighth-cat-death-in-six-weeks.

85. Berlant, *Cruel Optimism*, 6.

86. Berlant, 8.

6. 03-FLATS: ARCHITECTURE FILMMAKING, DISCIPLINARY QUESTIONS

1. Lei Yuan Bin and Lilian Chee, *03-FLATS*, Architectural essay film (Singapore: National University of Singapore and 13 Little Pictures, 2014); Lilian Chee, '*03-FLATS*: Domesticity, Home and Its Representations', *03-FLATS* Research Project, 2014, http://www.03-flats.com/; Lilian Chee, 'Reimagining Domesticity in *03-FLATS*: Entering Singaporean Domestic Space through the Essay Film', in *Making Visible: Architecture Filmmaking*, ed. Hugh Campbell and Igea Troiani (Bristol: Intellect Books, 2019), 117–38. For the trailers to *03-FLATS*, see: *03-FLATS*, 'Trailer', *03-FLATS*, accessed March 8, 2022, https://www.03-flats.com/trailer. *03-FLATS* was funded by MOE Tier 1 and MOE-NUS Start-up Grants. It received IRB clearance: NUS-IRB Reference code: 12-223; Approval number: NUS 1619.

2. See for example: Jason Pomeroy, 'Room at the Top—The Roof as an Alternative Habitable/Social Space in the Singapore Context', *Journal of Urban Design* 17, no. 3 (2012): 413–24, doi:10.1080/13574809.2012.666176; Tony Tan Keng Joo and Wong Tai-Chee, 'Public Housing in Singapore: A Sustainable Housing Form and Development', in *Spatial Planning for a Sustainable Singapore*, ed. Wong Tai-Chee, Belinda Yuen, and Charles Goldblum (Singapore: Springer, 2008), 135–50; Grace K. M. Wong, 'Vertical Cities as a Solution for Land Scarcity: The Tallest Public Housing Development in Singapore', *Urban Design International* 9, no. 1 (2004): 17–30, doi:10.1057/palgrave.udi.9000108; Sishir Chang, 'A High-Rise Vernacular in Singapore's Housing Development Board Housing', *Berkeley Planning Journal* 14, no. 1 (2000): 97–116, doi:10.5070/BP314112985.

3. Lilian Chee, 'Unhousing Sexuality: Sexuality and Singlehood in Singapore's Public Housing', in *Sexuality and Gender at Home: Experience, Politics, Transgression*, ed. Brent Carnell et al. (London; New York: Bloomsbury, 2017), 35–50; Lilian Chee, 'Domestic Digressions: Interrogating Singaporean Public Housing through Its Literary Forms', in *The Routledge Companion on Architecture, Literature and The City*, ed. Jonathan Charley (London: Routledge, 2018), 37–57.

4. Lilian Chee, 'Domesticity, Gender, Architecture: Locating an Expanded Field', in *The Routledge Companion to Contemporary Architectural History*, ed. Duanfang Lu (New York: Routledge, forthcoming), 185-201.

5. Filming started in September 2012 and was completed in July 2013. The editing process took almost a year, and a final cut was completed in May 2014. The film was officially released at the Busan International Film Festival in October 2014, where it was chosen to compete in the Wide Angle Documentary competition. *03-FLATS* won the Salaya International Documentary Prize in February 2015. The film was shown at the Singapore Film Festival in December 2015 to a sell-out audience. It has travelled to Kuala Lumpur, Seoul, Paris, Taipei, London, Delft, Zurich, Venice, Melbourne and Sydney, amongst other places.

6. For reviews of the film, see: 03-FLATS, 'Reviews', *03-FLATS*, accessed March 8, 2022, https://www.03-flats.com/reviews.

7. Housing and Development Board, *First Decade in Public Housing, 1960-1969* (Singapore: Housing and Development Board, 1970); Stephen. H. K. Yeh, *Public Housing in Singapore: A Multi-Disciplinary Study* (Singapore: Singapore University Press, 1975); Aline. K. Wong and Stephen H.K. Yeh, eds., *Housing a Nation: Twenty-Five Years of Public Housing in Singapore* (Singapore: Maruzen Asia, 1985); Housing and Development Board, 'About Us', HDB InfoWEB, accessed March 25, 2022, http://www.hdb.gov.sg/cs/infoweb/about-us.

8. Chua Beng-Huat, 'Not Depoliticized but Ideologically Successful: The Public Housing Programme in Singapore', *International Journal of Urban and Regional Research* 15, no. 1 (1991): 24–41, doi:10.1111/j.1468-2427.1991.tb00681.x; Chua Beng-Huat, *Political Legitimacy and Housing: Stakeholding in Singapore* (London: Routledge, 1997); Christopher Tremewan, 'Welfare and Governance: Public Housing under Singapore's Party-State', in *The East Asian Welfare Model: Welfare Orientalism and the State*, ed. Roger Goodman, Gordon White and Huck-Ju Kwon (London: Routledge, 1998), 77–105; Chua Beng-Huat, 'Public Housing Residents as Clients of the State', *Housing Studies* 15, no. 1 (2000): 45–60, doi:10.1080/02673030082469; Natalie Oswin, 'The Modern Model Family at Home in Singapore: A Queer Geography', *Transactions of the Institute of British Geographers* 35, no. 2 (2010): 256–68, doi:10.1111/j.1475-5661.2009.00379.x; Lilian Chee, 'Keeping Cats, Hoarding Things: Domestic Situations in the Public Spaces of the Singaporean Housing Block', *The Journal of Architecture* 22, no. 6 (2017): 1041–65, doi:10.1080/13602365.2017.1362024.

9. Jane M. Jacobs and Stephen Cairns, 'The Modern Touch: Interior Design and Modernisation in Post-Independence Singapore', *Environment and Planning A* 40, no. 3 (2008): 572–95, doi:10.1068/a39123; Lilian Chee, 'The Public Private Interior: Constructing the Modern Domestic Interior in Singapore's Public Housing', in *The Handbook of Interior Architecture and Design*, ed. Graeme Brooker and Lois Weinthal (London: Bloomsbury, 2013), 199–212.

10. The term 'heartlander' is used to denote a public housing neighbourhood-bound occupant. It was first introduced in 1991 but gained some controversial mileage in 1999 when during the National Day Rally speech, ex-Prime Minister Goh Chok Tong referenced it in opposition to the globalised 'cosmopolitan' citizen. For criticisms engaging the heartland or heartlander, see, for example: Yeo Wei-Wei, 'Of Trees and the Heartland: Singapore's Narratives', in *Beyond Description: Singapore Space Historicity*, ed. Ryan Bishop, John Phillips, and Yeo Wei-Wei (London: Routledge, 2004), 17–29; Kenneth Paul Tan, *Cinema and Television in Singapore: Resistance in One Dimension* (Boston: Brill, 2008), 185–218; Angelia Poon, 'Common Ground, Multiple Claims: Representing and Constructing Singapore's "Heartland"', *Asian Studies Review* 37, no. 4 (2013): 559–76, doi:10.1080/10357823.2013.844768.

11. The recent 2020 Singapore Household Survey records 22.6% of men aged 35-39 years and 19.3% of women aged 35-39 years as singles. The overall number of singles is also rising at 33.3% for men and 29.9% for women. 35 years is the threshold age for singles looking to purchase a Housing and Development Board (HDB) flat on their own. See: Department of Statistics Singapore, 'Singapore Census of Population 2020, Statistical Release 1: Demographic Characteristics, Education, Language and Religion' (Singapore: Department of Statistics, 2020), 26–27.

12. Charissa Yong, 'MND Rejects Petition to Reform Public Housing Policies for Single Parents', *Straits Times*, November 29, 2017, https://www.straitstimes.com/singapore/mnd-rejects-petition-to-reform-public-housing-policies-for-single-parents.

13. Lei and Chee, *03-FLATS*; Chee, '*03-FLATS*'; Chee, 'Reimagining Domesticity'. Various housing schemes, such as the Single Singapore Citizen Scheme and the Joint Singles Scheme, note that only unmarried or divorced Singaporean singles who are at least 35 years old, or Singaporeans who are widowed or an orphan who are at least 21 years old are eligible to purchase an HDB resale flat or apply for a 2-Room Flexi HDB Built-To-Order (BTO) flat. For debates on housing Singaporean singles, see, for example: Ng Jun Sen, 'New HDB flats: Why singles pay $15k premium', *Straits Times*, July 1, 2017, https://www.straitstimes.com/singapore/housing/new-hdb-flats-why-singles-pay-15k-premium; Kathleen F., 'MP Dr Lily Neo urges for HDB's Joint Singles Scheme to be reviewed to offer more privacy and flexibility', *Online Citizen Asia*, accessed March 5, 2022, https://www.onlinecitizenasia.com/2020/03/05/mp-dr-lily-neo-urges-for-hdbs-joint-singles-scheme-to-be-reviewed-to-offer-more-privacy-and-flexibility/.

14. Chee, 'Unhousing Sexuality', 35-50.

15. For a critique on biology and housing provision, see: Oswin, 'The Modern Model Family'.

16. The *Berita Singapura* series is a news magazine film commissioned and produced by the Singapore Ministry of Culture in the 1960s. Narrated in English, Malay, Mandarin, Tamil and also the Chinese dialects, Hokkien and Cantonese, these short films were screened in the cinemas before the movies of the day. They were also brought out to the rural hamlets (*kampongs*) to reach out to a wider audience. See: Ministry of Culture Singapore, *Berita Singapura: Housing Week*, Videocassette, News Magazines (Singapore: Ministry of Culture, Broadcasting Division, 1963); Ministry of Culture Singapore, *Berita Singapura: Celebration of New Housing*, Videocassette, News Magazines (Singapore: Ministry of Culture, Broadcasting Division, 1965); Ministry of Culture Singapore, *Berita Singapura: Visiting New Flats*, Videocassette, News Magazines (Singapore: Ministry of Culture, Broadcasting Division, 1967); Ministry of Culture Singapore, *Berita Singapura: A New Look At Housing*, Videocassette, News Magazines (Singapore: Ministry of Culture, Broadcasting Division, 1967).

17. Doreen Massey, *For Space* (London: SAGE Publications, 2005), 9.

18. Michael Herzfeld, 'Ethical and Epistemic Reflections on/of Anthropological Vision', in *Made to Be Seen: Perspectives on the History of Visual Anthropology*, ed. Marcus Banks and Jay Ruby (Chicago: University of Chicago Press, 2011), 316–17.

19. Jay Ruby, *Picturing Culture: Explorations of Film and Anthropology* (Chicago: University of Chicago Press, 2000); David MacDougall, *The Corporeal Image: Film, Ethnography, and the Senses* (Princeton: Princeton University Press, 2005); Sarah Pink, *Doing Visual Ethnography* (London: SAGE, 2006); Gillian Rose, *Visual Methodologies: An Introduction to the Interpretation of Visual Materials* (London: SAGE, 2007); Marcus Banks and Jay Ruby, eds., *Made to Be Seen: Perspectives on the History of Visual Anthropology* (Chicago: University of Chicago Press, 2012). The history of visual anthropology – arguably established after World War II with the formalisation of tertiary and institutional grants supporting visual outcomes from fieldwork – is incomplete without the inclusion of Margaret Mead and Gregory Bateson's seminal Balinese fieldwork films, or Jean Rouch's films on Nigerian Songhai rituals and ceremonies.

20. Roxana Waterson, 'Visual Anthropology and the Built Environment: Interpenetrations of the Visible and the Invisible', in *Made to Be Seen: Perspectives on the History of Visual Anthropology*, ed. Marcus Banks and Jay Ruby (Chicago: University of Chicago Press, 2011), 79.

21. Roxana Waterson, *Living House: An Anthropology of Architecture in South-East Asia* (Singapore: Oxford University Press, 1990).

22. Waterson, 'Visual Anthropology', 98–100. Waterson is describing the scenes from this film: Beate Engelbrecht, *Building Season in Tiébélé: A Royal Compound in Change (Kasena, Burkina Faso)* (Göttingen: Institut für den Wissenschaftlichen Film Göttingen, 1999).

23. Waterson, 'Visual Anthropology', 100.

24. Pink, *Doing Visual Ethnography*, 15.

25. Sarah Pink, *Doing Sensory Ethnography* (London: SAGE, 2009), 10, doi:10.4135/9781446249383.

26. Sarah Pink, *Home Truths: Gender, Domestic Objects and Everyday Life* (New York: Berg, 2004); Pink, *Doing Visual Ethnography*; Sarah Pink, 'Making the Sensory Home: Laundry Routes and Energy Flows', in *Situating Everyday Life: Practices and Places* (London: SAGE, 2012), 66–83; Sarah Pink and Kerstin Leder Mackley, 'Saturated and Situated: Expanding the Meaning of Media in the Routines of Everyday Life', *Media, Culture & Society* 35, no. 6 (2013): 677–91, doi:10.1177/0163443713491298.

27. Sarah Pink, 'Video and a Sense of the Invisible: Approaching Domestic Energy Consumption Through the Sensory Home', *Sociological Research Online* 17, no. 1 (2012): 87–105, doi:10.5153/sro.2583; Pink and Mackley, 'Saturated and Situated'; Energy & Digital Living, 'Sensory Ethnography', *Energy & Digital Living*, accessed March 16, 2022, http://energyanddigitalliving.com/doing-sensory-ethnography/.

28. Jay Ruby, 'Toward an Anthropological Cinema'. *Film Comment* 7, no. 1 (1971): 35–40; Herzfeld, 'Anthropological Vision'.

29. The filmmaker did not view research as antagonistic to aesthetic practice. Lei shared my admiration for Chantal Akerman's arthouse long-take domestic life of a housewife *Jeanne Dielman* (1975) as filmic precedent. He saw the film as an experiment of representational form including notions of durational cinema, essay film, still image film, and intermedial application of the essay film.

30. Suzanne Ewing, 'Field/Work', *Arq: Architectural Research Quarterly* 15, no. 4 (2011): 309, doi:10.1017/S1359135512000073.

31. Ewing 'Field/Work', 309. Emphasis mine.

32. Jane M. Jacob's ongoing research into the critical history of HDB's technological innovations is significant here.

33. Igea Troiani and Suzanne Ewing, 'Visual Research Methods in Architecture: New Forms of "Critical Visualization"', in *Visual Methodologies in Architectural Research*, ed. Suzanne Ewing and Igea Troiani (Bristol: Intellect Books, 2020), 1.

34. Troiani and Ewing, 28.

35. Igea Troiani and Alison Kahn, 'Beyond the Academic Book: New "Undisciplined" Corporeal Publication', *Architecture and Culture* 4, no. 1 (2016): 53, doi:10.1080/20507828.2015.1094228.

36. Hilar Stadler, Martino Stierli, and Peter Fischli, *Las Vegas Studio: Images from the Archive of Robert Venturi and Denise Scott Brown* (Zurich: Verlag Scheidegger & Spiess, 2015), 15, cited in Troiani and Ewing, 16.

37. Stadler, Stierli, and Fischli, 9, cited in Troiani and Ewing, 16.

38. Mark Linder, 'TRANSdisciplinarity', in *Hunch 9 – Disciplines*, ed. Penelope Dean (Rotterdam: Episode Publishers, 2005), 14.

39. Linder, 15.

40. Linder, 13.

41. Jane Rendell, 'Working Between and Across: Some Psychic Dimensions of Architecture's Inter- and Transdisciplinarity'. *Architecture and Culture* 1, no. 1 (2013): 131, doi:10.2752/175145213X13756908698685.

42. Rendell, 131.

43. For example: a dramatic plot versus an observational documentary; including music versus sitting with silence; allowing the camera to wander versus sticking with strict filming angles.

44. Ivone Margulies, *Nothing Happens: Chantal Akerman's Hyperrealist Everyday* (Durham: Duke University Press, 1996), 4.

45. For critiques on how domesticity has been denigrated see for example: Michel de Certeau, Luce Giard, and Pierre Mayol, *The Practice of Everyday Life, Volume 2: Living and Cooking*, trans. Timothy J. Tomasik (Minneapolis: University of Minnesota Press, 1998); Daniel Miller, ed., *Home Possessions: Material Culture Behind Closed Doors* (Oxford: Berg, 2001).

46. Igea Troiani, and Suzanne Ewing. 'Inside Architecture from the Outside: Architecture's Disciplinary Practices'. *Architecture and Culture* 2, no. 2 (2014): 163, doi:10.2752/205078214X14030010182308.

47. Troiani and Ewing, 'Architecture's Disciplinary Practices'.

48. Stan Allen, *Practice: Architecture, Technique and Representation* (New York: Routledge, 2009), xvii, cited in Troiani and Ewing, '"Critical Visualisation"', 155.

49. Gillian Rose and Divya P. Tolia-Kelly, 'Visuality/Materiality: Introducing a Manifesto for Practice', in *Visuality/Materiality: Images, Objects and Practices*, ed. Gillian Rose and Divya P. Tolia-Kelly (Surrey, UK: Ashgate, 2012), 2.

50. Rose and Tolia-Kelly, 3.

51. Rose and Tolia-Kelly, 4.

52. Timothy Corrigan, *The Essay Film: From Montaigne, After Marker* (New York: Oxford University Press, 2011).

53. Corrigan, 13, 14.

54. Corrigan, 6.

55. Nora M. Alter, 'The Political Im/Perceptible in the Essay Film: Farocki's "Images of the World and the Inscription of War"', *New German Critique*, no. 68 (1996): 171, doi:10.2307/3108669.

56. Laura Rascaroli, *The Personal Camera: Subjective Cinema and the Essay Film* (London: Wallflower, 2009); Corrigan, *The Essay Film*; Elizabeth Papazian and Caroline Eades, eds., *The Essay Film: Dialogue, Politics, Utopia* (London: Wallflower, 2016); Nora M. Alter and Timothy Corrigan, eds., *Essays on the Essay Film* (New York: Columbia University Press, 2017).

57. Nora M. Alter, *Projecting History: German Nonfiction Cinema, 1967-2000* (Ann Arbor, MI: University of Michigan Press, 2002), 7–8.

58. Laura Rascaroli, *How the Essay Film Thinks* (New York: Oxford University Press, 2017).

59. Laura Rascaroli, 'The Essay Film: Problems, Definitions, Textual Commitments', *Framework: The Journal of Cinema and Media* 49, no. 2 (2008): 35, doi:10.1353/frm.0.0019.

60. Corrigan, 23.

61. Theodor W. Adorno, *Notes to Literature*, vol. 1, ed. Rolf Tiedemann, trans. Shierry Weber Nicholson (New York: Columbia University Press, 1991), 160.

62. Penelope Haralambidou, 'The Architectural Essay Film', *Arq: Architectural Research Quarterly* 19, no. 3 (2015): 237, doi:10.1017/S1359135515000524.

63. Giuliana Bruno, *Public Intimacy: Architecture and the Visual Arts* (Cambridge, MA: MIT Press, 2007), 195; Peter Gidal, 'Films', in *Films and Painting*, ed. David Herbert (London: Studio Vista, 1971), 84.

64. Bruno, 200.

65. Margulies, 65–99.

66. Richard Brody, 'To Life: A Monthlong Retrospective of Chantal Akerman's Films', *New Yorker*, April 4, 2016, https://www.newyorker.com/magazine/2016/04/04/a-chantal-akerman-retrospective.

67. Margulies, 73.

68. Rosalind E. Krauss, *The Originality of the Avant Garde and Other Modernist Myths* (Cambridge, MA: MIT Press, 1985), 253, cited in Margulies, *Nothing Happens,* 73.

69. Margaret Mead, 'Visual Anthropology in a Discipline of Words', in *Principles of Visual Anthropology,* ed. Paul Hockings (Berlin: De Gruyter, 1995), 9.

70. Margulies, 37.

71. Margulies, 69.

72. Margulies, 69.

73. Doreen Massey, 'Living in Wythenshawe', in *The Unknown City: Contesting Architecture and Social Space,* ed. Iain Borden et al. (Cambridge, MA: MIT Press, 2001), 463.

74. Mads Farsø and Rikke Munck Petersen, 'Conceiving Landscape through Film: Filmic Explorations in Design Studio Teaching', *Architecture and Culture* 3, no. 1 (2015): 81, doi:10.2752/20507821 5X14236574273664.

75. Rose and Tolia-Kelly, 3.

76. For a good account of self-reflexivity, see: Jane M. Jacobs, Stephen Cairns, and Ignaz Strebel, 'Materializing Vision: Performing a High-Rise View', in *Visuality/Materiality: Images, Objects and Practices,* ed. Gillian Rose and Divya P. Tolia-Kelly (Surrey, UK: Ashgate, 2012), 133–52.

77. Sarah Pink and Kerstin Leder Mackley, 'Re-Enactment Methodologies for Everyday Life Research: Art Therapy Insights for Video Ethnography', *Visual Studies* 29, no. 2 (2014): 151, doi:10.1080/147258 6X.2014.887266.

78. Barbara Maria Stafford, 'Crystal and Smoke: Putting Image Back in Mind', in *A Field Guide to a New Meta-Field: Bridging the Humanities-Neurosciences Divide,* ed. Barbara Maria Stafford (Chicago: University of Chicago Press, 2011), 1, cited by Pink and Mackley, 'Re-Enactment Methodologies', 152.

79. Pink and Mackley, 'Re-Enactment Methodologies', 153.

80. The term 'disobedient domesticity' suggests insurgency and independence. See also: 'DISOBEDIENT BUILDINGS', accessed March 25, 2022, https://www.disobedientbuildings.com/; 'INSURGENT DOMESTICITIES', accessed March 10, 2023, https://www.socialdifference.columbia. edu/projects-/insurgent-domesticities.

81. Barbara Bolt, 'The Magic is in Handling', in *Practice as Research: Approaches to Creative Arts Enquiry,* ed. Estelle Barrett and Barbara Bolt (London: I. B. Tauris, 2010), 30.

82. Paul Carter, *Material Thinking: The Theory and Practice of Creative Research* (Carlton, Vic: Melbourne University Publishing, 2004).

83. Jane Rendell, 'Site-Writing: She Is Walking about in a Town Which She Does Not Know', *Home Cultures* 4, no. 2 (2007): 179, doi:10.2752/174063107X209019.

84. Dydia DeLyser et al., 'Forward Cultural Geographies', *Cultural Geographies* 25, no. 1 (2018): 5, doi:10.1177/1474474017745040.

85. Bolt, 33.

86. Bolt, 33.

87. Trinh T. Minh-Ha, and Eva Hohenberger, 'The Politics of Forms and Forces', in *D-Passage: The Digital Way,* by Trinh T. Minh-Ha (Durham: Duke University Press, 2013), 143.

88. Minh-Ha and Hogenberger, 143.

89. Igea Troiani and Hugh Campbell, 'Architecture Film: Framing the Discourse and Conditions of Production in Architectural Practice and Education', in *Architecture Filmmaking,* ed. Hugh Campbell and Igea Troiani (Bristol: Intellect Books, 2019), 14.

7. IN THE MIDST OF: FIELD NOTES AT A CEMETERY

1. Academic publications: Kevin Y. L. Tan, ed., *Spaces of the Dead: A Case from the Living* (Singapore: Singapore Heritage Society, 2011); Terence Chong and Chua Ai Lin, 'The Multiple Spaces of Bukit Brown', in *Public Space in Urban Asia*, ed. William S. W. Lim (Singapore: World Scientific, 2014), 26–55; Terence Chong, 'Bukit Brown Municipal Cemetery: Contestations of the Good Life in Singapore', in *Worlding Multiculturalisms: The Politics of Inter-Asian Dwelling*, ed. Daniel Goh Pei Siong and Wang Chih-Ming (Abingdon, Oxon: Routledge, 2015), 161–82.

2. World Monuments Fund, '2014 World Monuments Watch', World Monuments Fund, accessed April 8, 2022, https://www.wmf.org/downloads/press-kit/2014-Map-ENG.pdf.

3. Chong, 'Bukit Brown Municipal Cemetery', 161–2. Videos, blogs, websites: 'Ghosts of Bukit Brown Cemetery: GFS S03E03', YouTube video, 27:17, posted by 'Ghost Files Singapore', September 9, 2015, https://www.youtube.com/watch?v=LUKcrNEjQM8; All Things Bukit Brown, *All Things Bukit Brown - Heritage. Habitat. History* (blog), accessed April 12, 2022, http://bukitbrown.com/main/; 'The Rites and Rituals of Bukit Brown Cemetery', YouTube video, 14:22, posted by 'Roots Sg', September 28, 2015, https://www.youtube.com/watch?v=PckVbnTDSBQ; exhibitions e.g. *Bukit Brown: Documenting New Horizons of Knowledge* hosted by Singapore's National Library Board (NLB) in 2014, and seminars e.g. *An Inquiry into Singapore's Heritage, A Seminar by Educators and Students* (hosted September 8, 2012).

4. Nature Society (Singapore), 'Nature Society (Singapore)'s Position on Bukit Brown' (Position Paper, December 12, 2011), 5, https://www.nss.org.sg/documents/Nature%20Society%27s%20Position%20on%20Bukit%20Brown.pdf.

5. Asia Paranormal Investigators, 'Asia Paranormal Investigators', accessed April 10, 2022, http://api.sg/main/.

6. Kartini Saparudin, 'Bukit Brown Municipal Cemetery', Singapore Infopedia, accessed April 20, 2022, http://eresources.nlb.gov.sg/infopedia/articles/SIP_1358_2009-07-13.html; MBev 'Bukit Brown Cemetery: These Hills Are Alive', *UnTourist Singapore* (blog), accessed April 18, 2022, https://untouristsingapore.wordpress.com/2013/11/29/bukit-brown-cemetery-these-hills-are-alive/.

7. For historical information on Bukit Brown, see: Brenda S. A. Yeoh, *Contesting Space in Colonial Singapore: Power Relations and the Urban Built Environment* (Kuala Lumpur: Oxford University Press, 1996; Singapore: Singapore University Press, 2003); Elizabeth McKenzie, 'Bukit Brown: A Garden of History and Heritage', in *Spaces of the Dead: A Case from the Living*, ed. Kevin Y. L. Tan (Singapore: Singapore Heritage Society and Ethos Books, 2011), 54–99; Chong and Chua, Multiple Spaces'; Chong, 'Bukit Brown Municipal Cemetery'.

8. Brenda S. A. Yeoh, 'The Control of "Sacred" Space: Conflicts over the Chinese Burial Grounds in Colonial Singapore, 1880-1930', *Journal of Southeast Asian Studies* 22, no. 2 (1991): 290, doi:10.1017/S0022463400003891.

9. Chong, 180n1.

10. Chong, 164–6.

11. For a classification of Chinese burial grounds according to surname groups (*seh*), immigrants grouped according to particular districts of origin in China or dialect groups, see: Yeoh, 299; 288–90. Geographer Brenda Yeoh has done extensive historical research on the geographical and landed significance of Chinese burial grounds in the colonial period. Her work remains the most comprehensive and critical to date. Much of the historical background reiterated here draws on Yeoh's findings.

12. For example, in 1907, Cheng San Teng, the disused cemetery at Tanjong Pagar, was acquired to provide landfill material for ongoing land reclamation works at the Telok Ayer Basin. See: *Straits Times*, 'Untitled', November 19, 1906, https://eresources.nlb.gov.sg/newspapers/Digitised/Article/straitstimes19061119-1.2.27.

13. The cemetery at Tiong Bahru was the second largest clearance of graves. Seventy-odd acres of land was acquired by the Singapore Improvement Trust (SIT) for an Improvement and Housing Scheme—280 huts, 2000 squatters and their pigs were removed as were graves to make way for the building of roads and drainage. See: ARSM 1925, 'Appendix B: Singapore Improvement Trust [1925]', 23–4; ARSM 1926, 'Appendix B: Singapore Improvement Trust [1926]', 16–17; ARSM 1927, 'Appendix B: Singapore Improvement Trust [1927]', 18–19, all cited in Yeoh, 'The Control of "Sacred" Space', 306.

14. J. T. Thomson, Some Glimpses into Life in the Far East (London: Richardson, 1864), 280-1, cited in Yeoh, 'The Control of "Sacred" Space', 285.

15. Thomson, 282, cited in Yeoh, 'The Control of "Sacred" Space', 285.

16. *Malaya Tribune*, 'Chinese Burial Ground: Municipal By-Laws for Bukit Brown', October 27, 1917, cited in Chong, 'Bukit Brown Municipal Cemetery', 164–5.

17. Yeoh, 'The Control of "Sacred" Space', 309.

18. Such festivals include the Qing Ming Festival, involving the sweeping of ancestors' tombs, and the Winter Clothing Festival. The rites of both are documented in: 'The Rites and Rituals of Bukit Brown Cemetery'. The presence of such activities are further discussed in: Loh Kah Seng, 'History of the Dead, Heritage of the Living: Bukit Brown Cemetery in Singapore', *Kyoto Review of Southeast Asia* 12, September 2012, https://kyotoreview.org/issue-12/history-of-the-dead-heritage-of-the-living-bukit-brown-cemetery-in-singapore/; Roots SG, 'Qing Ming Festival', *Roots SG*, April 2018, accessed April 14, 2022, https://www.roots.sg/learn/resources/ich/qing-ming-festival.

19. See, for example: *Straits Times*, 'Caretaker's Death: Murder, Say Police', May 5, 1980, 11, https://eresources.nlb.gov.sg/newspapers/Digitised/Article/straitstimes19800505-1.2.62; *Straits Times*, 'Cemetery Blaze: Firemen's 45-Minute Battle', January 17, 1935, 12, https://eresources.nlb.gov.sg/newspapers/Digitised/Article/straitstimes19350117-1.2.79; Lau Fook Kong, 'Hardly a "Pleasant" Final Resting Place', *Straits Times*, November 27, 2004, 14; *Straits Times*, 'Faction Fight in a Cemetery', July 24, 1933, 12, https://eresources.nlb.gov.sg/newspapers/Digitised/Article/straitstimes19330724-1.2.69; *Straits Times*. 'Fight at a Funeral'. July 21, 1927, 8, https://eresources.nlb.gov.sg/newspapers/Digitised/Article/straitstimes19270721-1.2.28.

20. Mitch Rose, 'Gathering "Dreams of Presence": A Project for the Cultural Landscape', *Environment and Planning D: Society and Space* 24, no. 4 (2006): 543, doi:10.1068/d391t.

21. William J.T. Mitchell, 'Preface to the Second Edition of *Landscape and Power:* Space, Place and Landscape', in *Landscape and Power*, ed. William J.T. Mitchell, 2nd ed. (Chicago: University of Chicago Press, 2002), viii.

22. Henri Lefebvre, 'Plan of The Present Work', in *The Production of Space*, trans. Donald Nicholson-Smith (Malden, MA: Blackwell, 1991), 39.

23. Mitchell, x.

24. John Wylie, *Landscape* (Abingdon, Oxon: Routledge, 2007), 5.

25. John Berger, *About Looking* (London: Bloomsbury, 2009), 202.

26. Berger, 204.

27. Berger, 205.

28. Jacques Derrida and Bernard Stiegler, 'Spectrographies', in *The Spectralities Reader: Ghosts and Hauntings in Contemporary Cultural Theory*, ed. María del Pilar Blanco and Esther Peeren (New York: Bloomsbury, 2013), 38.

29. Jo Frances Maddern and Peter Adey, 'Editorial: Spectro-Geographies', *Cultural Geographies* 15, no. 3 (2008): 292, doi:10.1177/1474474008091328.

30. John Wylie, 'The Spectral Geographies of W.G. Sebald'. *Cultural Geographies* 14, no. 2 (2007): 172, doi:10.1177/1474474007075353.

31. Singapore Heritage Society, 'Bukit Brown Wayfinder: A Self-Guided Walking Trail' (Guide, October 2017), https://www.singaporeheritage.org/bukitbrownwayfinder/Bukit-Brown-Wayfinder-Guide.pdf. The *Wayfinder* was launched in October 2017 with the support of Singapore's Ministry of National Development and the National Heritage Board.

32. The description is based on material in the 'Wayfinder', a fully annotated and illustrated trail developed collaboratively between civil society groups, The Singapore Heritage Society and *All Things Bukit Brown*, who engaged state agencies including the Urban Redevelopment Authority, National Parks Board and the Land Transport Authority.

33. Singapore Heritage Society, 13.

34. Founded in 2005 by brothers Charles and Raymond Goh, API is, 'a paranormal research-based Society' that strives to systematically analyse strange and occult happenings in Singapore and Southeast Asia. For more on the API, see API's own website: Asia Paranormal Investigators.

35. See Map Version 4, updated 1 September 2013, at: Raymond Goh, 'Bukit Brown Map - DIY Walking Guide', *BUKIT BROWN: Living Museum of History and Heritage* (blog), May 10, 2015, https://www.bukitbrown.org/2015/05/bukit-brown-map-diy-walking-guide.html. The circled names are the only differences from Version 3 of the map, updated 25 February 2012, at: All Things Bukit Brown. 'Master Maps of Bukit Brown'. *All Things Bukit Brown - Heritage. Habitat. History* (blog), February 9, 2012, http://bukitbrown.com/main/?p=1170.

36. Stan Allen, *Practice: Architecture, Technique and Representation* (New York: Routledge, 2009), 49, 51.

37. Allen, 50.

38. James C. Scott, *Seeing like a State: How Certain Schemes to Improve the Human Condition Have Failed* (New Haven, CT: Yale University Press, 1998).

39. Paul Carter, *Dark Writing: Geography, Performance, Design* (Honolulu: University of Hawai'i Press, 2009), 28.

40. Carter, 19.

41. Michel Serres, *The Troubadour of Knowledge*, trans. S. F. Glaser and W. Pauldon (Ann Arbor, MI: University of Michigan Press, 1997), 69, cited in Carter, 28.

42. The number of stakes here is derived from the graves documented by the Bukit Brown Documentation Project, which shows a photo taken of each grave, taken with its respective stake. See: Bukit Brown Documentation Project, 'The Bukit Brown Cemetery Documentation Project', 2013, accessed April 12, 2022, http://www.bukitbrown.info/.

43. Land Transport Authority, 'Projects: Exhumation', 2012, https://www.lta.gov.sg/content/ltaweb/en/roads-and-motoring/projects/Exhumation.html.

44. Paul Emmons, 'Drawing Sites: Site Drawings', in *Architecture and Field/Work*, ed. Suzanne Ewing et al. (Abingdon, Oxon: Routledge, 2010), 120–22.

45. Leon Battista Alberti, *On the Art of Building in Ten Books*, trans. J. Rykwert, N. Leach and R. Tavernor (Cambridge, MA: MIT Press, 1988), 101, cited in Emmons, 121.

46. Carol Herselle Krinsky, 'Cesare Cesariano and the Como Vitruvius Edition of 1521' (PhD Diss., New York University, 1965), 100, cited in Emmons, 120.

47. Emmons, 119–22.

48. Alberti, *Art of Building*, 381, cited in Emmons, 121–22.

49. Von Braschler, *Confessions of a Reluctant Ghost Hunter: A Cautionary Tale of Encounters with Malevolent Entities and Other Disembodied Spirits* (Rochester, VT: Destiny Books, 2014), 52.

50. For a photo showing one such card with its associated tape, see 9th photo from the top in: Teo Siyang, 'Plant Walk @ Bukit Brown', *Urban Forest* (blog), September 25, 2011, https://uforest.blogspot.com/2011/09/plant-walk-bukit-brown.html.

51. National Archives of Singapore, 'Burial Registers of Bukit Brown Cemetery', National Archives of Singapore, accessed April 9, 2022, http://www.nas.gov.sg/BukitBrown_signage.htm.

52. In 2012, research exploring the digital mapping of the Bukit Brown graves and topography through GPS was undertaken, for example, see: Owen Noel Newton, Chamika Deshan, Natalie Pang, and Ryohei Nakatsu. 'Mobile Augmented Reality for Bukit Brown Cemetery Navigation', in *Proceedings of the 11th ACM SIGGRAPH International Conference on Virtual-Reality Continuum and Its Applications in Industry*, edited by Stephen N. Spencer (New York: Association for Computing Machinery, 2012), 59–62, doi:10.1145/2407516.2407534.

53. Terence Chong and Chua Ai Lin, 'Saving Bukit Brown', *Straits Times*, November 17, 2011, 32; Singapore Heritage Society, 'Position Paper on Bukit Brown' (Position Paper, January 2012), http://www.singaporeheritage.org/wp-content/uploads/2011/11/SHS_BB_Position_Paper.pdf.

54. Mitchell, ix.

55. Robin Dripps, 'Groundwork', in *Site Matters: Design Concepts, Histories, and Strategies*, ed. Carol J. Burns and Andrea Kahn (New York: Routledge, 2005), 61.

56. J. B. Harley, 'Deconstructing the Map', in *The New Nature of Maps: Essays in the History of Cartography*, ed. Paul Laxton, (Baltimore: Johns Hopkins Press, 2001), 163, cited in Carter, *Dark Writing*, 17.

57. Carter, 28.

58. In Tan Pin Pin's documentary film 'Gravediggers' Luck', the exhumation worker Ah Kow strongly believes that the work of gravedigging is taboo and implies diminishing fortunes for himself and his family, hence a winning number which he finds (at the gravesite, or around his neighbourhood) signifies an assurance that his fortunes are not yet over. See: Tan Pin Pin, *Gravediggers' Luck*, Documentary, 22:29 (Singapore: VHQ TV, 2003), posted September 17, 2015, https://www.tanpinpin.com/gravediggers-luck-2/.

59. Nicholas Yong, 'Buried: A life', *Straits Times*, April 10, 2011, Life!, 7.

60. See photograph of exhumation site reproduced in this newspaper article: Grace Chua, 'Exhumation at Bukit Brown Cemetery Begins', *Straits Times*, December 19, 2013, https://www.straitstimes.com/singapore/exhumation-at-bukit-brown-cemetery-begins.

61. Kuo Pau Kun, 'The Coffin Is Too Big For The Hole', in *Images at the Margins: A Collection of Kuo Pao Kun's Plays* (Singapore: Times Books International, 2000), 60–71; Lilian Chee, 'Domestic Digressions: Interrogating Singaporean Public Housing through Its Literary Forms', in *The Routledge Companion on Architecture, Literature and The City*, ed. Jonathan Charley (London: Routledge, 2018), 37–57.

62. Robert David Sack, *Conceptions of Space in Social Thought: A Geographic Perspective* (London: Macmillan Press, 1980), 155, cited in Brenda S. A. Yeoh and Tan Boon Hui, 'The Politics of Space: Changing Discourses on Chinese Burial Grounds in Post-War Singapore', *Journal of Historical Geography* 21, no. 2 (1995): 192. doi:10.1016/0305-7488(95)90035-7.

63. Robert D. Sack, 'Conceptions of Geographic Space', *Progress in Human Geography* 4, no. 3 (1980): 337, cited in Yeoh and Tan, 'Politics of Space', 192.

64. Yeoh and Tan, 192.

65. J. J. M. de Groot, *Chinese Geomancy* (Shaftesbury, UK: Element, 1989), 21, cited in Yeoh and Tan, 'Politics of Space', 191.

66. Yeoh, 'The Control of "Sacred" Space', 294.

67. de Groot, 28, cited in Yeoh and Tan, 'Politics of Space', 192; Freedman, 7, cited in Yeoh, 'Control of "Sacred" Space', 192.

68. Oswald Spengler, *The Decline of the West*, abridged version by Helmut Werner (London: George Allen and Unwin, 1961), 115.

69. In *Family Rituals*, which was written during the Song Dynasty by Neo-Confucian scholar Zhu Xi (1130-1200), death rites occupy 87% of the text. Park Chang-Won, 'Confucian Beliefs and Traditions', in *Encyclopedia of Death and the Human Experience*, edited by Clifton D. Bryant and Dennis L. Peck (Thousand Oaks, CA: SAGE, 2009), 222, doi:10.4135/9781412972031.n73. According to Sinologist Patricia Ebrey Buckley, the *Family Rituals* prescribes rituals that 'expressed and reproduced the key principles underlying the family system, such as the relationships between parents and children, and ancestors and descendants. These rituals are modelled on revered, older Confucian sources. She further argues that the *Family Rituals*' overall scheme of writes 'can be described in terms of the shifting relations of the living and the dead'. For more on the *Family Rituals*, see: Patricia Buckley Ebrey, 'Introduction', in *Chu Hsi's 'Family Rituals'*, trans. Patricia Ebrey Buckley (Princeton: Princeton University Press, 1991), xiii–xxxi.

70. Park, 'Confucian Beliefs', 222.

71. 'Spirit money' refers to the paper money offerings burnt by the Chinese to provide for their ancestors in the afterlife, usually at festivals such as the Hungry Ghost Festival, or the Qingming Festival. For more on Chinese paper offerings and their associated customs, see: C. Fred Blake, *Burning Money: The Material Spirit of the Chinese Lifeworld*, (Honolulu: University of Hawai'i Press, 2011), 25–52; Chung Sheng Kuan and Li Dan, 'An Artistic and Spiritual Exploration of Chinese Joss Paper', *Art Education* 70, no. 6 (2017): 28–35, doi:10.1080/00043125.2017.1361770.

72. Blake, 50.

73. Xiao Difei 萧涤非, ed., *Tangshi Jianshang Cidian* 唐诗鉴赏辞典 [Dictionary for the Appreciation of Tang Poetry]. Shanghai: Shanghai Cishu Chubanshe 上海辞书出版社, 1983.

74. Robert Macfarlane, *The Old Ways: A Journey on Foot* (New York: Penguin Books, 2012), 17.

75. Macfarlane, 17.

76. Macfarlane, 17.

77. Elizabeth Helsinger, 'Blindness and Insights', in *Landscape Theory*, ed. Rachel Ziady DeLue and James Elkins (New York: Routledge, 2008), 323.

78. Macfarlane, 17.

79. Rebecca Solnit, *A Field Guide to Getting Lost* (New York: Penguin Books, 2005), 50–51. Robert Macfarlane also explains 'Shul' in a similar manner, see: Macfarlane, *The Old Ways*, 50-51.

80. Solnit, 50-51.

81. Brenda S. A. Yeoh, 'Bones of Contention: Chinese Burial Grounds in Colonial and Post-Colonial Singapore', in *Spaces of the Dead: A Case from the Living*, ed. Kevin YL Tan (Singapore: Singapore Heritage Society and Ethos Books, 2011), 291.

82. Macfarlane, 21.

83. Still led by knowledgeable and passionate volunteers who call themselves 'Brownies', the free-of-charge walks today continue to focus on thematic routes associated with the cemetery's history, including the cemetery's relationship with the two World Wars, the architectural history of tombstone designs, stories surrounding the lives of those buried here, and the natural history of Bukit Brown focusing on its exceptional flora and fauna. See: Chantal Sajan, 'Heritage Activists: Bukit Brown More than Just a Cemetery; It's a "Living Museum" of Singapore's Pioneers', *Straits Times*, September 12, 2020, https://www.straitstimes.com/lifestyle/home-design/heritage-activists-bukit-brown-more-than-just-a-cemetery-its-a-living-museum.

84. Rebecca Solnit, *Wanderlust: A History of Walking* (New York: Viking Penguin, 2000), 229.

85. Wylie, 'Spectral Geographies'; Peter Merriman, George Revill, Tim Cresswell, Hayden Lorimer, David Matless, Gillian Rose, and John Wylie, 'Landscape, Mobility, Practice', *Social & Cultural Geography* 9, no. 2 (2008): 191–212, doi:10.1080/14649360701856136; Sarah Pink, Phil Hubbard, Maggie O'Neill, and Alan Radley, 'Walking Across Disciplines: From Ethnography to Arts Practice', *Visual Studies* 25, no. 1 (2010): 1–7, doi:10.1080/14725861003606670; Tim Edensor, 'Walking in Rhythms: Place, Regulation, Style and the Flow of Experience', *Visual Studies* 25, no. 1 (2010): 69–79, doi:10.1080/14725861003606902; Jennie Middleton, 'Sense and the City: Exploring the Embodied Geographies of Urban Walking', *Social & Cultural Geography* 11, no. 6 (2010): 575–96, doi:10.1080/14649365.2010.497913; David Pinder, 'Errant Paths: The Poetics and Politics of Walking', *Environment and Planning D: Society and Space* 29, no. 4 (2011): 672–92, doi:10.1068/d10808; Goh, 'Walking the Global City'; Hannah Macpherson, 'Walking Methods in Landscape Research: Moving Bodies, Spaces of Disclosure and Rapport', *Landscape Research* 41, no. 4 (2016): 425–32, doi:10.1080/01426397.2016.1156065.

86. Tim Edensor, 'Sensing the Ruin', *Senses and Society* 2, no. 2 (2007): 217–32, doi:10.2752/174589307X203100.

87. Edensor, 225.

88. Carter, *Dark Writing*, 93. Carter also mentions this relationship of 'traits' to 'traces/tracks' in: Carter, *Dark Writing*, 257n13.

89. Carter, 93.

90. Carter, 93.

91. John Wylie, 'A Single Day's Walking: Narrating Self and Landscape on the South West Coast Path', *Transactions of the Institute of British Geographers* 30, no. 2 (2005): 245.

92. Wylie, 240.

93. Paul Virilio and Sylvère Lotringer, *Pure War: Twenty-Five Years Later*, trans. Mark Polizzotti (Los Angeles: Semiotext(e), 2008), 48, cited in Allen, *Practice*, 72.

94. Liz Wells, *Land Matters: Landscape Photography, Culture and Identity* (London: I.B. Tauris, 2011), 12.

95. Matthew Gandy, 'Unintentional Landscapes', *Landscape Research* 41, no. 4 (2016): 434, doi:10.1080/01426397.2016.1156069.

96. Peg Rawes, 'Introduction', in *Relational Architectural Ecologies: Architecture, Nature and Subjectivity*, ed. Peg Rawes (London: Routledge, 2013), 1–18.

97. Rawes, 10.

98. Peg Rawes, 'Architectural Ecologies of Care', in *Relational Architectural Ecologies: Architecture, Nature and Subjectivity*, ed. Peg Rawes (London: Routledge, 2013), 40.

99. Rawes, 'Architectural Ecologies of Care', 51.

100. Marcus O'Donnell, 'Walking, Writing and Dreaming: Rebecca Solnit's Polyphonic Voices', *Journalism* 16, no. 7 (2014): 944, doi:10.1177/1464884914553078.

9. THE SEA, AND THE SEA: INFRASTRUCTURE AND THE DIALECTICAL IMAGE

1. In 2019, Singapore's total coastline length was recorded at 203km. About 70 per cent of the coastline is protected from erosion by waves and storms, with structures such as concrete seawalls and stone revetments. The rest of the coast consists of natural areas such as beaches and mangroves. With rising sea levels threatening the low-lying coastlines, current measures taken include coastal reservoirs, pumps and tidal gates at mouths of waterways. To safeguard Singapore's coastline, the Coastal and Flood Protection Fund (CFPF) of $5 billion was set up in 2020 to finance Singapore's capital expenditure for coastal protection and drainage infrastructure. See: Department of Statistics Singapore, 'Yearbook of Statistics Singapore 2019' (Singapore: Department of Statistics, 2019), 3; Public Utilities Board Singapore, 'PUB to lead Singapore's coastal protection efforts', March 4, 2020, https://www.pub.gov.sg/news/pressreleases/PUBtoleadSingaporescoastalprotectionefforts.

2. Since 2010, all Singapore school students must undergo the SwimSafer programme to learn survival and safety skills in open waters. When the programme was first implemented in 2010, it was amidst reports that more than 65 per cent of drowning incidents in Singapore between 2005 and 2008 occurred in seas, rivers and reservoirs. The minimum threshold is to stay afloat for at least 15 minutes and swim 15 metres. The Singapore school curriculum was revamped to accommodate this year-long programme. See: Monica Kotwani, 'Further enhancements needed to improve water safety teaching: Dr Teo Ho Pin', *Channel News Asia*, April 2, 2018, https://www.channelnewsasia.com/news/singapore/swimsafer-programme-revamp-needed-survival-expert-10089340.

3. Stephen Graham, *Disrupted Cities: When Infrastructures Fail* (New York: Routledge, 2010), 1.

4. For example, Mark Revinder Frost, 'Emporium in Imperio: Nanyang Networks and the Straits Chinese in Singapore, 1819–1914', *Journal of Southeast Asian Studies* 36, no. 1 (February 2005): 29–66, doi: 10.1017/S0022463405000020; John N. Miksic, *Singapore and the Silk Road of the Sea, 1300–1800* (Singapore: NUS Press, 2013).

5. Lisa Björkman and Andrew Harris, 'Engineering Cities: Mediating Materialities, Infrastructural Imaginaries and Shifting Regimes of Urban Expertise', *International Journal of Urban and Regional Research* 42, no. 2 (2018): 245. See also: Graham, 'When Infrastructures Fail', 1–2.

6. Rosalind E. Krauss, *The Optical Unconscious* (Cambridge, MA: MIT Press, 1993), 2.

7. Michel Foucault, 'Of Other Spaces', *Diacritics* 16, no. 1 (Spring 1986): 27.

8. Johann Wolfgang von Goethe, 'Italian Journey 1786–1787', cited in Michel Taussig, 'The Beach (A Fantasy)', in *Landscape and Power*, ed. W. J. T. Mitchell, 2nd ed. (Chicago: University of Chicago Press, 2002), 330.

9. Sylvia Plath, 'Ocean W-1212', in *Johnny Panic and the Bible of Dreams* (New York: Harper Perennial, 2000), 26.

10. Plath, 32.

11. Taussig, 320.

12. *Singapore's Integrated Urban Coastal Management* (Singapore: Technical Committee on the Coastal and Marine Environment, 2013), 5.

13. Amitav Ghosh, *The Great Derangement: Climate Change and the Unthinkable,* (Chicago: Chicago University Press, 2016), 35.

14. Ghosh, 36.

15. Stephen Graham and Nigel Thrift, 'Out of Order: Understanding Repair and Maintenance', *Theory Culture & Society* 24, no. 3 (2007): 10.

16. Graham and Thrift, 8.

17. Art Radar, 'Sand and Sea: Charles Lim and Shabbir Hussain Mustafa on the Singapore Pavilion in Venice – Interview | Art Radar', accessed April 19, 2022, http://artradarjournal.com/2015/05/29/singapore-pavilion-interview-venice-biennale/. See also Charles Lim Yi Yong, accessed November 3, 2021, https://www.seastate.sg/seastate. Most of Lim's SEA STATE videos can be viewed here.

18. Art Radar.

19. See: Ahmad Mashadi, 'Drifting Conversations', in *SEA STATE: Charles Lim Yi Yong,* ed. Shabbir Hussain Mustafa, (Singapore: National Arts Council, 2015), 116–21.

20. Shabbir Hussain Mustafa, 'Sea State 0 / Internal Waters. "Mr" Anthrobalanus', in *SEA STATE*, 27. The art historian Kevin Chua also discusses the appearance and significance of infrastructure in Lim's SEA STATE in his essay in the same volume; see: Kevin Chua, 'Sea State 2 / as Evil Disappears. The Sajahat Buoy: Notes on a Disappearance', in *SEA STATE,* 56–69.

21. David Teh, 'Charles Lim's Informatic Naturalism: Notes on SEA STATE 2', in *Charles Lim Sea State 2: As Evil Disappears* (Singapore: Future Perfect, 2012), 5.

22. Art Radar.

23. May Wong, 'M'sia Cannot Prove Ownership of Pedra Branca: S'pore Lawyers', *Today,* November 8, 2007, 4. See also: *The Singapore Free Press and Mercantile Advertiser,* 'The Horsburgh Light-House', March 20, 1902, Weekly edition, 3.

24. 'Pulau' is the Malay term for 'island'. *Straits Times*, 'Photographing Singapore Defences', November 24, 1900, 2; *Straits Times*, 'Photographing the Forts', July 31, 1901, 2; *Morning Tribute*, 'Sketching In Prohibited Area Alleged', January 9, 1941, 7; *Straits Times*, 'Did Not Know A War Was Going On', February 5, 1941, 11.

25. Keller Easterling, *Extrastatecraft: The Power of Infrastructure Space* (New York: Verso, 2014), 138.

26. Easterling, 15.

27. Easterling, 21.

28. Easterling, 73.

29. Easterling, 72.

30. Michel Foucault, *Power/Knowledge: Selected Interviews and Other Writings*, ed. Colin Gordon (New York: Vintage Books, 1980), 194, cited in Easterling, 92.

31. Easterling, 83.

32. Easterling, 73.

33. Easterling, 74.

34. Easterling, 75–76.

35. Easterling, 80.

36. Easterling, 90.

37. Easterling, 90.

38. Easterling, 92.

39. Easterling, 215.

40. Easterling, 238.

41. Easterling, 225.

42. Michael Fischer, 'Nomadic Video in Turbulent Sea States: How Art Becomes Critique', in *Probing Arts and Emergent Forms of Life* (Durham: Duke University Press, 2023), 100.

43. Fischer, 131.

44. Walter Benjamin, 'On the Theory of Knowledge, Theory of Progress', in *The Arcades Project*, trans. Howard Eiland and Kevin McLaughlin (Cambridge, MA: Harvard University Press, 2002), (N2a, 3), 462.

45. Ross Lipton, 'Benjamin's Dialectical Image and the Textuality of the Built Landscape', *Footprint Constellation of Awakening: Benjamin and Architecture*, Spring/Summer (2016): 76–77. doi: 10.7480/footprint.10.1.961.

46. Anthony Auerbach, 'Image No Metaphors: The Dialectical Image of Walter Benjamin', *Image [&] Narrative e-Journal* 18, 2007, http://www.imageandnarrative.be/inarchive/thinking_pictures/auerbach.htm.

47. Max Pensky, 'Method and Time: Benjamin's dialectical images', in *The Cambridge Companion to Walter Benjamin*, ed. David S. Ferris (New York: Cambridge University Press, 2004), 184.

48. 'Thinking involves not only the flow of thoughts, but their arrest as well. Where thinking suddenly stops in a configuration pregnant with tensions, it gives that configuration a shock'. Walter Benjamin, 'Theses on the philosophy of history' (completed in 1940 and translated by Harry Zohn), in Hannah Arendt (ed.) *Illuminations* (London: Fontana, 1992) 245–55. Cited in Jane Rendell, *Art and Architecture: A Place Between* (London: I. B. Tauris, 2006), 90.

49. Art Radar.

50. Pensky, 192.

51. Peter Burger and Jochen Schulte-Sasse, *Theory of the Avant-Garde*, trans. Michael Shaw (Minneapolis: University of Minnesota Press, 1984), 79, cited in Rendell, *Art and Architecture*, 91.

52. Walter Benjamin, *The Origin of German Tragic Drama*, trans. John Osborn (London: New Left Books, 1997), 214, cited in Jennifer Bloomer, 'In The Museyroom', *Assemblage*, no. 5 (February 1988): 60.

53. Taussig, 322.

54. Taussig, 326.

55. Taussig, 335.

56. Walter Benjamin, 'On the Theory of Knowledge, Theory of Progress', (N1, 1), 456.

57. Manuel Mogato, 'Manila Finds Marker With Chinese Writing, Buoys In Disputed Waters-Sources', *Reuters*, July 7, 2015, https://www.reuters.com/article/southchinasea-philippines-buoys-idUSL3N0ZG1VT20150706.

58. Mogato.

59. Aqil Haziq Mahmud, 'Malaysian Vessel Parked in Singapore Waters Is Used to Mark Territory', *Channel News Asia*, December 6, 2018, https://www.channelnewsasia.com/news/singapore/malaysian-port-limits-vessel-polaris-buoy-singapore-waters-11007250.

60. Mahmud, 'Malaysian Vessel'.; *Channel News Asia*, 'Singapore, Malaysia Maritime Dispute: A Timeline', December 6, 2018, https://www.channelnewsasia.com/news/singapore/singapore-malaysia-maritime-dispute-port-limits-timeline-11006762.

61. Lone Bertelsen and Andrew Murphie, 'An Ethics of Everyday Infinities and Powers: Félix Guattari on Affect and the Refrain', in *The Affect Theory Reader*, ed. Melissa Gregg and Gregory J. Seigworth (Durham: Duke University Press, 2010), 138–57; Gilles Deleuze and Felix Guattari, '1837: Of the Refrain' in *A Thousand Plateaus* (London: Athlone Press, 1988), 323.

62. Eric Tagliacozzo, 'The Lit Archipelago: Coast Lighting and the Imperial Optic in Insular Southeast Asia, 1860-1910', *Technology and Culture* 46, no. 2 (April 2005): 317.

63. Tagliacozzo describes the system of maritime buoys, beacons and lighthouses in the 1870s Dutch and English race to light the archipelago seas as the European colonial 'shaping of the frontier', and in so doing, 'concretizing' and 'making rigid' of 'imaginary lines', alongside their greater project of map making. See: Eric Tagliacozzo, 'Mapping the Frontier', in *Secret Trades, Porous Borders: Smuggling and States Along a Southeast Asian Frontier, 1865-1915* (New Haven: Yale University Press, 2005), 46.

64. Linda Chee, 'Ships Advised to Keep Clear of H Lighthouse', *Business Times*, July 21, 1989, 2; *Business Times*, 'Johor Has Proof That Island Belongs to Malaysia', July 25, 1989, 22.

65. Fischer, 111.

66. Mustafa, 30.

67. Chua, 'Sea State 2', 62–67.

68. Colin Fournier, 'Oceanic Commons', in *Our Ocean Guide*, ed. MAP Office (Cannaregio, Venice: Lightbox Publishing, 2017), 48.

69. United Nations Office of Legal Affairs Division for Ocean Affairs and the Law of the Sea, *Commission on the Limits of the Continental Shelf*, accessed April 21, 2022, https://www.un.org/Depts/los/clcs_new/continental_shelf_description.htm.

70. Tagliacozzo, 'The Lit Archipelago', 312–13.

71. Tagliacozzo, 313.

72. Tung Shi Yun, 'Explore Three Singapore Lighthouses on the National Heritage Board's Lighthouse Trail', *Straits Times*, June 26, 2014, https://www.straitstimes.com/singapore/explore-three-singapore-lighthouses-on-the-national-heritage-boards-lighthouse-trail. See: Peter Ooi, 'Beacon of Hope', *Today*, December 1, 2008. Additional: Yu-Mei Balasingamchow, 'Lighthouse', *Monoa, University of Hawai'i Press* 26, no. 1 (Summer 2014): 31–41.

73. Chua, 'Sea State 2', 62.

74. S. Jayakumar and Tommy T. B. Koh. *Pedra Branca: The Road to the World Court* (Singapore: NUS Press, 2009), 2.

75. Jayakumar and Koh, 8–9.

76. *The Singapore Free Press and Mercantile Advertiser*, 'The Horsburgh Light-House'.

77. Gilbert Murray, 'Violent Seas Marooned Men Who Built Horsburgh Light', *Straits Times*, December 31, 1955, 9. The article reports, "In the 27 years from 1824 to 1851, no less than 25 ships reported hitting reefs or went aground on shoals in the area, 16 of which became total wrecks."

78. Murray.

79. Measurements are given as: At low water mark, 450 feet in its greatest length and its greatest breadth, 250 feet and at high water 140 and 100 feet respectively. The highest point of the rock is 26 feet, 11 inches above high water spring tides … . See: *Singapore Free Press and Mercantile Advertiser*, 'The Horsburgh Light-House'; Derrick A. Paulo, 'White Rock of Contention Goes Before World Court', *Today*, November 6, 2007, 2.

80. *The Singapore Free Press and Mercantile Advertiser*, 'The Horsburgh Light-House'.

81. 'The Horsburgh Light-House'; Paulo, 'White Rock of Contention'.

82. Example of lamp replacements: In 1887, see: *The Singapore Free Press and Mercantile Advertiser*, 'The Horsburgh Light-House'. Lamp replacements in 1931: *Malaya Tribune*, 'Light To Be Established', April 27, 1931, 7.

83. Ng Boon Yian, 'Pedra Branca, An Island of Two Stories', *Today*, December 27, 2002, 2nd ed., 1.

84. *Straits Times*, 'Fishing Now Allowed Near Horsburgh Lighthouse', August 26, 1989.

85. For a similar argument of affective refrain held in the persisting media image, see Bertelsen and Murphie, 'An Ethics of Everyday Infinities and Powers'.

86. Tan Wooi Leong, 'Jurong Island: What It Takes to Achieve a World-Class Petrochemicals Hub', *Surbana Jurong Private Limited*, April 25, 2017, https://surbanajurong.com/resources/perspectives/jurong-island-takes-achieve-world-class-petrochemicals-hub/. See: Grace Chua, 'PUB Looks at Jurong Island Water Needs', *Straits Times*, January 3, 2013, https://www.asiaone.com/print/News/Latest+News/Singapore/Story/A1Story20130101-392832.html.

87. Tan, 'Jurong Island'. See also: Charles Yi Yong Lim, 'SEASTATE 6', *seastories.sg*, accessed April 16, 2022, https://www.seastories.sg/6; Samanth Subramanian, 'How Singapore Is Creating More Land for Itself', *New York Times*, April 20, 2017, https://www.nytimes.com/2017/04/20/magazine/how-singapore-is-creating-more-land-for-itself.html.

88. Jolene Loh, 'The Abstractions Of Geopolitics, Singapore Pavilion at the 56th Venice Biennale. Charles Lim: SEA STATE', *ArtAsiaPacific*, accessed April 20, 2022, http://artasiapacific.com/Magazine/WebExclusives/TheAbstractionsOfGeopolitics. Also see: Lim, 'SEASTATE 6'.

89. Rachael Squire, '"Do You Dive?": Methodological Considerations for Engaging with "Volume"', *Geography Compass* 11, no. 7 (2017): 3.

90. Squire, 1.

91. Squire, 3.

92. Fischer, 121.

93. Art Radar.

94. For the bilateral relations and provisions between Singapore and Malaysia, see Chapter 3, 'After the Last Train'.

95. Fischer, 110.

96. Cashin House will be restored as part of Lim Chu Kang Nature Park, a new nature area located at the west of Sungei Buloh Wetland Reserve. The building will be 'enhanced sensitively' for both natural and built heritage, and will include an exhibition space, seminar rooms and a seaview terrace for visitors. See: National Parks Board, 'New Sungei Buloh Nature Park Network to expand Singapore's natural capital along our northern coasts as part of efforts to make Singapore a City in Nature', August 19, 2020, https://www.nparks.gov.sg/news/2020/8/new-sungei-buloh-nature-park-network-to-expand-singapore%E2%80%99s-natural-capital-along-our-northern-coasts-as-part-of-efforts-to-make-singapore-a-city-in-nature.

97. The Long and Winding Road, 'Cashin's Pier, lost and found', August 19, 2020, https://thelongnwindingroad.wordpress.com/2020/08/19/cashins-pier-lost-and-found/; Our Vanishing Backyard, 'Lim Chu Kang Jetty', accessed April 14, 2022, https://ljkj2000.wixsite.com/ourvanishingbackyard/lim-chu-kang-jetty-and-pier.

98. *Straits Times Weekly*, 'Muar News', June 29, 1892, cited in Timothy P. Barnard, ed., 'A Whale's Tale: Excerpts from the Annual Reports of the Raffles Museum and Library, Newspaper Accounts and Visitors' Memories', in *Nature Contained: Environmental Histories of Singapore* (Singapore: NUS Press, 2014), 179–80.

99. *Straits Times Weekly*.

100. R. Hanitsch, *Annual Report on the Raffles Library and Museum for the Year 1905* (Singapore: Government Printing Office, 1906), 3–4, in Barnard, 'A Whale's Tale', 180.

101. *Straits Times Weekly*.

102. *Straits Times Weekly*.

103. Hanitsch, *Raffles Library and Museum*, 3–4.

104. *Straits Times Weekly*.

105. Kate Pocklington and Iffah Iesa, *A Whale Out of Water: The Salvage of Singapore's Sperm Whale* (Singapore: Lee Kong Chian Natural History Museum, 2016), 5.

106. Pocklington and Iesa, 11.

107. Pocklington and Iesa, 16.

108. Pocklington and Iesa, 14.

109. Pocklington and Iesa, 16.

110. Pocklington and Iesa, 22.

111. Pocklington and Iesa, 26.

112. The duration is mentioned in Pocklington and Iesa, 130.

113. Pocklington and Iesa, 23.

114. Pocklington and Iesa, 32.

115. Pocklington and Iesa, 32.

116. Pocklington and Iesa, 33.

117. Pocklington and Iesa, 12.

118. Barbara Watson Andaya, 'Seas, Oceans and Cosmologies in Southeast Asia', *Journal of Southeast Asian Studies* 48, no. 3 (October 2017): 358.

119. Andaya, 358.

120. Taussig, 326.

121. Kathleen Stewart, 'Afterword: Worlding Refrains', in *The Affect Theory Reader*, ed. Melissa Gregg and Gregory J Seigworth (Durham: Duke University Press, 2010), 340.

122. Stewart, 340.

123. Ben Anderson and Paul Harrison, 'The Promise of Non-Representational Theories', in *Taking-Place: Non-Representational Theories and Geography*, ed. Ben Anderson and Paul Harrison (Farnham: Ashgate, 2010), 8.

124. Donna Haraway, *Staying with the Trouble: Making Kin in the Chthulucene* (Durham: Duke University Press, 2016), 12.

BIBLIOGRAPHY

'DISOBEDIENT BUILDINGS'. Accessed March 25, 2022. https://www.disobedientbuildings.com/.

'Ghosts of Bukit Brown Cemetery: GFS S03E03'. YouTube video, 27:17. Posted by 'Ghost Files Singapore', September 9, 2015. https://www.youtube.com/watch?v=LUKcrNEjQM8.

'INSURGENT DOMESTICITIES', accessed March 10, 2023, https://www.socialdifference.columbia.edu/projects-/insurgent-domesticities.

'The Rites and Rituals of Bukit Brown Cemetery'. YouTube video, 14:22. Posted by 'Roots Sg', September 28, 2015. https://www.youtube.com/watch?v=PckVbnTDSBQ.

03-FLATS. 'Reviews'. *03-FLATS*. Accessed March 8, 2022. https://www.03-flats.com/reviews.

———. 'Trailer'. *03-FLATS*. Accessed March 8, 2022. https://www.03-flats.com/trailer.

Adorno, Theodor W. *Notes to Literature*. Vol. 1. Edited by Rolf Tiedemann, translated by Shierry Weber Nicholson. New York: Columbia University Press, 1991.

Ahmed, Sara. 'Happy Objects'. In *The Affect Theory Reader*, edited by Melissa Gregg and Gregory J. Seigworth, 29–51. Durham: Duke University Press, 2010.

———. 'Travelling with Strangers'. *Journal of Intercultural Studies* 42, no. 1 (2 January 2021): 8–23. https://doi.org/10.1080/07256868.2020.1859204.

Alberti, Leon Battista. *On the Art of Building in Ten Books*. Translated by J. Rykwert, N. Leach and R. Tavernor. Cambridge, MA: MIT Press, 1988.

All Things Bukit Brown. 'Bukit Brown: Documenting New Horizons of Knowledge'. *All Things Bukit Brown - Heritage. Habitat. History* (blog). Accessed April 12, 2022. http://bukitbrown.com/main/?p=9202.

———. 'Master Maps of Bukit Brown'. *All Things Bukit Brown - Heritage. Habitat. History* (blog), February 9, 2012. http://bukitbrown.com/main/?p=1170.

———. *All Things Bukit Brown - Heritage. Habitat. History* (blog). Accessed April 12, 2022. http://bukitbrown.com/main/.

Allen, R. *The New Penguin English Dictionary*. London: Penguin Books, 2001.

Allen, Stan. *Practice: Architecture, Technique and Representation*. New York: Routledge, 2009.

Allewaert, Monique. *Ariel's Ecology: Plantations, Personhood, and Colonialism in the American Tropics.* Minneapolis: University of Minnesota Press, 2013.

Alter, Nora M. 'The Political Im/Perceptible in the Essay Film: Farocki's "Images of the World and the Inscription of War"'. *New German Critique*, no. 68 (1996): 165–92. doi:10.2307/3108669.

———. *Projecting History: German Nonfiction Cinema, 1967-2000.* Ann Arbor, MI: University of Michigan Press, 2002.

Alter, Nora M., and Timothy Corrigan, eds. *Essays on the Essay Film.* New York: Columbia University Press, 2017.

Andaya, Barbara Watson. 'Seas, Oceans and Cosmologies in Southeast Asia'. *Journal of Southeast Asian Studies 48,* no. 3 (October 2017): 349–371.

Anderson, Ben, and Paul Harrison. 'The Promise of Non-Representational Theories'. In *Taking-Place: Non-Representational Theories and Geography*, edited by Ben Anderson and Paul Harrison, 1–36. Farnham: Ashgate, 2010.

Anstey, Tim, Katja Grillner, and Rolf Hughes, eds. *Architecture and Authorship.* London: Black Dog Publishing, 2007.

Armstrong, Isobel. *The Radical Aesthetic.* Oxford: Blackwell Publishing, 2000.

Art Radar. 'Sand and Sea: Charles Lim and Shabbir Hussain Mustafa on the Singapore Pavilion in Venice – Interview | Art Radar'. Accessed April 19, 2022. http://artradarjournal.com/2015/05/29/singapore-pavilion-interview-venice-biennale/.

Asia Paranormal Investigators. 'Asia Paranormal Investigators'. Accessed April 10, 2022. http://api.sg/main/.

AsiaOne. 'Man, 76, Found Dead in Trash-Filled Flat'. August 15, 2012. https://www.asiaone.com/print/News/Latest%2BNews/Singapore/Story/A1Story20120815-365468.html.

Auerbach, Anthony. 'Image No Metaphors: The Dialectical Image of Walter Benjamin'. *Image [&] Narrative e-Journal* 18, (September 2007). http://www.imageandnarrative.be/inarchive/thinking_pictures/auerbach.htm.

Baker, Steve. *Picturing the Beast: Animals, Identity and Representation.* Manchester: Manchester University Press, 1993.

Balasingamchow, Yu-Mei. 'Lighthouse'. *Monoa, University of Hawai'i Press 26,* no. 1 (Summer 2014): 31–41.

Banks, Marcus, and Jay Ruby, eds. *Made to Be Seen: Perspectives on the History of Visual Anthropology.* Chicago: University of Chicago Press, 2012.

Barnard, Timothy P., ed. 'A Whale's Tale: Excerpts from the Annual Reports of the Raffles Museum and Library, Newspaper Accounts and Visitors' Memories'. In *Nature Contained: Environmental Histories of Singapore*, 179–183. Singapore: NUS Press, 2014.

Barr, Michael D. *Lee Kuan Yew: The Beliefs behind the Man.* Washington: Georgetown University Press, 2000.

Barthes, Roland. *A Roland Barthes Reader*. Edited by Susan Sontag. London: Vintage, 2000.

———. *Camera Lucida.* Translated by Richard Howard. London: Vintage, 2000.

Benjamin, Walter. 'On the Theory of Knowledge, Theory of Progress'. In *The Arcades Project*, translated by Howard Eiland and Kevin McLaughlin, 456–88. Cambridge, MA: Harvard University Press, 2002.

———. 'The Storyteller: Reflections on the Works of Nikolai Leskov'. In *Illuminations: Essays and Reflections*, edited by Hannah Arendt, 83–107. London: Pimlico, 1999.

———. 'Theses on the Philosophy of History'. In *Illuminations: Essays and Reflections,* edited by Hannah Arendt, translated by Harry Zohn, 245–55. London: Pimlico, 1992.

———. 'The Work of Art in the Age of Mechanical Reproduction'. In *Illuminations: Essays and Reflections,* edited by Hannah Arendt, translated by Harry Zohn, 211–44. London: Pimlico, 1999.

———. *The Origin of German Tragic Drama*. Translated by John Osborn. London: New Left Books, 1997.

Berger, John. *About Looking*. London: Bloomsbury, 2009.

Bergson, Henri. *An Introduction to Metaphysics*. Translated by T. E. Hulme. New York: G. P. Putnam's Sons, 1912.

Berlant, Lauren. 'Austerity, Precarity, Awkwardness'. November 2011. https://supervalentthought.files.wordpress.com/2011/12/berlant-aaa-2011final.pdf.

———. *Cruel Optimism*. Durham: Duke University Press Books, 2011.

———. *The Queen of America Goes to Washington City: Essays on Sex and Citizenship*. Durham: Duke University Press, 1997.

Berlant, Lauren, and Kathleen Stewart. *The Hundreds*. Durham: Duke University Press Books, 2019.

Bermann, Karen. 'The House Behind'. In *Places Through the Body*, edited by Heidi J. Nast and Steve Pile, 165–80. London, New York: Routledge, 1998.

Bertelsen, Lone, and Andrew Murphie. 'An Ethics of Everyday Infinities and Powers: Félix Guattari on Affect and the Refrain'. In *The Affect Theory Reader*, edited by Melissa Gregg and Gregory J. Seigworth, 138–57. Durham: Duke University Press, 2010.

Björkman, Lisa, and Andrew Harris. 'Engineering Cities: Mediating Materialities, Infrastructural Imaginaries and Shifting Regimes of Urban Expertise'. *International Journal of Urban and Regional Research* 42, no. 2 (2018): 244–62.

Blackman, Lisa. 'Immaterial Bodies: Affect, Embodiment, Mediation'. London: SAGE Publications, 2012. https://doi.org/10.4135/9781446288153.

Blackman, Lisa, and Couze Venn. 'Affect'. *Body & Society* 16, no. 1 (1 March 2010): 7–28. https://doi.org/10.1177/1357034X09354769.

Blake, C. Fred. *Burning Money: The Material Spirit of the Chinese Lifeworld*. Honolulu: University of Hawai'i Press, 2011.

Bloomer, Jennifer. 'Abodes of Theory and Flesh: Tabbles of Bower'. *Assemblage*, no. 17 (April 1992): 7–29. https://doi.org/10.2307/3171221.

———. 'Hold It (Meditations upon a Gorgonzola Cheese)'. *Journal of Architectural Education (1984-)* 40, no. 2 (1987): 8–9.

———. 'In The Museyroom'. *Assemblage*, no. 5 (February 1988): 58–65.

———. *Architecture and the Text: The (S)Crypts of Joyce and Piranesi*. New Haven: Yale University Press, 1995.

Bolt, Barbara. 'The Magic is in Handling'. In *Practice as Research: Approaches to Creative Arts Enquiry,* edited by Estelle Barrett and Barbara Bolt, 27–34. London: I.B. Tauris, 2010.

Boomgaard, Peter. *Frontiers of Fear: Tigers and People in the Malay World, 1600–1950*. New Haven, CT: Yale University Press, 2001.

Boym, Svetlana. 'Ruins of the Avant-Garde: From Tatlin's Tower to Paper Architecture'. In *Ruins of Modernity,* edited by Julia Hell and Andreas Schönle, 58–85. Durham: Duke University Press, 2010.

Brinkema, Eugenie. *The Forms of the Affects*. Durham: Duke University Press, 2014.

Brody, Richard. 'To Life: A Monthlong Retrospective of Chantal Akerman's Films'. *New Yorker*. April 4, 2016. https://www.newyorker.com/magazine/2016/04/04/a-chantal-akerman-retrospective.

Brophy, Philip. 'Singapore Biennale 2006'. *Philip Brophy*. October 2006. http://www.philipbrophy.com/projects/rstff/SingaporeBiennale2006_A.html.

Brown, C. C., trans. *Sejarah Melayu or Malay Annals: An Annotated Translation*. Kuala Lumpur: Oxford University Press, 1970.

Bruno, Giuliana. *Public Intimacy: Architecture and the Visual Arts*. Cambridge, MA: MIT Press, 2007.

Buck, Frank. *Wild Cargo*. New York: Simon and Schuster, 1932.

Buckley, Charles Burton. *An Anecdotal History of Old Times in Singapore*. Singapore: Fraser & Neave, 1902. New edition with introduction by Constance Mary Turnbull. Singapore: Oxford University Press, 1984.

Bukit Brown Documentation Project. 'The Bukit Brown Cemetery Documentation Project'. 2013. Accessed April 12, 2022. http://www.bukitbrown.info/.

Burger, Peter, and Jochen Schulte-Sasse. *Theory of the Avante-Garde*. Translated by Michael Shaw. Minneapolis: University of Minnesota Press, 1984.

Burgess, Anthony. *The Malayan Trilogy*. London: Vintage Books, 2000.

Burns, Karen. 'Architecture/Discipline/Bondage'. In *Desiring Practices: Architecture, Gender and the Interdisciplinary*, edited by K. Ruedi, S. Wigglesworth, and D. McCorquodale, 73–86. London: Black Dog Publishing, 1996.

Business Times. 'Johor Has Proof That Island Belongs to Malaysia'. July 25, 1989, 22.

Cairns, Stephen, and Jane M. Jacobs. *Buildings Must Die: A Perverse View of Architecture*. Cambridge, MA: MIT Press, 2014.

Cairns, Stephen, Jane M. Jacobs, Jiang Yingying, Rita Padawangi, Sharon Siddique, and Eugene Tan. 'Void Decks – Singapore Void Decks'. In *Public Space in Urban Asia*, edited by William S. W. Lim, 80-89. Singapore: World Scientific, 2014.

Cameron, John. *Our Tropical Possessions in Malayan India*. London: Smith, Elder, 1865. Facsimile of the first edition, with an introduction by Wang Gungwu. Kuala Lumpur: Oxford University Press, 1965.

Carter, Paul. *Dark Writing: Geography, Performance, Design*. Honolulu: University of Hawai'i Press, 2009.

———. *Material Thinking: The Theory and Practice of Creative Research*. Carlton, Vic: Melbourne University Publishing, 2004.

Cat Welfare Society. 'Frequently Asked Questions | Cat Welfare Society'. Accessed March 23, 2022. http://www.catwelfare.org/faq.

Chan, Ying-kit. 'No Room to Swing a Cat? Animal Treatment and Urban Space in Singapore'. *Southeast Asian Studies* 5, no. 2 (2016): 305–29. https://doi.org/10.20495/seas.5.2_305.

Chang, Jiat-Hwee. *A Genealogy of Tropical Architecture: Colonial Networks, Nature and Technoscience*. London: Routledge, 2016.

Chang, Li Lin. 'Singapore's Troubled Relations with Malaysia: A Singapore Perspective'. *Southeast Asian Affairs*, no. 1 (2003): 259–74.

Chang, Sishir. 'A High-Rise Vernacular in Singapore's Housing Development Board Housing'. *Berkeley Planning Journal* 14, no. 1 (2000): 97–116. doi:10.5070/BP314112985.

Channel News Asia. 'Singapore, Malaysia Maritime Dispute: A Timeline'. December 6, 2018. https://www.channelnewsasia.com/news/singapore/singapore-malaysia-maritime-dispute-port-limits-timeline-11006762.

Chee, Lilian. '*03-FLATS:* Domesticity, Home and Its Representations'. *03-FLATS* Research Project, 2014. http://www.03-flats.com/.

———. 'After the Last Train: Narrating the Tentative Monument through Simryn Gill's Tanjong Pagar Railway Station'. In *Proceedings of the Society of Architectural Historians, Australia and New Zealand: 33, Gold*, edited by AnnMarie Brennan and Philip Goad, 122–32. Melbourne: SAHANZ, 2016.

———. 'Amidst…: Afterimage, Affect, Architecture'. In *Remote Practices: Architecture at a Distance*, edited by Matthew Mindrup and Lilian Chee, 137–43. London: Lund Humphries, 2022.

———. 'Domestic Digressions: Interrogating Singaporean Public Housing through Its Literary Forms'. In *The Routledge Companion on Architecture, Literature and The City*, edited by Jonathan Charley, 37–57. London: Routledge, 2018.

———. 'Domesticity, Gender, Architecture: Locating an Expanded Field'. In *The Routledge Companion to Contemporary Architectural History*, edited by Duanfang Lu, forthcoming. New York: Routledge, 2021.

———. 'Keeping Cats, Hoarding Things: Domestic Situations in the Public Spaces of the Singaporean Housing Block'. *The Journal of Architecture* 22, no. 6 (2017): 1041–65, doi:10.1080/13602365.2017.1362024.

———. 'Materializing the Tiger in the Archive: Creative Research and Architectural History'. In *Feminist Practices: Interdisciplinary Approaches to Women in Architecture*, edited by Lori Brown, 155–65. Surrey: Ashgate, 2011.

———. 'Reimagining Domesticity in 03-FLATS: Entering Singaporean Domestic Space through the Essay Film'. In *Making Visible: Architecture Filmmaking*, edited by Hugh Campbell and Igea Troiani, 117–38. Bristol: Intellect Books, 2019.

———. 'The Domestic Residue: Feminist Mobility and Space in Simryn Gill's Art'. *Gender, Place & Culture 19*, no. 6 (2012): 750–70. doi: 10.1080/0966369X.2012.674924.

———. 'The Public Private Interior: Constructing the Modern Domestic Interior in Singapore's Public Housing'. In *The Handbook of Interior Architecture and Design*, edited by Graeme Brooker and Lois Weinthal, 199–212. London: Bloomsbury, 2013.

———. 'Under the billiard table: Animality, anecdote and the tiger's subversive significance at the Raffles Hotel'. *Singapore Journal of Tropical Geography* 32, no. 3 (2011): 350–64, https://doi.org/10.1111/j.1467-9493.2011.00437.x.

———. 'Unhousing Sexuality: Sexuality and Singlehood in Singapore's Public Housing'. In *Sexuality and Gender at Home: Experience, Politics, Transgression*, edited by Brent Pilkey, Rachel Scicluna, Ben Campkin and Barbara Penner, 35–50. London: Bloomsbury, 2017.

Chee, Linda. 'Ships Advised to Keep Clear of H Lighthouse'. *Business Times*, July 21, 1989, 2.

Chew, Hui Min. 'Another Cat Found Dead at Foot of Yishun Block; Eighth Cat Death in Six Weeks'. *Straits Times*. October 30, 2015. http://www.straitstimes.com/singapore/another-cat-found-dead-at-foot-of-yishun-block-eighth-cat-death-in-six-weeks.

Chong, Terence, and Chua Ai Lin. 'Saving Bukit Brown'. *Straits Times*. November 17, 2011.

———. 'The Multiple Spaces of Bukit Brown'. In *Public Space in Urban Asia*, edited by William S. W. Lim, 26–55. Singapore: World Scientific, 2014.

Chong, Terence. 'Bukit Brown Municipal Cemetery: Contestations of the Good Life in Singapore'. In *Worlding Multiculturalisms: The Politics of Inter-Asian Dwelling*, edited by Daniel Goh Pei Siong and Wang Chih-Ming, 161–82. Abingdon, Oxon: Routledge, 2015.

Choo, Yun Ting. 'Cats Injured by Glue Traps Meant for Rats in Redhill'. *Straits Times*. January 23, 2019. https://www.straitstimes.com/singapore/environment/cats-injured-by-glue-traps-meant-for-rats-in-redhill.

Chopard, Kelly. *Terry's Raffles Adventures*. Illustrated by Patrick Yee. Singapore: Landmark Books, 1996.

Chua, Beng-Huat. 'Not Depoliticized but Ideologically Successful: The Public Housing Programme in Singapore'. *International Journal of Urban and Regional Research* 15, no. 1 (1991): 24–41. doi:10.1111/j.1468-2427.1991.tb00681.x.

———. 'Public Housing Residents as Clients of the State'. *Housing Studies* 15, no. 1 (2000): 45–60, doi:10.1080/02673030082469.

———. *Communitarian Ideology and Democracy in Singapore*. New York: Routledge, 1996.

———. *Political Legitimacy and Housing: Stakeholding in Singapore*. London: Routledge, 1997.

Chua, Grace. 'Exhumation at Bukit Brown Cemetery Begins'. *Straits Times*. December 19, 2013. https://www.straitstimes.com/singapore/exhumation-at-bukit-brown-cemetery-begins.

———. 'PUB looks at Jurong Island water needs'. *Straits Times*. January 3, 2013. https://www.asiaone.com/print/News/Latest+News/Singapore/Story/A1Story20130101-392832.html.

Chua, Kevin. 'Sea State 2 / as Evil Disappears. The Sajahat Buoy: Notes on a Disappearance'. In *SEA STATE: Charles Lim Yi Yong*, edited by Shabbir Hussain Mustafa, 56–69. Singapore: National Arts Council, 2015.

———. 'The Tiger and the Theodolite: George Coleman's Dream of Extinction'. *FOCAS* 6 (2007): 124–49.

Chung, Sheng Kuan, and Li Dan. 'An Artistic and Spiritual Exploration of Chinese Joss Paper'. *Art Education* 70, no. 6 (2017): 28–35. doi:10.1080/00043125.2017.1361770.

Collett, Anne. *Tracking the Literature of Tropical Weather: Typhoons, Hurricanes, and Cyclones*. New York: Springer, 2016.

Corbusier, Le. *L'art décoratif dáujourd'hui*. Paris: Éditions G. Grés et Cie, 1925.

———. *Precisions: On the Present State of Architecture and City Planning*. Translated by Edith Schreiber Aujame. Cambridge, MA: MIT Press, 1991.

———. *The Decorative Art of Today*. Translated by James Dunnet. In Cambridge, MA: MIT Press, 1987.

Corrigan, Timothy. *The Essay Film: From Montaigne, After Marker*. New York: Oxford University Press, 2011.

Cowell, F. R. *The Athenaeum: Club and Social Life in London, 1824-1973*. London: Heinemann, 1975.

Creamer, Kathy. *The Tiger Who Came to Tea*. Singapore: Oxford University Press, 1996.

Cuthbertson, Debbie. 'Australia's First Venice Biennale Pavilion Finds Second Life as New Building Progresses'. *Sydney Morning Herald*. October 18, 2014. http://www.smh.com.au/entertainment/art-and-design/australias-first-venice-biennale-pavilion-finds-second-life-as-new-building-progresses-20141017-117imr.html.

Davis, Lucy. 'Zones of Contagion: The Singapore Body Politic and the Body of the Street-Cat'. In *Considering Animals: Contemporary Studies in Human-Animal Relations*, edited by Carol Freeman, Elizabeth Leane and Yvette Watt, 279–302. Abingdon, Oxon: Ashgate, 2011.

De Certeau, Michel, Luce Giard, and Pierre Mayol. *The Practice of Everyday Life, Volume 2: Living and Cooking*. Translated by Timothy J. Tomasik. Minneapolis: University of Minnesota Press, 1998.

De Certeau, Michel. *The Practice of Everyday Life*. Translated by Steven Rendall. Berkeley: University of California Press, 1988.

de Groot, J. J. M. *Chinese Geomancy*. Shaftesbury, UK: Element, 1989.

Delany, Samuel R. *The Motion of Light in Water: Sex and Science Fiction Writing in the East Village, 1957-1965*. New York: Arbor House Pub Co, 1988.

Deleuze, Gilles, and Felix Guattari. '1837: Of the Refrain'. In *A Thousand Plateaus*, 310–50. London: Athlone Press, 1988.

DeLyser, Dydia, Harriet Hawkins, Anna Secor, and John Wylie. 'Forward Cultural Geographies'. *Cultural Geographies* 25, no. 1 (2018): 3–5. doi:10.1177/1474474017745040.

Denis, Armand. *Cats of the World*. Boston: Houghton Mifflin Co., 1954.

Department of Statistics Singapore. 'Singapore Census of Population 2020, Statistical Release 1: Demographic Characteristics, Education, Language and Religion'. Singapore: Department of Statistics Singapore, 2020.

———. 'Yearbook of Statistics Singapore 2019'. Singapore: Department of Statistics Singapore, 2019.

Derrida, Jacques, and Bernard Stiegler. 'Spectrographies'. *In The Spectralities Reader: Ghosts and Hauntings in Contemporary Cultural Theory*, edited by María del Pilar Blanco and Esther Peeren, 37–51. New York: Bloomsbury, 2013.

Derrida, Jacques. 'From "The Deaths of Roland Barthes"'. In *Critical Essays on Roland Barthes*, edited by Diana Knight, 129–37. New York: G. K. Hall & Co., 2000.

———. 'The Animal That Therefore I Am (More to Follow)'. In *Animal Philosophy*, edited by Matthew Callarco and Peter Atterton, 113–28. London: Continuum, 2004.

———. *Archive Fever: A Freudian Impression*. Translated by Eric Prenowitz. Chicago: University of Chicago Press, 1996.

———. *Dissemination*. Translated by Barbara Johnson. London: Athlone Press, 1981.

DeSilvey, Caitlin. 'Making Sense of Transience: An Anticipatory History'. *Cultural Geographies* 19, no. 1 (2012): 31–54. doi:10.1177/1474474010397599.

———. 'Observed Decay: Telling Stories with Mutable Things'. *Journal of Material Culture* 11, no. 3 (2006): 318–38. doi:10.1177/1359183506068808.

Douglas, Mary. *Purity and Danger*. London: Routledge, 1966.

Dripps, Robin. 'Groundwork'. In *Site Matters: Design Concepts, Histories, and Strategies*, edited by Carol J. Burns and Andrea Kahn, 59–91. New York: Routledge, 2005.

Driver, Felix, and Brenda S.A Yeoh. 'Constructing the Tropics: Introduction'. *Singapore Journal of Tropical Geography* 21, no. 1 (2000): 1–5. doi: 10.1111/1467-9493.00059.

Duncan, James S. *The City as Text: The Politics of Landscape Interpretation in the Kandyan Kingdom*. Cambridge, UK: Cambridge University Press, 1989.

Easterling, Keller. *Extrastatecraft: The Power of Infrastructure Space*. New York: Verso, 2014.

Ebrey, Patricia Buckley. 'Introduction'. In Chu Hsi's *'Family Rituals'*, translated by Patricia Ebrey Buckley, xiii–xxxi. Princeton: Princeton University Press, 1991.

Edensor, Tim. 'Sensing the Ruin'. *Senses and Society* 2, no. 2 (2007): 217–32. doi:10.2752/17458930 7X203100.

———. 'Walking in Rhythms: Place, Regulation, Style and the Flow of Experience'. *Visual Studies* 25, no. 1 (2010): 69–79. doi:10.1080/14725861003606902.

———. 'Waste Matter - The Debris of Industrial Ruins and the Disordering of the Material World'. *Journal of Material Culture* 10, no. 3 (2005): 311–32. doi:10.1177/1359183505057346.

———. *Industrial Ruins: Space, Aesthetics and Materiality*. New York; Oxford: Berg, 2005.

Emmons, Paul. 'Drawing Sites: Site Drawings'. In *Architecture and Field Work*, 119–29. Abingdon, Oxon: Routledge, 2010.

Energy & Digital Living. 'Sensory Ethnography'. *Energy & Digital Living*. Accessed March 16, 2022. http://energyanddigitalliving.com/doing-sensory-ethnography/.

Engelbrecht, Beate. *Building Season in Tiébélé: A Royal Compound in Change (Kasena, Burkina Faso)*. Göttingen: Institut für den Wissenschaftlichen Film Göttingen, 1999.

Ewing, Suzanne. 'Field/Work'. *Arq: Architectural Research Quarterly* 15, no. 4 (2011): 309–11. doi:10.1017/S1359135512000073.

F., Kathleen. 'MP Dr Lily Neo urges for HDB's Joint Singles Scheme to be reviewed to offer more privacy and flexibility'. *Online Citizen Asia*. Accessed March 5, 2022. https://www.onlinecitizenasia.com/2020/03/05/mp-dr-lily-neo-urges-for-hdbs-joint-singles-scheme-to-be-reviewed-to-offer-more-privacy-and-flexibility/.

Farsø, Mads, and Rikke Munck Petersen. 'Conceiving Landscape through Film: Filmic Explorations in Design Studio Teaching'. *Architecture and Culture* 3, no. 1 (2015): 65–86. doi:10.2752/205078215X14236574273664.

Fischer, Michael. 'Nomadic Video in Turbulent Sea States: How Art Becomes Critique'. In *Probing Arts and Emergent Forms of Life*. Durham: Duke University Press, 2023.

Fitzgerald, Michael. 'Against Blankness: The Inhabiting Spaces of Simryn Gill'. *Art Asia Pacific*. Accessed February 15, 2022. http://artasiapacific.com/Magazine/82/AgainstBlanknessTheInhabitingSpacesOfSimrynGill.

Flower, Raymond. *Year of the Tiger*. Singapore: Times Books International, 1986.

Forty, Adrian. 'Primitive: The Word and Concept'. In *Primitive: Original Matters in Architecture*, edited by Jo Ogders, Flora Samuel, and Adam Sharr, 3–14. Abingdon, Oxon: Routledge, 2006.

Foucault, Michel. 'Of Other Spaces'. *Diacritics* 16, no. 1 (Spring 1986): 22–27.

———. *Ethics: Subjectivity and Truth*. Edited by Paul Rainbow. New York: The New Press, 1997.

———. *Power/Knowledge: Selected Interviews and Other Writings*. Edited by Colin Gordon. New York: Vintage Books, 1990.

Fournier, Colin. 'Oceanic Commons'. In *Our Ocean Guide*, edited by MAP Office, 46-53. Cannaregio, Venice: Lightbox Publishing, 2017.

Frampton, Kenneth. *Modern Architecture: A Critical History*. London: Thames and Hudson, 1980.

Frichot, Hélène. 'Following Hélène Cixous' Steps Towards A Writing Architecture'. *Architectural Theory Review* 15, no. 3 (1 December 2010): 312–23. https://doi.org/10.1080/13264826.2010.524310.

———. 'Infrastructural Affects: Challenging the Autonomy of Architecture'. In *Architectural Affects after Deleuze and Guattari*, edited by Marko Jobst and Hélène Frichot, 33–55. Oxon: Routledge, 2021.

———. *Creative Ecologies: Theorizing the Practice of Architecture*. New York: Bloomsbury Visual Arts, 2018.

Frichot, Hélène, and Naomi Stead. 'Waking Ideas from Their Sleep: An Introduction to Ficto-Critical Writing in and of Architecture'. In *Writing Architectures: Ficto-Critical Approaches*, edited by Hélène Frichot and Naomi Stead, 11–24. London: Bloomsbury, 2020.

Frost, Mark Revinder. 'Emporium in Imperio: Nanyang Networks and the Straits Chinese in Singapore, 1819–1914'. *Journal of Southeast Asian Studies* 36, no. 1 (February 2005): 29–66. doi: 10.1017/S0022463405000020.

Gallop, Jane. *Anecdotal Theory*. Durham: Duke University Press, 2002.

Gan, Ling Kai. 'Little India Residents' Woes'. *New Paper*. June 20, 2011. https://www.asiaone.com/print/News/AsiaOne%2BNews/Singapore/Story/A1Story20110620-285089.html.

Gandy, Matthew. 'Unintentional Landscapes'. *Landscape Research* 41, no. 4 (2016): 433–40. doi:10.1080/01426397.2016.1156069.

George, Cherian. *Singapore: The Air-Conditioned Nation; Essays on the Politics of Comfort and Control, 1990–2000*. Singapore: Landmark Books, 2000.

Ghosh, Amitav. 'Wild Fictions'. Essay. Accessed February 3, 2022. https://www.amitavghosh.com/docs/Wild%20Fictions.pdf.

———. *The Great Derangement: Climate Change and the Unthinkable*. Chicago: University of Chicago Press, 2016.

Gibbs, Anna. 'Writing and Danger: The Intercorporeality of Affect'. In *Creative Writing: Theory beyond Practice*, edited by Nigel Krauth and Tess Brady, 157–68. Teneriffe: Post Pressed, 2016.

Gidal, Peter. 'Films'. In *Films and Painting*, edited by David Herbert, 80–151. London: Studio Vista, 1971.

Gill, Jaipal Singh. 'Commentary: Do We Need to Do More to Protect Our Community Cats?'. *Channel News Asia*. June 2, 2021. https://www.channelnewsasia.com/commentary/community-cats-cruelty-ang-mo-kio-spca-cases-animal-abuse-1823451.

Gill, Simryn. *Guide to the Murals of the Tanjong Pagar Railway Station*. Singapore: Singapore Biennale 2006 and the Singapore National Arts Council, 2006.

Gillett, Sidney. *The Earlier History of Billiard Tables and Accessories as Seen from the Sales Journals of John Thurston, 1818–1843*. London: Thurston and Co., 1996.

Gilloch, Graeme. *Myth and Metropolis: Walter Benjamin and the City*. Cambridge: Polity Press, 1996.

Glendinning, Miles. 'From European Welfare State to Asian Capitalism: The Transformation of "British Public Housing" in Hong Kong and Singapore'. In *Architecture and the Welfare State*, edited by Mark Swenarton, Tom Avermaete, and Dirk van den Heuvel, 299-320. London: Routledge, 2015.

Goethe, Johann Wolfgang von. 'Italian Journey 1786–1787', cited in Michel Taussig, 'The Beach (A Fantasy)', in *Landscape and Power*, ed. W. J. T. Mitchell, 2nd ed., 330. Chicago: University of Chicago Press, 2002.

Goh, Raymond. 'Bukit Brown Map - DIY Walking Guide'. *BUKIT BROWN: Living Museum of History and Heritage* (blog). May 10, 2015. https://www.bukitbrown.org/2015/05/bukit-brown-map-diy-walking-guide.html.

Goh, Robbie B. H. 'Things to a Void Utopian Discourse, Communality and Constructed Interstices in Singapore Public Housing'. In *Theorizing the Southeast Asian City as Text*, edited by Brenda S. A. Yeoh and Robbie B. H. Goh, 51–75. Singapore: World Scientific, 2003.

Graham, Stephen, and Nigel Thrift. 'Out of order: Understanding repair and maintenance'. *Theory, Culture & Society* 24, no. 3 (2007): 1–25.

Graham, Stephen. 'When Infrastructure Fail'. In *Disrupted Cities: When Infrastructures Fail*, 1–26. New York: Routledge, 2010.

Griffiths, Huw, Ingrid Poulter, and David Sibley. 'Feral Cats in the City'. In *Animal Spaces, Beastly Places: New Geographies of Human-Animal Relations*, edited by Chris Philo and Chris Wilbert, 56–70. London: Routledge, 2000.

Grillner, Katja, and Klaske Havik. 'Between Sites and Stories, Between Text and Times: A Dialogue about Fiction and Narrative in Architectural Inquiries'. In *Writing Place: Investigations in Architecture and Literature*, edited by Klaske Havik, Jorge Mejia Hernandez, Oliveira Susana, Mark Proosten, and Mike Schafer, 158–72. Rotterdam: nai010, 2016.

Hanitsch, R. *Annual Report on the Raffles Library and Museum for the Year 1905*. Singapore: Government Printing Office, 1906.

Haralambidou, Penelope. 'The Architectural Essay Film'. *Arq: Architectural Research Quarterly* 19, no. 3 (2015): 234–48. doi:10.1017/S1359135515000524.

Haraway, Donna. *Staying with the Trouble: Making Kin in the Chthulucene*. Durham: Duke University Press, 2016.

Harley, J. B. 'Deconstructing the Map'. In *The New Nature of Maps: Essays in the History of Cartography*, edited by Paul Laxton, 149–68. Baltimore: Johns Hopkins Press, 2001.

Hartman, Saidiya. 'Intimate History, Radical Narrative'. AAIHS, May 22, 2020. https://www.aaihs.org/intimate-history-radical-narrative/.

Hassan, Rizal. 'Some Sociological Implications of Public Housing in Singapore'. *South-East Asian Journal of Sociology* 2 (1969): 23-26.

———. 'Social and Psychological Implications of High Population Density'. *University of Singapore, Department of Sociology Working Paper*, no. 47 (1975).

Hauser, Kitty. 'Artist Simryn Gill Becomes Australia's Accidental Ambassador at Venice Biennale'. *The Australian*. May 25, 2013. http://www.theaustralian.com.au/arts/review/artists-simryn-gill-becomes-australias-accidental-ambassador-at-venice-biennale/story-fn9n8gph-1226649130185.

Hee, Limin. *Constructing Singapore Public Space*. Singapore: Springer Science+Business Media, 2017.

Helsinger, Elizabeth. 'Blindness and Insights'. In *Landscape Theory*, edited by Rachel Ziady DeLue and James Elkins, 323–42. New York: Routledge, 2008.

Heng, Janice. 'Quota for HDB Subletting to Foreigners: 8% Cent for Neighbourhoods, 11% for Blocks'. *Straits Times*. January 16, 2014. http://www.straitstimes.com/singapore/quota-for-hdb-subletting-to-foreigners-8-cent-for-neighbourhoods-11-for-blocks.

Herzfeld, Michael. 'Ethical and Epistemic Reflections on/of Anthropological Vision'. In *Made to Be Seen: Perspectives on the History of Visual Anthropology*, edited by Marcus Banks and Jay Ruby, 313–33. Chicago: University of Chicago Press, 2011.

Hetherington, Kevin. 'Phantasmagoria/Phantasm Agora: Materialities, Spatialities and Ghosts'. In 'Spatial Hauntings', edited by Kevin Hetherington and Monica Degan. Special issue, *Space and Culture* 4, no. 11–12 (2001): 24–41.

Highmore, Ben. 'Formations of Feeling, Constellation of Things'. *Cultural Studies Review* 22, no. 1 (4 April 2016): 144–167. https://doi.org/10.5130/csr.v22i1.4413.

Hill, Jonathan. *Occupying Architecture: Between the Architect and the User*. London: Routledge, 1998.

HistorySG. 'Singapore Separates from Malaysia and Becomes Independent'. Accessed February 13, 2022. https://eresources.nlb.gov.sg/history/events/dc1efe7a-8159-40b2-9244-cdb078755013.

Ho, Tzu Nyen. 'Every Cat in History Is I'. *FOCAS* 6 (2007): 150–67.

———. *One or Several Tigers*. 2017. Shadow puppets, LED lights, smoke machine, show control system, and video (2-channel, 10-channel sound, 16:9 format, colour), 33 min. National Gallery Singapore, Singapore. https://www.nationalgallery.sg/see-do/programme-detail/740/light-to-night--one-or-several-tigers-by-ho-tzu-nyen.

Hollis, Edward. *The Secret Lives of Buildings: From the Parthenon to the Vegas Strip in Thirteen Stories*. London: Portobello Books Ltd, 2010.

Housing and Development Board. 'About Us'. *HDB InfoWEB*. Accessed March 25, 2022. http://www.hdb.gov.sg/cs/infoweb/about-us.

———. 'Annual Report 1977/1978'. Singapore: Housing and Development Board, 1978.

———. 'Eligibility to Buy New HDB Flat'. *HDB InfoWEB*. Accessed March 22, 2022. http://www.hdb.gov.sg/cs/infoweb/residential/buying-a-flat/new/hdb-flat.

———. 'Keeping Pets'. *HDB InfoWEB*. Accessed March 22, 2022. http://www.hdb.gov.sg/cs/infoweb/community/neighbourliness/being-a-good-neighbour/keeping-pets.

———. *50,000 Up: Homes for the People*. Singapore: Housing Development Board, 1966.

———. *Annual report 2020/2021*. 2021. https://www20.hdb.gov.sg/fi10/fi10221p.nsf/hdb/2021/assets/ebooks/annual-report-2021.pdf.

———. *First Decade in Public Housing, 1960-1969*. Singapore: Housing and Development Board, 1970.

Huxley, Tim. 'Singapore and Malaysia: A Precarious Balance?' *The Pacific Review* 4, no. 3 (1991): 204–13. doi.org:10.1080/09512749108718919.

Huxtable, Ada L. 'Russian Engineers Take on Problem of Mass Housing'. *Chicago Tribune*. October 19, 1967. http://archives.chicagotribune.com/1967/10/19/page/10/article/russian-engineers-take-on-problem-of-mass-housing.

Ibrahim, Iskandar. 'Why I Drove the Last Train out'. *ST.Blogs, Straits Times*. July 1, 2011. http://blogs.straitstimes.com/2011/7/1/royal-on-the-rails.

Ingraham, Catherine. *Architecture and the Burdens of Linearity*. New Haven, CT: Yale University Press, 1998.

———. *Architecture, Animal, Human: The Asymmetrical Condition*. London: Routledge, 2006.

Jacobs, Jane M. 'The Only Way is Up'. Panel discussion, National Library Board, Singapore, November 6, 2015.

Jacobs, Jane M., and Stephen Cairns. 'The Modern Touch: Interior Design and Modernisation in Post-Independence Singapore'. *Environment and Planning* A 40, no. 3 (2008): 572–95, doi:10.1068/a39123.

Jacobs, Jane M., Stephen Cairns, and Ignaz Strebel. 'Materializing Vision: Performing a High-Rise View'. In *Visuality/Materiality: Images, Objects and Practices*, edited by Gillian Rose and Divya P. Tolia-Kelly, 133–52. Surrey, UK: Ashgate, 2012.

Jayakumar, S., and Tommy T. B. Koh. *Pedra Branca: The Road to the World Court*. Singapore: NUS Press, 2009.

Kadir, Mohamad. *Malaysia-Singapore: Fifty Years of Contentions, 1965-2015*. Kuala Lumpur: The Other Press, 2015.

Kartini, Saparudin. 'Bukit Brown Municipal Cemetery'. *Singapore Infopedia*. Accessed April 20, 2022. http://eresources.nlb.gov.sg/infopedia/articles/SIP_1358_2009-07-13.html.

Kipling, Rudyard. *The Jungle Book*. New York: Signet, 1961.

Kong, Lily, and Brenda S. A. Yeoh. 'Urban Conservation in Singapore: A Survey of State Policies and Popular Attitudes'. *Urban Studies (Edinburgh, Scotland)* 31, no. 2 (1994): 247–65. doi.org:10.1080/00420989420080231.

Kostof, Spiro, ed. *The Architect: Chapters in the History of a Profession*. New York: Oxford University Press, 1977.

Kotwani, Monica. 'Further enhancements needed to improve water safety teaching: Dr Teo Ho Pin'. *Channel News Asia*. April 2, 2018. https://www.channelnewsasia.com/news/singapore/swimsafer-programme-revamp-needed-survival-expert-1008934.

Krauss, Rosalind E. *The Optical Unconscious*. Cambridge, MA: MIT Press, 1993.

———. *The Originality of the Avant Garde and Other Modernist Myths*. Cambridge, MA: MIT Press, 1985.

Krinsky, Carol Herselle. 'Cesare Cesariano and the Como Vitruvius Edition of 1521'. PhD Diss., New York University, 1965.

Kuo, Pao Kun. 'The Coffin Is Too Big For The Hole'. In *Images at the Margins: A Collection of Kuo Pao Kun's Plays*, 60–71. Singapore: Times Books International, 2000.

Land Transport Authority. 'Projects: Exhumation'. 2012. https://www.lta.gov.sg/content/ltaweb/en/roads-and-motoring/projects/Exhumation.html.

Lau, Fook Kong. 'Hardly a "Pleasant" Final Resting Place'. *Straits Times*, November 27, 2004, 14.

Lau, Jean. 'Woman's 20-Year Hoarding Habit Irritates Neighbours'. *New Paper*. May 1, 2015. http://news.asiaone.com/news/singapore/womans-20-year-hoarding-habit-irritates-neighbours-0.

Lavin, Sylvia. 'Architecture in Extremis'. *Log 22* (2011): 51–61.

Le Guin, Ursula K. *Buffalo Gals and Other Animal Presences*. London: Victor Gollancz, 1990.

Lee, Hsien Loong. 'Prime Minister Lee Hsien Loong's National Day Rally 2011 (Speech in English), Sunday, 14 August 2011, at University Cultural Centre, National University of Singapore'. Transcript, Prime Minister's Office Singapore. August 14, 2011. http://www.pmo.gov.sg/mediacentre/prime-minister-lee-hsien-loongs-national-day-rally-2011-speech-english-sunday-14-august.

———. 'Speech to Parliament on Reading of Penal Code (Amendment) Bill, 22 October' Speech. Parliament of Singapore, Singapore, October 22, 2007.

Lee, Kuan Yew. 'Text of a Talk by the Prime Minister Lee Kuan Yew, Received as a Voicecast from London and Broadcast by Radio Singapore on 9th July 1963'. Transcript, Radio Singapore. September 7, 1963. National Archives of Singapore. http://www.nas.gov.sg/archivesonline/speeches/record-details/73dd0b4c-115d-11e3-83d5-0050568939ad.

———. *From Third World to First: The Singapore Story 1965-2000*. Singapore: Singapore Press Holdings, 2000.

———. *The Singapore Story: Memoirs of Lee Kuan Yew*. Singapore: Prentice Hall, 1998.

Lefebvre, Henri. *The Production of Space*. Translated by Donald Nicholson-Smith. Malden, MA: Blackwell, 1991.

Lei, Yuan Bin (Dir.) and Lilian Chee (PI). *03-FLATS*. Architectural Essay Film. Singapore: National University of Singapore and 13 Little Pictures, 2014.

Lejeune, Anthony. *The Gentlemen's Clubs of London*. London: Parkgate Books, 1984.

Leutemann, Heinrich. *Unterbrochene Strassenmessung auf Singapore* [Road Surveying Interrupted in Singapore]. 1865-85. Wood engraving. Collection of National Museum of Singapore, National Heritage Board, Singapore.

Lim, Charles Yi Yong. 'SEASTATE 6'. *seastories.sg*. Accessed April 16, 2022. https://www.seastories.sg/6.

———. 'Charles Lim Yi Yong'. Accessed November 3, 2021. https://www.seastate.sg/seastate.

Lim, Jessica. 'Tanjong Pagar Station a National Monument'. *Straits Times*. April 9, 2011. https://www.straitstimes.com/singapore/tanjong-pagar-station-a-national-monument.

Lim, Yan Liang. 'Singapore and Malaysia Hold Ceremony to Resume Work on Woodlands-Johor Rail Link; Operations Expected to Begin at End-2026'. *Straits Times*. July 30, 2020. https://www.straitstimes.com/singapore/singapore-and-malaysia-sign-deal-to-resume-work-on-woodlands-johor-rail-link.

Linder, Mark. 'TRANSdisciplinarity'. In *Hunch 9 – Disciplines*, edited by Penelope Dean, 12–15. Rotterdam: Episode Publishers, 2005.

Lipton, Ross. 'Benjamin's Dialectical Image and the Textuality of the Built Landscape'. *Footprint Constellation of Awakening: Benjamin and Architecture*, Spring/Summer (2016): 75–90. doi: 10.7480/footprint.10.1.961.

Liu, Gretchen. *Raffles Hotel*. Singapore: Editions Didier Millet, 2006.

Liu, Thai Ker. 'Design for Better Living Conditions'. In *Public Housing in Singapore: A Multi-Disciplinary Study*, edited by Stephen H. K. Yeh, 117–84. Singapore: Singapore University Press and HDB, 1975.

———. 'Overview'. In *Housing a Nation: Twenty-Five Years of Public Housing in Singapore*, edited by Aline K. Wong and Stephen H. K. Yeh, 1–29. Singapore: Maruzen Asia, 1985.

Livingstone, David N. 'Tropical Hermeneutics: Fragments for a Historical Narrative; An Afterword'. *Singapore Journal of Tropical Geography* 21, no. 1 (2000): 76–91. doi:10.1111/1467-9493.00066.

Loh, Jolene. 'The Abstractions Of Geopolitics, Singapore Pavilion at the 56th Venice Biennale. Charles Lim: SEA STATE'. *ArtAsiaPacific*. Accessed April 20, 2022. http://artasiapacific.com/Magazine/WebExclusives/TheAbstractionsOfGeopolitics.

Loh, Kah Seng. 'History of the Dead, Heritage of the Living: Bukit Brown Cemetery in Singapore'. *Kyoto Review of Southeast Asia* 12, September 2012. https://kyotoreview.org/issue-12/history-of-the-dead-heritage-of-the-living-bukit-brown-cemetery-in-singapore/.

Loh, Kin Kit. 'Domestica(ated) Landscapes: Contesting the Open Spaces of Public Housing in Singapore'. Masters of Architecture Diss., National University of Singapore, 2014.

Lokko, Lesley Naa Norle. 'Black Matter(s): Such as? Does It?' In *Architecture: The Subject Is Matter*, edited by Jonathan Hill, 175–92. London: Routledge, 2001.

———. *White Papers, Black Marks: Architecture, Race, Culture*. Minneapolis: University of Minnesota Press, 2000.

MacDougall, David. *The Corporeal Image: Film, Ethnography, and the Senses*. Princeton: Princeton University Press, 2005.

Macfarlane, Robert. *The Old Ways: A Journey on Foot*. New York: Penguin Books, 2012.

Macpherson, Hannah. 'Walking Methods in Landscape Research: Moving Bodies, Spaces of Disclosure and Rapport'. *Landscape Research* 41, no. 4 (2016): 425–32. doi:10.1080/01426397.2016.1156065.

Maddern, Jo Frances, and Peter Adey. 'Editorial: Spectro-Geographies'. *Cultural Geographies* 15, no. 3 (2008): 291–95. doi:10.1177/1474474008091328.

Mahmud, Aqil Haziq. 'Malaysian Vessel Parked in Singapore Waters Is Used to Mark Territory'. *Channel News Asia*. December 6, 2018. https://www.channelnewsasia.com/news/singapore/malaysian-port-limits-vessel-polaris-buoy-singapore-waters-11007250.

Malaya Tribune. 'Chinese Burial Ground: Municipal By-Laws for Bukit Brown'. October 27, 1917. https://eresources.nlb.gov.sg/newspapers/Digitised/Article/maltribune19210830-1.2.74.

———. 'Light To Be Established'. April 27, 1931, 7.

Mann, Thomas. 'Death in Venice'. In *Death in Venice: Complete, Authoritative Text with Biographical and Historical Contexts, Critical History, and Essays from Five Contemporary Critical Perspectives*, edited by Naomi Ritter, 23–88. Boston: Bedford Books, 1998.

Manning, Erin. *The Minor Gesture*. Durham: Duke University Press, 2016.

Margulies, Ivone. *Nothing Happens: Chantal Akerman's Hyperrealist Everyday*. Durham: Duke University Press, 1996.

Mashadi, Ahmad. 'Drifting Conversations'. In *SEA STATE: Charles Lim Yi Yong*, edited by Shabbir Hussain Mustafa, 116–21. Singapore: National Arts Council, 2015.

Massey, Doreen. 'Living in Wythenshawe'. In *The Unknown City: Contesting Architecture and Social Space*, edited by Iain Borden, Joe Kerr, Jane Rendell, and Alicia Pivaro, 458–74. Cambridge, MA: MIT Press, 2001.

———. *For Space*. London: SAGE Publications, 2005.

Massumi, Brian, Jacob Ferrington, Alina Hechler, and Jannell Parsons. 'Affect and Immediation: An Interview with Brian Massumi', 2019. https://doi.org/10.13023/DISCLOSURE.28.09.

Massumi, Brian. *Architectures of the Unforeseen: Essays in the Occurrent Arts*. Minneapolis: University of Minnesota Press, 2019.

———. *Politics of Affect*. Cambridge: Polity Press, 2015.

———. *Semblance and Event: Activist Philosophy and the Occurrent Arts*. Technologies of Lived Abstraction. Cambridge, MA: MIT Press, 2011.

Matthews, Sean. 'Change and Theory in Raymond Williams's Structure of Feeling'. *Pretexts: Literary and Cultural Studies* 10, no. 2 (1 November 2001): 179–94. https://doi.org/10.1080/10155490120106032.

Maycroft, Neil. 'Not Moving Things Along: Hoarding, Clutter and Other Ambiguous Matter'. *Journal of Consumer Behaviour* 8, no. 6 (November 2009): 354–364. doi:10.1002/cb.298.

MBev. 'Bukit Brown Cemetery: These Hills Are Alive'. *UnTourist Singapore* (blog). Accessed April 18, 2022. https://untouristsingapore.wordpress.com/2013/11/29/bukit-brown-cemetery-these-hills-are-alive/.

McKenzie, Elizabeth. 'Bukit Brown: A Garden of History and Heritage'. In *Spaces of the Dead: A Case from the Living*, edited by Kevin Y. L. Tan, 54–99. Singapore: Singapore Heritage Society and Ethos Books, 2011.

Mead, Margaret. 'Visual Anthropology in a Discipline of Words'. In *Principles of Visual Anthropology*, edited by Paul Hockings, 3–10. Berlin: De Gruyter, 1995.

Merriman, Peter, George Revill, Tim Cresswell, Hayden Lorimer, David Matless, Gillian Rose, and John Wylie. 'Landscape, Mobility, Practice'. *Social & Cultural Geography* 9, no. 2 (2008): 191–212. doi:10.1080/14649360701856136.

Middleton, Jennie. 'Sense and the City: Exploring the Embodied Geographies of Urban Walking'. *Social & Cultural Geography* 11, no. 6 (2010): 575–96. doi:10.1080/14649365.2010.497913.

Miksic, John N. *Singapore and the Silk Road of the Sea, 1300–1800*. Singapore: NUS Press, 2013.

Miller, Daniel, ed. *Home Possessions: Material Culture Behind Closed Doors*. Oxford: Berg, 2001.

Miller, Nancy K. *Getting Personal: Feminist Occasions and Other Autobiographical Acts*. New York: Routledge, 1991.

Mills, J. V. G., ed. *Ma Huan: Ying-yai sheng-lan; The Overall Survey of the Ocean's Shores; [1433]*. Cambridge: Cambridge University Press, 1970.

Minh-Ha, Trinh T. 'The Plural Void: Barthes and Asia'. In *Critical Essays on Roland Barthes*, edited by Diana Knight, 209-18. New York: G. K. Hall & Co., 2000.

Minh-Ha, Trinh T., and Eva Hohenberger. 'The Politics of Forms and Forces'. In *D-Passage: The Digital Way*, by Trinh T. Minh-Ha, 141–68. Durham: Duke University Press, 2013.

Ministry of Culture Singapore. *Berita Singapura: A New Look at Housing*, Videocassette, News Magazines. Singapore: Ministry of Culture, Broadcasting Division, 1967.

———. *Berita Singapura: Celebration of New Housing*, Videocassette, News Magazines. Singapore: Ministry of Culture, Broadcasting Division, 1965.

———. *Berita Singapura: Housing Week*, Videocassette, News Magazines. Singapore: Ministry of Culture, Broadcasting Division, 1963.

———. *Berita Singapura: Visiting New Flats*, Videocassette, News Magazines. Singapore: Ministry of Culture, Broadcasting Division, 1967.

Mitchell, William J.T. 'Preface to the Second Edition of *Landscape and Power:* Space, Place and Landscape'. In *Landscape and Power*, edited by William J.T. Mitchell, vii–xii. 2nd ed. Chicago: University of Chicago Press, 2002.

Mogato, Manuel. 'Manila Finds Marker With Chinese Writing, Buoys In Disputed Waters-Sources'. *Reuters*. July 7, 2015. https://www.reuters.com/article/southchinasea-philippines-buoys-idUSL3N0ZG1VT20150706.

Morning Tribute. 'Sketching In Prohibited Area Alleged'. January 9, 1941, 7.

Morris, Meaghan. 'Banality in Cultural Studies'. *Discourse* 10, no. 2 (1988): 3–29.

Murray, Gilbert. 'Violent Seas Marooned Men Who Built Horsburgh Light'. *Straits Times*, December 31, 1955, 9.

Mustafa, Shabbir Hussain. 'Sea State 0 / Internal Waters. "Mr" Anthrobalanus'. In *SEA STATE: Charles Lim Yi Yong*, edited by Shabbir Hussain Mustafa, 24–31. Singapore: National Arts Council, 2015.

National Archives of Singapore. 'Burial Registers of Bukit Brown Cemetery'. National Archives of Singapore. Accessed April 9, 2022. http://www.nas.gov.sg/BukitBrown_signage.htm.

National Heritage Board. *Community Heritage Series III: Void Decks*. Singapore: National Heritage Board, 2013.

National Parks Board. 'New Sungei Buloh Nature Park Network to expand Singapore's natural capital along our northern coasts as part of efforts to make Singapore a City in Nature'. August 19, 2020. https://www.nparks.gov.sg/news/2020/8/new-sungei-buloh-nature-park-network-to-expand-singapore%E2%80%99s-natural-capital-along-our-northern-coasts-as-part-of-efforts-to-make-singapore-a-city-in-nature.

Nature Society (Singapore). 'Nature Society (Singapore)'s Position on Bukit Brown'. Position Paper. December 12, 2011. https://www.nss.org.sg/documents/Nature%20Society%27s%20Position%20on%20Bukit%20Brown.pdf.

Navaro-Yashin, Yael. *The Make-Believe Space: Affective Geography in a Postwar Polity*. Durham: Duke University Press Books, 2012.

Nemet-Nejat, Murat. *The Peripheral Space of Photography*. Kobenhavn: Green Integer, 2003.

Newbold, Thomas John. *Political and Statistical Account of the British Settlements in the Straits of Malacca...with a History of the Malayan States on the Peninsula of Malacca*. Vol. 2. London: Murray, 1839.

Newton, Owen Noel, Chamika Deshan, Natalie Pang, and Ryohei Nakatsu. 'Mobile Augmented Reality for Bukit Brown Cemetery Navigation'. In *Proceedings of the 11th ACM SIGGRAPH International Conference on Virtual-Reality Continuum and Its Applications in Industry*, edited by Stephen N. Spencer, 59–62. New York: Association for Computing Machinery, 2012. doi:10.1145/2407516.2407534.

Ng, Boon Yian. 'Pedra Branca, An Island of Two Stories'. *Today*. December 27, 2002, 2nd ed., 1.

Ng, Jun Sen. 'New HDB flats: Why singles pay $15k premium'. *Straits Times*, July 1, 2017, https://www.straitstimes.com/singapore/housing/new-hdb-flats-why-singles-pay-15k-premium.

Ng, Wei Kai. '"Every Few Steps There Was a Cat Lying on the Floor": How SCDF Officers Saved 13 Cats from Fire'. *Straits Times*. April 30, 2022. https://www.straitstimes.com/singapore/every-few-steps-there-was-a-cat-lying-on-the-floor-how-scdf-officers-saved-13-cats-from-fire.

O'Donnell, Marcus. 'Walking, Writing and Dreaming: Rebecca Solnit's Polyphonic Voices'. *Journalism* 16, no. 7 (2014): 936–52. doi:10.1177/1464884914553078.

Ooi, Giok Ling, and Thomas T. W. Tan. 'The Social Significance of Public Spaces in Public Housing Estates'. In *Public Space: Design, Use and Management*, edited by Chua Beng-Huat and Norman Edwards, 69–81. Singapore: Singapore University Press for Centre for Advanced Studies, Faculty of Social Sciences, 1992.

Ooi, Peter. 'Beacon of Hope'. *Today*. December 1, 2008.

Orwell, George. 'A Hanging'. In *George Orwell: Essays*, 14–18. London: Penguin Books, c.1968, 2000.

———. 'Shooting an Elephant'. In *George Orwell: Essays*, 18–25. London: Penguin Books, c.1968, 2000.

———. *Burmese Days*. London: Penguin Books, c.1934, 2014.

———. *George Orwell: A Life in Letters*. Edited by Peter Davison. London: Harvill Secker, 2010.

Oswin, Natalie. 'The Modern Model Family at Home in Singapore: A Queer Geography'. *Transactions of the Institute of British Geographers* 35, no. 2 (2010): 256–68. doi:10.1111/j.1475-5661.2009.00379.x.

Our Vanishing Backyard. 'Lim Chu Kang Jetty'. Accessed April 14, 2022. https://ljkj2000.wixsite.com/ourvanishingbackyard/lim-chu-kang-jetty-and-pier.

Panagia, Davide. *Ten Theses for an Aesthetics of Politics*. Forerunners: Ideas First 16. Minneapolis: Minnesota University Press, 2016. https://manifold.umn.edu/read/ten-theses-for-an-aesthetics-of-politics/section/efa1c1e5-8b4f-4264-947c-06e4361137f6#ch01.

———. *The Political Life of Sensation*. Durham: Duke University Press, 2009.

Papazian, Elizabeth, and Caroline Eades, eds. *The Essay Film: Dialogue, Politics, Utopia*. London: Wallflower, 2016.

Pardon, George Frederick [Captain Crawley]. *The Billiard Book*. London: Longman, Green, and Co., 1866.

Park, Chang-Won. 'Confucian Beliefs and Traditions'. In *Encyclopedia of Death and the Human Experience*, edited by Clifton D. Bryant and Dennis L. Peck, 222–23. Thousand Oaks, CA: SAGE, 2009. doi:10.4135/9781412972031.n73.

Paulo, Derrick A. 'White Rock of Contention Goes Before World Court'. *Today*. November 6, 2007, 2.

Penner, Barbara, and Charles Rice. 'Detecting Architecture: Questions of Evidence in Architectural History'. In *Detecting Architecture*. New York, 2009.

Penner, Barbara. 'Researching Female Public Toilets: Gendered Spaces, Disciplinary Limits'. *Journal of International Women's Studies* 6, no. 2 (2005): 81–98.

Pensky, Max. 'Method and Time: Benjamin's dialectical images'. In *The Cambridge Companion to Walter Benjamin*, edited by David S. Ferris, 177–98. New York: Cambridge University Press, 2004.

Perry, Martin, Lily Kong, and Brenda S. A. Yeoh. *Singapore: A Developmental City State*. New York: Wiley, 1997.

Perry, Richard. *The World of the Tiger*. London: Cassell, 1964.

Pevsner, Nikolaus. *An Outline of European Architecture*. Harmondsworth: Penguin, 1977.

———. *Pioneers of the Modern Movement: From William Morris to Walter Gropius*. London: Faber, 1936.

Picon, Antoine, and Karen Bates. 'Anxious Landscapes: From the Ruin to Rust'. *Grey Room*, no. 1 (2000): 65–83.

Pieris, Anoma. 'Dwelling in Ruins: Affective Materialities of the Sri Lankan Civil War'. *The Journal of Architecture* 22, no. 6 (18 August 2017): 1001–20. https://doi.org/10.1080/13602365.2017.1363265.

Pinder, David. 'Errant Paths: The Poetics and Politics of Walking'. *Environment and Planning D: Society and Space* 29, no. 4 (2011): 672–92. doi:10.1068/d10808.

Pink, Sarah. 'Making the Sensory Home: Laundry Routes and Energy Flows'. In *Situating Everyday Life: Practices and Places*, 66–83. London: SAGE, 2012.

———. 'Video and a Sense of the Invisible: Approaching Domestic Energy Consumption Through the Sensory Home'. *Sociological Research Online* 17, no. 1 (2012): 87–106, doi:10.5153/sro.2583.

———. *Doing Sensory Ethnography*. London: SAGE, 2009. doi:10.4135/9781446249383.

———. *Doing Visual Ethnography*. London: SAGE, 2006.

———. *Home Truths: Gender, Domestic Objects and Everyday Life*. New York: Berg, 2004.

Pink, Sarah, and Kerstin Leder Mackley. 'Re-Enactment Methodologies for Everyday Life Research: Art Therapy Insights for Video Ethnography'. *Visual Studies* 29, no. 2 (2014): 146–54. doi:10.1080/1472586X.2014.887266.

———. 'Saturated and Situated: Expanding the Meaning of Media in the Routines of Everyday Life'. *Media, Culture & Society* 35, no. 6 (2013): 677–91. doi:10.1177/0163443713491298.

Pink, Sarah, Phil Hubbard, Maggie O'Neill, and Alan Radley. 'Walking Across Disciplines: From Ethnography to Arts Practice'. Visual Studies 25, no. 1 (2010): 1–7. doi:10.1080/14725861003606670.

Plath, Sylvia. 'Ocean W-1212'. In *Johnny Panic and the Bible of Dreams*, 26–32. New York: Harper Perennial, 2000.

Pocklington, Kate, and Iffah Iesa. *A Whale Out of Water: The Salvage of Singapore's Sperm Whale*. Singapore: Lee Kong Chian Natural History Museum, 2016.

Pomeroy, Jason. 'Room at the Top—The Roof as an Alternative Habitable/Social Space in the Singapore Context'. *Journal of Urban Design* 17, no. 3 (2012): 413–24. doi:10.1080/13574809.2012.666176.

Poon, Angelia. 'Common Ground, Multiple Claims: Representing and Constructing Singapore's "Heartland"'. *Asian Studies Review* 37, no. 4 (2013): 559–76. doi:10.1080/10357823.2013.844768.

Pregarz, Roberto. *Memories of Raffles: 22 Years with a Grand Old Hotel*. Singapore: Treasury Publishing, 1990.

Preservations of Monuments Board and Urban Redevelopment Authority of Singapore. 'Historic Railway Stations to be kept for Future Generations'. Media Release. April 8, 2011. https://www.nas.gov.sg/archivesonline/data/data/pdfdoc/20110415003/joint_pmb-ura_release_on_tprs_and_btrs__final_.pdf.

Public Utilities Board Singapore. 'PUB to lead Singapore's coastal protection efforts'. March 4, 2020. https://www.pub.gov.sg/news/pressreleases/PUBtoleadSingaporescoastalprotectionefforts.

Quammen, David. *Monster of God: The Man-Eating Predator in the Jungles of History and of the Mind.* New York: W.W. Norton, 2003.

Rae, Paul. 'Cat's Entertainment: Feline Performance in the Lion City'. *The Drama Review* 51, no. 1 (2007): 119–37.

Rahim, Lily Zubaidah. 'Singapore-Malaysia Relations: Deep-Seated Tensions and Self-Fulfilling Prophecies'. *Journal of Contemporary Asia 29*, no. 1 (1999): 38–55. doi:10.1080/00472339980000031.

Rascaroli, Laura. 'The Essay Film: Problems, Definitions, Textual Commitments'. *Framework: The Journal of Cinema and Media* 49, no. 2 (2008): 24–47. doi:10.1353/frm.0.0019.

———. *The Personal Camera: Subjective Cinema and the Essay Film.* London: Wallflower, 2009.

———. *How the Essay Film Thinks.* New York: Oxford University Press, 2017.

Rawes, Peg. 'Architectural Ecologies of Care'. In *Relational Architectural Ecologies: Architecture, Nature and Subjectivity*, edited by Peg Rawes, 40–55. London: Routledge, 2013.

———. 'Introduction'. In *Relational Architectural Ecologies: Architecture, Nature and Subjectivity*, edited by Peg Rawes, 1–18. London: Routledge, 2013.

Rendell, Jane. 'An Atlas of the Welsh dresser'. In *Surface Tension: Problematics of Site*, edited by Ken Ehrlich and Brandon LaBelle, 283–92. New York: Errant Bodies, 2003.

———. 'Prelude: The Ways in Which We Write'. In *Writing Architectures: Ficto-Critical Approaches*, edited by Hélène Frichot and Naomi Stead, 1–10. London: Bloomsbury, 2020.

———. 'Site-Writing: She Is Walking about in a Town Which She Does Not Know'. *Home Cultures* 4, no. 2 (2007): 177–99. doi:10.2752/174063107X209019.

———. 'Working Between and Across: Some Psychic Dimensions of Architecture's Inter- and Transdisciplinarity'. *Architecture and Culture* 1, no. 1 (2013): 128–40. doi:10.2752/175145213X13756908698685.

———. *Art and Architecture: A Place Between.* London: I.B. Tauris, 2006.

———. *Site-Writing: The Architecture of Art Criticism.* London: I.B. Tauris, 2010.

———. *The Pursuit of Pleasure: Gender, Space and Architecture in Regency London.* London: Athlone Press, 2002.

Rendell, Jane, and Pamela Wells. 'The Place of Prepositions: A Space Inhabited by Angels'. In *Architecture: The Subject Is Matter*, 131–158. London: New York, 2001.

Report of Survey: Colony of Singapore Masterplan. Singapore: Government Print Office. 1955; 1958.

Rogers, Barbara. *Men Only: An Investigation into Men's Organizations.* London: Pandora, 1988.

Roots SG. 'Qing Ming Festival'. *Roots* SG, April 2018. Accessed April 14, 2022. https://www.roots.sg/learn/resources/ich/qing-ming-festival.

Rose, Gillian, and Divya P. Tolia-Kelly. 'Visuality/Materiality: Introducing a Manifesto for Practice'. In *Visuality/Materiality: Images, Objects and Practices*, edited by Gillian Rose and Divya P. Tolia-Kelly, 1–11. Surrey, UK: Ashgate, 2012.

Rose, Gillian. *Visual Methodologies: An Introduction to the Interpretation of Visual Materials.* London: SAGE, 2007.

Rose, Mitch. 'Gathering "Dreams of Presence": A Project for the Cultural Landscape'. *Environment and Planning D: Society and Space* 24, no. 4 (2006): 537–54. doi:10.1068/d391t.

Roy, Arundhati. *The Ministry of Utmost Happiness*. London: Hamish Hamilton, 2017.

Ruby, Jay. 'Toward an Anthropological Cinema'. *Film Comment* 7, no. 1 (1971): 35–40.

———. *Picturing Culture: Explorations of Film and Anthropology*. Chicago: University of Chicago Press, 2000.

Sack, Robert David. 'Conceptions of Geographic Space'. *Progress in Human Geography* 4, no. 3 (1980): 313–45. doi:10.1177/030913258000400301.

———. *Conceptions of Space in Social Thought: A Geographic Perspective*. London: Macmillan Press, 1980.

Sajan, Chantal. 'Heritage Activists: Bukit Brown More than Just a Cemetery; It's a "Living Museum" of Singapore's Pioneers'. *Straits Times*. September 12, 2020. https://www.straitstimes.com/lifestyle/home-design/heritage-activists-bukit-brown-more-than-just-a-cemetery-its-a-living-museum.

Scanlan, John A. *On Garbage*. London: Reaktion, 2005.

Schor, Naomi. *Reading in Detail: Aesthetics and the Feminine*. London: Methuen, 1987.

Scott, James C. *Seeing like a State: How Certain Schemes to Improve the Human Condition Have Failed*. New Haven, CT: Yale University Press, 1998.

Scott, Joan W. 'The Evidence of Experience'. *Critical Inquiry* 17, no. 4 (1991): 773–97. http://www.jstor.org/stable/1343743.

Segrest, Robert. 'The Perimeter Projects: Notes for Design'. *Assemblage*, no. 1 (1986): 25–35.

Seigworth, Gregory J, and Melissa Gregg. 'An Inventory of Shimmers'. In *The Affect Theory Reader*, edited by Melissa Gregg and Gregory J Seigworth, 1–25. Durham: Duke University Press, 2010.

Seng, Eunice. 'Habitation and the Invention of a Nation, Singapore 1926-1979'. PhD Diss., Columbia University, 2015.

Serres, Michel. *Hermes*. Edited by Josue V. Harari and David F. Bell. Baltimore: Johns Hopkins University Press, 1982.

———. *The Troubadour of Knowledge*. Translated by S. F. Glaser and W. Pauldon. Ann Arbor, MI: University of Michigan Press, 1997.

Sharp, Ilsa. *There is Only One Raffles: The Story of a Grand Hotel*. London: Souvenir Press, 1981.

Sim, Fann. 'Plight of "Homeless" Woman in Bedok Captures Netizens' Attention'. *Yahoo! News*. January 8, 2013. https://sg.news.yahoo.com/plight-of--homeless--woman-in-bedok-captures-attention-of-netizens-153857867.html.

Simpson, David. 'Raymond Williams: Feeling for Structures, Voicing "History"'. *Social Text*, no. 30 (1992): 9–26. https://doi.org/10.2307/466464.

Singapore Heritage Society. 'Bukit Brown Wayfinder: A Self-Guided Walking Trail'. Guide, October 2017. https://www.singaporeheritage.org/bukitbrownwayfinder/Bukit-Brown-Wayfinder-Guide.pdf.

———. 'Position Paper on Bukit Brown'. Position Paper, January 2012. http://www.singaporeheritage.org/wp-content/uploads/2011/11/SHS_BB_Position_Paper.pdf.

Singapore Institute of Architects Magazine. 'Park Road Development'. *Singapore Institute of Architects* 28–29 (October 1968): 9–19.

Singapore's Integrated Urban Coastal Management. Singapore: Technical Committee on the Coastal and Marine Environment, 2013.

Singapore Sunday Times. 'Speech at UMNO General Assembly by Tunku Abdul Rahman'. May 16, 1965.

Smith, Barbara Herrnstein. *Contingencies of Value: Alternative Perspectives for Critical Theory*. Cambridge, MA: Harvard University Press, 1988.

Solnit, Rebecca. *A Field Guide to Getting Lost*. New York: Penguin Books, 2005.

———. *Wanderlust: A History of Walking*. New York: Viking Penguin, 2000.

Spengler, Oswald. *The Decline of the West*. Abridged version by Helmut Werner. London: George Allen and Unwin, 1961.

Squire, Rachael. '"Do You Dive?": Methodological Considerations for Engaging with "Volume"'. *Geography Compass* 11, no. 7 (2017): 1–11.

Stadler, Hilar, Martino Stierli, and Peter Fischli. *Las Vegas Studio: Images from the Archive of Robert Venturi and Denise Scott Brown*. Zurich: Verlag Scheidegger & Spiess, 2015.

Stafford, Barbara Maria. 'Crystal and Smoke: Putting Image Back in Mind'. In *A Field Guide to a New Meta-Field: Bridging the Humanities-Neurosciences Divide*, edited by Barbara Maria Stafford, 1–63. Chicago: University of Chicago Press, 2011.

Stallybrass, Peter, and Allon White. 'Introduction'. In *The Politics and Poetics of Transgression*, 1–26. Ithaca, NY: Cornell University Press, 1986.

Stead, Naomi. 'Architecture and Memory in W. G. Sebald's Austerlitz'. *Arq: Architectural Research Quarterly* 19, no. 1 (March 2015): 41–48. https://doi.org/10.1017/S1359135515000263.

Stewart, Kathleen. 'Afterword: Worlding Refrains'. In *The Affect Theory Reader*, edited by Melissa Gregg and Gregory J. Seigworth, 339–53. Durham: Duke University Press, 2010.

———. 'Atmospheric Attunements'. *Society and Space*, D, 29 (2011): 445–53. https://doi.org/10.1068/d9109.

———. 'In the World That Affect Proposed'. *Cultural Anthropology* 32, no. 2 (n.d.): 192–98. https://doi.org/10.14506/ca32.2.03.

———. 'Pockets'. *Communication and Critical Cultural Studies* 9, no. 4 (2012): 365.

———. 'Precarity's Forms'. *Cultural Anthropology* 27, no. 3 (2012): 518–25.

———. *Ordinary Affects*. Durham: Duke University Press, 2007.

Straits Times Weekly. 'Muar News'. June 29, 1892.

Straits Times. 'A Legend Roars Back to Life'. February 11, 1986.

———. 'A New Hotel in Singapore'. September 19, 1887.

———. 'A Tiger Alarm'. September 17, 1902.

———. 'A Tiger in Town: Shot at Raffles Hotel under the Billiard Room'. August 13, 1902.

———. 'Caretaker's Death: Murder, Say Police'. May 5, 1980, 11. https://eresources.nlb.gov.sg/newspapers/Digitised/Article/straitstimes19800505-1.2.62.

———. 'Cemetery Blaze: Firemen's 45-Minute Battle'. January 17, 1935, 12. https://eresources.nlb.gov.sg/newspapers/Digitised/Article/straitstimes19350117-1.2.79.

———. 'Did Not Know A War Was Going On'. February 5, 1941, 11.

———. 'Faction Fight in a Cemetery'. July 24, 1933, 12. https://eresources.nlb.gov.sg/newspapers/Digitised/Article/straitstimes19330724-1.2.69.

———. 'Fight at a Funeral'. July 21, 1927, 8. https://eresources.nlb.gov.sg/newspapers/Digitised/Article/straitstimes19270721-1.2.28.

———. 'Fishing Now Allowed Near Horsburgh Lighthouse'. August 26, 1989.

———. 'Multi-Use Buildings: First Goes Up in Chinatown'. April 21, 1967. https://eresources.nlb.gov.sg/newspapers/Digitised/Article/straitstimes19670421-1.2.24?ST=1&AT=search&k=Multi-Use%20Buildings:%20First%20Goes%20Up%20in%20Chinatown&QT=multi-use,buildings,first,goes,up,in,chinatown&oref=article.

———. 'Photographing Singapore Defences'. November 24, 1900, 2.

———. 'Photographing The Forts'. July 31, 1901, 2.

———. 'Raffles Hotel: New Billiard Room To Be Constructed and Other Improvements Made'. October 9, 1906.

———. 'Untitled'. November 19, 1906. https://eresources.nlb.gov.sg/newspapers/Digitised/Article/straitstimes19061119-1.2.27.

Strasser, Susan. *Waste and Want: A Social History of Trash*. New York: Metropolitan Books, 1999.

Subramanian, Samanth. 'How Singapore Is Creating More Land for Itself'. *New York Times*, April 20, 2017, https://www.nytimes.com/2017/04/20/magazine/how-singapore-is-creating-more-land-for-itself.html.

Sumartojo, Shanti, and Sarah Pink. *Atmospheres and the Experiential World: Theory and Methods*. Oxon: Routledge, 2019.

Tagliacozzo, Eric. 'Mapping the Frontier'. In *Secret Trades, Porous Borders: Smuggling and States Along a Southeast Asian Frontier, 1865-1915*, 27–52. New Haven: Yale University Press, 2005.

———. 'The Lit Archipelago: Coast Lighting and the Imperial Optic in Insular Southeast Asia, 1860-1910'. *Technology and Culture* 46, no. 2 (April 2005): 306–28.

Tai, Ching Ling. *Housing and High-Rise Living: A Study of Singapore's Public Housing*. Singapore: Chopmen, 1988.

Tan, Kenneth Paul. *Cinema and Television in Singapore: Resistance in One Dimension*. Boston: Brill, 2008.

Tan, Kevin Y. L., ed. *Spaces of the Dead: A Case from the Living*. Singapore: Singapore Heritage Society, 2011.

Tan, Pin Pin. *Gravediggers' Luck*. Documentary, 22:29. Singapore: VHQ TV, 2003. Posted September 17, 2015. https://www.tanpinpin.com/gravediggers-luck-2/.

———. *Moving House (2001)*. 16mm, Documentary, 22:26. Discovery Channel International, 2001. https://www.tanpinpin.com/about/moving-house/.

Tan, Tony Keng Joo, and Wong Tai-Chee. 'Public Housing in Singapore: A Sustainable Housing Form and Development'. In *Spatial Planning for a Sustainable Singapore,* edited by Wong Tai-Chee, Belinda Yuen, and Charles Goldblum, 135–50. Singapore: Springer, 2008.

Tan, Tony Keng Joo, Loh Choon Tong, Tan Sioe An, and Lau Woh Cheong. 'Physical Planning and Design'. In *Housing a Nation: Twenty-Five Years of Public Housing in Singapore,* edited by Aline K. Wong and Stephen H. K. Yeh, 56–112. Singapore: Maruzen Asia, 1985.

Tan, Wooi Leong. 'Jurong Island: What It Takes to Achieve a World-Class Petrochemicals Hub'. *Surbana Jurong Private Limited.* April 25, 2017. https://surbanajurong.com/resources/perspectives/jurong-island-takes-achieve-world-class-petrochemicals-hub/.

Taussig, Michael. 'The Beach (A Fantasy)'. In *Landscape and Power,* edited by W. J. T. Mitchell, 2nd ed., 316–347. Chicago: University of Chicago Press, 2002.

———. *I Swear I Saw This: Drawings in Fieldwork Notebooks, Namely My Own*. Chicago: Chicago University Press, 2011.

———. *Mastery of Non-Mastery in the Age of Meltdown.* Chicago: Chicago University Press, 2020.

———. *The Corn Wolf.* Chicago: University of Chicago Press, 2015.

Teh, David. 'Charles Lim's Informatic Naturalism: Notes on SEA STATE 2'. In *Charles Lim Sea State 2: As Evil Disappears*, 2–8. Singapore: Future Perfect, 2012.

Teo, Siyang. 'Plant Walk @ Bukit Brown'. *Urban Forest* (blog). September 25, 2011. https://uforest.blogspot.com/2011/09/plant-walk-bukit-brown.html.

Teyssot, Georges. *A Topology of Everyday Constellations*. Cambridge, MA: MIT Press, 2013.

The Long and Winding Road. 'Cashin's Pier, lost and found'. August 19, 2020. https://thelongnwindingroad.wordpress.com/2020/08/19/cashins-pier-lost-and-found/.

The Malayan Saturday Post. 'Opening of the new F.M.S.R. terminal station'. May 7, 1932.

The Singapore Free Press and Mercantile Advertiser. 'Opening of Singapore's New Station'. May 3, 1932.

———. 'The Horsburgh Light-House'. March 20, 1902, Weekly edition, 3.

Thomson, J. T. *Some Glimpses into Life in the Far East*. London: Richardson, 1864.

Torri, Davide. 'Her Majesty's Servants: The Tamed and the Wild Under the British Raj'. In *Charming Beauties and Frightful Beasts: Non-Human Animals in South Asian Myth, Ritual and Folklore*, edited by Fabrizio M. Ferrari and Thomas Dähnhardt, 32–48. Sheffield, UK: Equinox, 2013.

Tremewan, Christopher. 'Public Housing: The Working Class Barracks'. In *The Political Economy of Social Control in Singapore*, 45–73. London: St Martin's Press, 1994.

———. 'Welfare and Governance: Public Housing under Singapore's Party-State'. In *The East Asian Welfare Model: Welfare Orientalism and the State*, edited by Roger Goodman, Gordon White and Huck-Ju Kwon, 77–105. London: Routledge, 1998.

Troiani, Igea, and Alison Kahn. 'Beyond the Academic Book: New 'Undisciplined' Corporeal Publication'. *Architecture and Culture* 4, no. 1 (2016): 51–71. doi:10.1080/20507828.2015.1094228.

Troiani, Igea, and Hugh Campbell. 'Architecture Film: Framing the Discourse and Conditions of Production in Architectural Practice and Education'. In *Architecture Filmmaking*, edited by Hugh Campbell and Igea Troiani, 1–20. Bristol: Intellect Books, 2019.

Troiani, Igea, and Suzanne Ewing. 'Inside Architecture from the Outside: Architecture's Disciplinary Practices'. *Architecture and Culture* 2, no. 2 (2014): 151–66. doi:10.2752/205078214X14030010182308.

———. 'Visual Research Methods in Architecture: New Forms of "Critical Visualisation"'. In *Visual Methodologies in Architectural Research*, edited by Suzanne Ewing and Igea Troiani, 1–29. Bristol: Intellect Books, 2020.

Tung, Shi Yun. 'Explore Three Singapore Lighthouses on the National Heritage Board's Lighthouse Trail'. *Straits Times*. June 26, 2014. https://www.straitstimes.com/singapore/explore-three-singapore-lighthouses-on-the-national-heritage-boards-lighthouse-trail.

Tunku, Abdul Rahman Putran. *Looking Back: The Historic Years of Malaya and Malaysia*. Kuala Lumpur: Pustaka Antara, 1977.

United Nations Office of Legal Affairs Division for Ocean Affairs and the Law of the Sea. *Commission on the Limits of the Continental Shelf*. April 21, 2022. https://www.un.org/Depts/los/clcs_new/continental_shelf_description.htm.

Urban Redevelopment Authority of Singapore. 'Citation for Adaptive Reuse of the Former Tanjong Pagar Railway Station'. *Rail Corridor*. Accessed February 13, 2022. https://www.ura.gov.sg/MS/railcorridor/rfp/awards/Winners.aspx.

———. 'Overview'. *Rail Corridor*. Accessed February 13, 2022. https://www.ura.gov.sg/MS/railcorridor/rfp/overview.aspx.

Vinegar, Aron, and Jorge Otero-Pailos. 'On Preserving the Openness of the Monument'. *Future Anterior Journal of Historic Preservation, History, Theory, and Criticism* 9, no. 2 (2013): iii-vi.

Virilio, Paul, and Sylvère Lotringer. *Pure War: Twenty-Five Years Later*. Translated by Mark Polizzotti. Los Angeles: Semiotext(e), 2008.

Von Braschler. *Confessions of a Reluctant Ghost Hunter: A Cautionary Tale of Encounters with Malevolent Entities and Other Disembodied Spirits*. Rochester, VT: Destiny Books, 2014.

Waterson, Roxana. 'Visual Anthropology and the Built Environment: Interpenetrations of the Visible and the Invisible'. In *Made to Be Seen: Perspectives on the History of Visual Anthropology*, edited by Marcus Banks and Jay Ruby, 74–107. Chicago: University of Chicago Press, 2011.

———. *Living House: An Anthropology of Architecture in South-East Asia*. Singapore: Oxford University Press, 1990.

Wells, Liz. *Land Matters: Landscape Photography, Culture and Identity*. London: I.B. Tauris, 2011.

Wessing, Robert. *The Soul of Ambiguity: The Tiger in Southeast Asia*. DeKalb, IL: Centre for Southeast Asian Studies, Northern Illinois University, 1986.

Williams, Raymond. *Marxism and Literature*. New York: Oxford University Press, 1977.

Wong, Aline K., and Stephen H.K. Yeh, eds. *Housing a Nation: Twenty-Five Years of Public Housing in Singapore*. Singapore: Maruzen Asia, 1985.

Wong, Grace K. M. 'Vertical Cities as a Solution for Land Scarcity: The Tallest Public Housing Development in Singapore'. *Urban Design International* 9, no. 1 (2004): 17–30. doi:10.1057/palgrave.udi.9000108.

Wong, May. 'M'sia Cannot Prove Ownership of Pedra Branca: S'pore Lawyers'. *Today*. November 8, 2007, 4.

Woodcock, George. *The Crystal Spirit: A Study of George Orwell*. London: Fourth Estate, 1967.

Woon, Wallace. 'ST Explains: Why Cats Are Banned from HDB Flats, and What Lifting the Ban Would Take'. The Straits Times. September 5, 2022. https://www.straitstimes.com/singapore/environment/st-explains-why-cats-are-banned-from-hdb-flats-and-what-is-being-done-to-change-it.

World Monuments Fund. '2014 World Monuments Watch'. World Monuments Fund. Accessed April 8, 2022. https://www.wmf.org/downloads/press-kit/2014-Map-ENG.pdf.

Wylie, John. 'A Single Day's Walking: Narrating Self and Landscape on the South West Coast Path'. *Transactions of the Institute of British Geographers* 30, no. 2 (2005): 234–47.

———. 'The Spectral Geographies of W.G. Sebald'. *Cultural Geographies* 14, no. 2 (2007): 171–88. doi:10.1177/1474474007075353.

———. *Landscape*. Abingdon, Oxon: Routledge, 2007.

Wylie, Lesley. *Colonial Tropes and Postcolonial Tricks: Rewriting the Tropics in the Novela de La Selva*. Liverpool: Liverpool University Press, 2011. doi:10.5949/UPO9781846315220.

Xiao Difei 萧涤非, ed. *Tangshi Jianshang Cidian*唐诗鉴赏辞典 [Dictionary for the Appreciation of Tang Poetry]. Shanghai: Shanghai Cishu Chubanshe, 1983.

Yao, Souchou. *Singapore: The State and the Culture of Excess*. New York: Routledge, 2007.

Yeh, Stephen H. K. *Public Housing in Singapore: A Multi-Disciplinary Study*. Singapore: Singapore University Press, 1975.

Yeo, Wei-Wei. 'Of Trees and the Heartland: Singapore's Narratives'. In *Beyond Description: Singapore Space Historicity*, edited by Ryan Bishop, John Phillips, and Yeo Wei-Wei, 17–29. London: Routledge, 2004.

Yeoh, Brenda S. A. 'Bones of Contention: Chinese Burial Grounds in Colonial and Post-Colonial Singapore'. *In Spaces of the Dead: A Case from the Living*, edited by Kevin Y. L. Tan, 280–94. Singapore: Singapore Heritage Society and Ethos Books, 2011.

———. 'The Control of "Sacred" Space: Conflicts over the Chinese Burial Grounds in Colonial Singapore, 1880-1930'. *Journal of Southeast Asian Studies* 22, no. 2 (1991): 282–311, doi: 10.1017/S0022463400003891.

———. *Contesting Space in Colonial Singapore: Power Relations and the Urban Built Environment*. Kuala Lumpur: Oxford University Press, 1996; Singapore: NUS Press, 2003.

Yeoh, Brenda S. A., and L. Y. Ho. 'Popular Cognition of National Monuments in Singapore'. *Environmental Monitoring and Assessment* 44, no. 1 (1997): 21–31. doi:10.1023/A:1005755717141.

Yeoh, Brenda S. A., and Lily Kong. 'Reading Landscape Meanings: State Constructions and Lived Experiences in Singapore's Chinatown'. *Habitat International* 18, no. 4 (1994): 17–35. doi:10.1016/0197-3975(94)90015-9.

———. 'The Notion of Place in the Construction of History, Nostalgia and Heritage in Singapore'. *Singapore Journal of Tropical Geography* 17, no. 1 (1996): 52–65. doi:10.1111/j.1467-9493.1996.tb00084.x.

Yeoh, Brenda S. A., and Tan Boon Hui. 'The Politics of Space: Changing Discourses on Chinese Burial Grounds in Post-War Singapore'. *Journal of Historical Geography* 21, no. 2 (1995): 184–201. doi:10.1016/0305-7488(95)90035-7.

Yong, Charissa. 'MND Rejects Petition to Reform Public Housing Policies for Single Parents'. *Straits Times*. November 29, 2017. https://www.straitstimes.com/singapore/mnd-rejects-petition-to-reform-public-housing-policies-for-single-parents.

Yong, Chun Yuan. 'Tanjong Pagar Railway Station'. *Singapore Infopedia*. Accessed February 10, 2022. https://eresources.nlb.gov.sg/infopedia/articles/SIP_954_2005-01-10.html.

Yong, Nicholas. 'Buried: A Life'. *Straits Times*. April 10, 2011, Life!, 7.

Zarecor, Kimberly E. 'Architecture in Eastern Europe and the Former Soviet Union'. In *A Critical History of Contemporary Architecture, 1960–2010*, edited by Elie G. Haddad and David Rifkind, 255–74. Surrey, England: Ashgate, 2014.

INDEX

Note: Page numbers in *italic* refer to figure captions.

For Product Safety Concerns and Information please contact our EU representative GPSR@taylorandfrancis.com Taylor & Francis Verlag GmbH, Kaufingerstraße 24, 80331 München, Germany

Batch number: 10733633

Printed by Printforce, the Netherlands